# VOLCANISM AND GLOBAL ENVIRONMENTAL CHANGE

The 2010 eruption of Eyjafjallajökull resulted in unprecedented disruption to global air travel and caused major flooding in Iceland, highlighting the importance of understanding how volcanic processes affect the Earth's surface and atmosphere.

Covering a key connection between geological processes and life on Earth, this multidisciplinary volume describes the effects of volcanism on the environment by combining present-day observations of volcanism and environmental changes with information from past eruptions preserved in the geologic record. The book discusses the origins, features and timing of volumetrically large volcanic eruptions; methods for assessing gas and tephra release in the modern day and the palaeo-record; and the impacts of volcanic gases and aerosols on the environment, from ozone depletion to mass extinctions. The significant advances that have been made in recent years in quantifying and understanding the impacts of present and past volcanic eruptions are presented and review chapters are included, making this a valuable book for academic researchers and graduate students in volcanology, climate science, palaeontology, atmospheric chemistry and igneous petrology.

DR ANJA SCHMIDT is an Academic Research Fellow at the School of Earth and Environment, University of Leeds, quantifying the effects of volcanism on the atmosphere, the climate system and society by combining volcanological datasets and atmospheric modelling. Dr Schmidt has been awarded a University of Leeds Research Scholarship, as well as a Springer Thesis Prize for her Ph.D work on modelling tropospheric volcanic aerosols.

DR KIRSTEN E. FRISTAD is a NASA Postdoctoral Fellow at the NASA Ames Research Center, investigating the role of volcanism and hydrothermal activity on life and environmental change. Active in field-based research, she spent two seasons field-testing Mars Curiosity Rover instruments in Svalbard, and received a Fulbright Fellowship to study in Norway.

DR LINDA T. ELKINS-TANTON is Director of the School of Earth and Space Exploration at Arizona State University. Her research interests include silicate melting and solidification processes, planetary formation and early evolution, and the formation of large volcanic provinces. She is a two-time National Academy of Sciences Kavli Frontiers of Science Fellow, and now sits on the National Academy Committee on Astrobiology and Planetary Science.

# VOLCANISM AND GLOBAL ENVIRONMENTAL CHANGE

*Edited by*

ANJA SCHMIDT
*University of Leeds*

KIRSTEN E. FRISTAD
*NASA Ames Research Center*

LINDA T. ELKINS-TANTON
*Arizona State University*

CAMBRIDGE
UNIVERSITY PRESS

# CAMBRIDGE
## UNIVERSITY PRESS

University Printing House, Cambridge CB2 8BS, United Kingdom

One Liberty Plaza, 20th Floor, New York, NY 10006, USA

477 Williamstown Road, Port Melbourne, VIC 3207, Australia

4843/24, 2nd Floor, Ansari Road, Daryaganj, Delhi - 110002, India

79 Anson Road, #06-04/06, Singapore 079906

Cambridge University Press is part of the University of Cambridge.

It furthers the University's mission by disseminating knowledge in the pursuit of education, learning and research at the highest international levels of excellence.

www.cambridge.org
Information on this title: www.cambridge.org/9781107633544

© Cambridge University Press 2015

First published 2015
First paperback edition 2017

*A catalogue record for this publication is available from the British Library*

*Library of Congress Cataloging in Publication data*
Volcanism and global environmental change / edited by Anja Schmidt, University of Leeds, Kirsten Fristad, NASA Ames Research Center, Linda Elkins-Tanton, Arizona State University.
pages cm
ISBN 978-1-107-05837-8 (Hardback)
1. Volcanism–Environmental aspects. 2. Global environmental change.
3. Paleogeography. 4. Paleoclimatology. I. Schmidt, Anja, 1980– editor.
II. Fristad, Kirsten, editor. III. Elkins-Tanton, Linda T., editor.
QE522.V64 2015
551.21–dc23 2014021444

ISBN 978-1-107-05837-8 Hardback
ISBN 978-1-107-63354-4 Paperback

..................................................................................................................

# Table of contents

*Colour plates are to be found between pp. 176 and 177.*

# List of contributors

**Alessandro Aiuppa**
*Università di Palermo, Palermo, Italy; Istituto Nazionale di Geofisica e Vulcanologia, Sezione di Palermo, Palermo, Italy*

**Nick T. Arndt**
*Université Joseph Fourier, Grenoble, France*

**Jean Besse**
*Institut de Physique du Globe de Paris, Paris, France; Université Paris Diderot, Paris, France*

**Benjamin A. Black**
*University of California, Berkeley, Berkeley CA, USA*

**Terrence J. Blackburn**
*Carnegie Institution for Science, Washington, DC, USA*

**Nicole Bobrowski**
*Institut für Umweltphysik, Heidelberg, Germany*

**Samuel A. Bowring**
*Massachusetts Institute of Technology, Cambridge, MA, USA*

**Seth D. Burgess**
*Massachusetts Institute of Technology, Cambridge, MA, USA*

**Kevin Burke**
*University of Houston, Texas, TX, USA*

**Ying Cui**
*The Pennsylvania State University, University Park, PA, USA*

**Vincent Courtillot**
*Institut de Physique du Globe de Paris, Paris, France; Université Paris Diderot, Paris, France*

**Amy Donovan**
*University of Cambridge, Cambridge, UK*

**Linda T. Elkins-Tanton**
*School of Earth and Space Exploration, Arizona State University, Tempe, AZ, USA*

**Anna Fetisova**
*Moscow State University, Moscow, Russia; Russian Academy of Sciences, Moscow, Russia*

**Frédéric Fluteau**
*Institut de Physique du Globe de Paris, Paris, France; Université Paris Diderot, Paris, France*

**Kirsten E. Fristad**
*ORAU/NASA Ames Research Center, Mountain View, Moffett Field, CA, USA*

**Lori S. Glaze**
*NASA Goddard Space Flight Center, Greenbelt, MD, USA*

**Thor H. Hansteen**
*GEOMAR Helmholtz Centre for Ocean Research Kiel, Kiel, Germany*

**Morgan T. Jones**
*Centre for Earth Evolution and Dynamics (CEED), University of Oslo, Oslo, Norway*

**Jeffrey T. Kiehl**
*National Center for Atmospheric Research, Boulder, CO, USA*

**Nadezhda A. Krivolutskaya**
*Vernadsky Institute of Geochemistry, Moscow, Russia*

**Kirstin Krüger**
*University of Oslo, Oslo, Norway; GEOMAR Helmholtz Centre for Ocean Research Kiel, Kiel, Germany*

**Lee R. Kump**
*The Pennsylvania State University, University Park, PA, USA*

**Steffen Kutterolf**
*GEOMAR Helmholtz Centre for Ocean Research Kiel, Kiel, Germany*

**Dimitry V. Kuzmin**
*V. S. Sobolev Institute of Geology and Mineralogy, Novosibirsk, Russia; Novosibirsk State University, Novosibirsk, Russia*

**Jean-François Lamarque**
*National Center for Atmospheric Research, Boulder, CO, USA*

**A. Latyshev**
*Moscow State University, Moscow, Russia; Russian Academy of Sciences, Moscow, Russia*

**Kimberly V. Lau**
*Stanford University, Stanford, CA, USA*

**Tamsin A. Mather**
*University of Oxford, Oxford, UK*

**Katja M. Meyer**
*Department of Environmental and Earth Sciences, Willamette University, Salem, OR, USA*

**Clive Oppenheimer**
*University of Cambridge, Cambridge, UK*

**Vladimir Pavlov**
*Russian Academy of Sciences, Moscow, Russia; Kazan Federal University, Kazan, Tatarstan, Russia*

**Jonathan L. Payne**
*Stanford University, Stanford, CA, USA*

**Ingrid Ukstins Peate**
*University of Iowa, Iowa City, IA, USA*

**David Pieri**
*Jet Propulsion Laboratory, California Institute of Technology, Pasadena, CA, USA*

**Sverre Planke**
*Volcanic Basin Petroleum Research (VBPR), Oslo Science Park, Oslo, Norway; Centre for Earth Evolution and Dynamics (CEED), University of Oslo, Oslo, Norway*

**Ulrich Platt**
*Institut für Umweltphysik, Heidelberg, Germany*

**Alexander Polozov**
*Russian Academy of Sciences, Moscow, Russia*

**Fred Prata**
*Norwegian Institute for Air Research, Kjeller, Norway*

**Gemma Prata**
*University of Oxford, Oxford, UK*

**David M. Pyle**
*University of Oxford, Oxford, UK*

**Andy Ridgwell**
*University of Bristol, Bristol, UK*

**Alan Robock**
*Rutgers University, New Brunswick, NJ, USA*

**Ellen K. Schaal**
*Department of Geology, Lawrence University, Appleton, WI, USA*

**Anja Schmidt**
*University of Leeds, Leeds, UK*

**Stephen Self**
*The Open University, Milton Keynes, UK; Department of Earth and Planetary Science, University of California, Berkeley, CA, USA*

**Christine Shields**
*National Center for Atmospheric Research, Boulder, CO, USA*

**Juan Carlos Silva-Tamayo**
*Department of Earth and Environment, University of Leeds, Leeds, UK*

**Alexander V. Sobolev**
*Université Joseph Fourier, Grenoble, France*

**Stephan V. Sobolev**
*German Research Center for Geosciences, Potsdam, Germany*

**Henrik Svensen**
*Centre for Earth Evolution and Dynamics (CEED), University of Oslo, Oslo, Norway*

**Trond H. Torsvik**
*University of Oslo, Oslo, Norway*

**Roman Veselovskiy**
*Moscow State University, Moscow, Russia; Russian Academy of Sciences, Moscow, Russia*

# Preface

*Linda T. Elkins-Tanton*
*Anja Schmidt*
*Kirsten E. Fristad*

On the 8th June in 1783 CE a fissure on Iceland opened and the devastating Laki eruption began. Seemingly a simple basaltic fire-fountaining event, it defied common assumptions about basaltic volcanism by emitting vast amounts of halogens and sulfur species to the atmosphere. The eruption caused severe environmental and climatic changes in the northern hemisphere that lasted for several years. Even with the fidelity of human recordkeeping at the time, scientists today are still investigating why and how the Laki eruption affected the environment.

The geologic record reveals that volcanism has occurred on a wide range of scales throughout Earth history, from the formation of small cinder cones to giant flood basalt provinces. Coeval sedimentary records indicate that some of these past eruptions, continental flood basalts in particular, may have caused dramatic changes to the global environment, affecting climate, environmental chemistry, and perhaps triggering mass extinctions. One of the largest of these continental flood basalt eruptions occurred 252 million years ago in present-day Siberia. Much of the lava is thought to have been produced in fissure eruptions, such as Laki in Iceland, and death ensued, not only from starvation. The coeval end-Permian extinction of species was global and came close to eliminating multicellular life in the oceans and, to a lesser extent, on land.

Not every volcanic eruption causes significant environmental change, however, and the mechanisms driving different modes of volcanism and their variable environmental impact are areas of ongoing research. Eruption volume, magnitude and explosivity are obvious indicators of the potential for environmental effects. Large erupted volumes and high explosivities indicate large volatile mass fluxes and increase the chance of volatile species being lofted high into the atmosphere. Once in the stratosphere, gases and aerosol particles can circulate the globe, affecting climate and the environment on global scales. Beyond this simple beginning, however, is a landscape of complex physical and chemical interactions that still remain to be explored and understood.

Methods for assessing the effects of volcanism on the environment are increasingly diverse as new technology and techniques enable measurements that were previously unattainable. Today, scientists can actively monitor eruptions with instrumentation on the ground and on satellites to measures plume sizes, dispersion rates, and plume compositions, including sulfur, carbon and halogen compounds. We can measure global temperature changes in the years following an eruption as well as changes to surface water chemistry and primary productivity due to ash deposition and ash-leachate dissolution. Although advances have been made in recent years, many questions remain regarding issues such as the production and dispersion of ash and its effect on airplanes. Understanding the effects of volcanism on climate and environment is limited, however, in the small slices of time and styles of volcanism experienced during human history.

To understand the full range of styles and impacts of volcanism, we must look to the geologic record. In the palaeo-record, traditional petrography and physical volcanology are supplemented with pressure and chemistry estimates from fluid inclusions, with nanoprobe measurements of volatiles in melt inclusions and with disaggregation and grain-size measurements and magnetic conglomerate tests in volcaniclastics, to determine palaeo-eruption dynamics, explosivity and volatile content. A variety of isotope and geochemical proxies are used to understand the impact of volcanism, in the geologic past, on global temperature and other environmental conditions.

Previously, the voluminous flood basalts in the geologic record were thought to contain low levels of climate-changing volatiles. Recent work on the Siberian flood basalts, however, indicates that the eruptions mobilized vast amounts of carbon and sulfur-bearing species, along with ozone-depleting chlorofluorocarbons. Similarly, recent work provides estimates of the volatiles released by the Deccan Traps and the Central Atlantic Igneous Province, indicating their potential to severely affect the environment and life on Earth. Given the impacts of volcanic activity in the past and the similarity in composition between volcanic volatiles and anthropogenic emissions, a better understanding of how volcanic volatiles contributed to past global environmental change has direct application to both volcanic and anthropogenic climate change today.

The richest discoveries and most important advances in science can often be made in interdisciplinary work. This volume brings together geologists, atmospheric scientists, climate scientists, volcanologists, palaeobiologists and modelers, to find a fruitful path forward in better understanding how solid Earth processes affect the atmosphere, and thus result in global environmental change affecting habitability on Earth, both in the present day and the geological past.

The chapters in *Volcanism and Global Environmental Change* are divided into three sections. In Part One, 'Large volume volcanism: origins, features, and timing', Ukstins Peate and Elkins-Tanton highlight a specific aspect of large igneous provinces, their common inclusion of explosive basaltic volcanism producing in some cases tremendous volumes of volcaniclastic deposits. In some cases, these volcaniclastics were produced by interaction with ground water but, in others, they were driven from depth with an explosive force that indicates a far greater potential for environmental change than has been supposed. Oppenheimer and Donovan discuss the poorly understood phenomena dubbed 'super-eruptions', perhaps the greatest single volcanic threat to modern humankind.

Large igneous provinces hold our fascination not only because of their immense size and their lack of modern analogs, but also because their physical origins are still debated (discussed here by Torsvik and Burke). The apparent link between large igneous provinces and global extinction events is being demonstrated with smaller and smaller errors as laboratory geochronology techniques improve and as fieldwork continues; the state of this art is described by Burgess, Blackburn and Bowring. With increasing geochronology fidelity comes better knowledge of the duration of these eruptions (which in many cases may have been far less than the commonly quoted million years). Palaeomagnetism offers a method for examining the rapidity of emplacement of packets of flows, as they record the continuous gradual movement of the magnetic pole, and, in conjunction with ages for the whole province, for assessing the total length of non-eruptive intervals. This technique is demonstrated for the Siberian Traps by Pavlov and co-authors.

In Part Two, 'Assessing gas and tephra release in the present day and palaeorecord', the state of the art of present-day gas and ash measurements by ground-based, satellite and aircraft instruments are discussed by Aiuppa, Prata, Platt and Bobrowski, and Pieri. Sophisticated ground-based instrumentation is becoming more common and will continue to expand in coming years as population growth puts more and more people at risk of volcanic hazards; gas monitoring is becoming as standard as seismometers on volcanoes around the world. Meanwhile, instrumentation on satellites and aircraft can target both volcanoes near population centers and those that are more remote.

Measuring the gas release rates, degrees of explosivity, and climatic effects of eruptions from the geologic record is more challenging. The following four chapters concern gases emitted by flood basalts and their link to extinctions. Sobolev and co-authors describe a model for the source of volatiles in the Siberian flood basalts, while Self and co-authors discuss plume heights, composition and timescales of flood basalt eruptions, and Svensen and co-authors discuss the potential for volatiles baked from country rocks by the heat and mass transfer of the flood basalts.

The final part of the book, 'Modes of volcanically induced global environmental change', contains chapters on atmospheric and climate change caused by volcanism by Schmidt and Robock and by Mather and Pyle. Courtillot *et al.* summarize the decades of work done at the Institut de Physique du Globe on defining and addressing this broad topic. Finally, the specific effects of halogens emitted to the atmosphere are discussed by Krueger and co-authors, while Jones covers the environmental effects of ash deposition. Schaal and co-authors describe the evidence for ocean anoxia and its relation to volcanism, specifically during the end-Permian, and, Cui, Kump and Ridgwell present models of ocean acidification induced by carbon release from various sources including volcanic $CO_2$, and the link to the end-Permian extinction. The book is completed by Black and co-authors, who evaluate proposed environmental effects using a global model of atmospheric chemistry and climate for the end-Permian.

The interactions that occur between volcanism and the environment on global scales are numerous and complex. Past and present eruptions provide natural experiments on the environmental impact of volcanism that could never be created in the laboratory. Flood basalt volcanism associated with mass extinctions, in particular, offers end-member constraints on the extent to which Earth's ecosystems can adapt to an abrupt shift in atmospheric composition. The compositions of some past volcanically released gases are frighteningly evocative of anthropogenic emissions, and these episodes of past volcanism may offer clues to the ways that humankind is currently affecting the Earth's ecosystems. Ultimately, volcanic processes from the present day and geologic record provide scenarios through which we may begin to understand the implications of anthropogenic activities, such as fossil-fuel burning, natural-resource utilization and landscape modification.

# Part One

Large volume volcanism: origins, features and timing

# 1

# Large igneous provinces and explosive basaltic volcanism

INGRID UKSTINS PEATE AND LINDA T. ELKINS-TANTON

## 1.1 Introduction

Large igneous provinces are recognized from the Precambrian at 3.79 Ga (Ernst, 2013), and extend through well-preserved examples from the Mesozoic and Cenozoic (Ross *et al.*, 2005; Bryan and Ferrari, 2013, and references therein). While originally inferred to consist of a layer-cake sequence of massive and laterally continuous effusive basaltic lava flows, detailed volcanostratigraphy studies have generated a more nuanced view of province architecture, highlighting that many provinces include a significant component of clastic material derived from volcanic and sedimentary formation mechanisms. Conversely, some of the volumetrically largest basaltic volcaniclastic deposits appear to be associated with large igneous provinces (Ross *et al.*, 2005).

The importance of volcaniclastic deposits – and the implications for paleoenvironmental reconstructions, eruption dynamics, and climate impact – is one of the key concepts to emerge from scientific studies of large igneous provinces over the last 25 years. Ross *et al.* (2005) recognized, and highlighted, the near-ubiquitous occurrence of mafic volcaniclastic deposits as an integral component in large igneous provinces. These deposits contain information – some unique – on primary fragmentation mechanisms, eruptive processes, and depositional environments. Mafic volcaniclastic deposits provide a record of what we now recognize as complex temporal and spatial volcanic heterogeneity in large igneous provinces, and allow us to reconstruct their tectonic and physical evolution as an equally significant and complementary story to that of the geochemical evolution of magmatism. We provide a brief overview of mafic volcaniclastic deposits and formation mechanisms, and spotlight recent work highlighting their utility for interpreting large-scale tectonic evolution and climate impact issues related to large igneous province emplacement.

*Volcanism and Global Environmental Change*, eds. Anja Schmidt, Kirsten E. Fristad and Linda T. Elkins-Tanton.
Published by Cambridge University Press. © Cambridge University Press 2015.

## 1.2 Mafic volcanic-derived clastic deposits

Clastic deposits composed of mafic volcanic particles – in any proportion from partly to entirely – can be generated by a wide variety of mechanisms spanning the full range of volcanic to sedimentary processes, and the resultant textures and morphologies are likewise highly variable. Three genetic categories of mafic volcanic-derived clastic deposits, based on formation mechanisms, are *primary* and *reworked volcaniclastic deposits*, and *epiclastic deposits* (White and Houghton, 2006).

### 1.2.1 Primary volcaniclastic deposits

Primary volcaniclastic deposits are formed from fragmental material deposited as a direct result of explosive or effusive eruptions. There are four end-member types of deposits: autoclastic, pyroclastic, hyaloclastic, and peperitic (White and Houghton, 2006; White *et al.*, 2009). The main factors controlling formation of these deposits are magma eruption rates, concentration of magmatic volatiles, and presence and relative abundance of external water, either as freestanding bodies or in saturated sediments. Mobilization of magma-generated particles is unique compared to other sedimentation processes that depend exclusively on gravity, because primary volcanic particles may acquire transport energy from their source – e.g. explosive expansion, lava flow velocity – and may initially be independent of slope or depositional base level (Fisher and Smith, 1991).

*Autoclastic* deposits are products of auto brecciation, and are generated as effusive lavas that exceed the viscosity-strain rate threshold and fragment (Peterson and Tilling, 1980); rapid cooling promotes groundmass crystallization and crust disruption (Cashman *et al.*, 1999). A'a lava flow morphology is characterized by brecciated upper and lower surfaces; slabby or rubbly pahoehoe have broken or brecciated upper crusts and are transitional between pahoehoe and a'a (e.g. Guilbaud *et al.*, 2005).

*Pyroclastic* deposits form from explosive volcanic eruption plumes and jets or pyroclastic density currents and can be generated by magmatic or phreatomagmatic fragmentation mechanisms, or a complicated interplay of both (e.g. Graettinger *et al.*, 2013). Magmatic fragmentation represents a minor mechanism for generating mafic volcaniclastic deposits in large igneous provinces, and documented examples are rare. The Columbia River Basalt vent sites in the Roza Member have densely agglutinated and welded spatter and highly vesicular scoria fall deposits (Brown *et al.*, 2014).

Involvement of aquifer or surface water in mafic eruptions leads to large-scale phreatomagmatic volcanism, and can generate deposits with volumes of up to

$10^2$ to $10^5$ km$^3$ (Ross *et al.*, 2005). Phreatomagmatic pyroclastic and hyaloclastic deposits, along with peperite, represent the full spectrum of products from magma–water interaction (Wohletz, 2002). External water is integral for formation, but magmatic volatiles are not precluded, and in the case of phreatomagmatic eruptions, may also play a role in fragmentation. Clast-forming processes during hydromagmatism include four primary mechanisms: magmatic explosivity, steam explosivity, cooling-contraction granulation, and dynamic stressing; all are dependent on the magma to water ratio (Wohletz, 1983; Kokelaar, 1986). Phreatomagmatic pyroclastic deposits are produced by the optimal fuel (magma) to coolant (water or sediment-laden water) mixture to generate explosivity (magma to pure water mass ratio of ~ 0.33: White, 1996), whereas hyaloclastic deposits are volumetrically dominated by water and peperite dominated by wet sediment (wet sediment to magma mass ratios > 1: Wohletz, 2002).

*Hyaloclastic* deposits are solely generated by quench fragmentation during magma–water interaction, and result from effusive magma contacting abundant water, in either marine or continental settings. Pillow lavas, pillow–palagonite breccias, and hyaloclastites are the most typical products of mafic magma quenching and spalling in a subaqueous environment.

*Peperite* deposits result from magma interaction with unconsolidated, water-bearing clastic deposits in shallow intrusions, subaqueous or surface environments (White *et al.*, 2000; Skilling *et al.*, 2002). Experimental and theoretical studies suggest that mechanisms of magma–water interaction and magma–sediment–water interaction may be similar (Kokelaar, 1986), and peperites rely on fluidization and vigorous injection and mixing of water-saturated sediments and lava (Kokelaar, 1982). Recognition of peperite indicates contemporaneity of sedimentation and volcanism (Busby-Spera and White, 1987). Given the ubiquity of environments that could generate peperite, there are relatively few documented examples in large igneous provinces. However, this may be a function of identification rather than absence. Even in the predominantly arid desert environment that the Paraná–Etendeka Large Igneous Province was emplaced into, peperites formed where pahoehoe lava flows interacted with wet lacustrine silts and clays, which collected in low-lying topography of lava flow surfaces (Waichel *et al.*, 2007).

### 1.2.2 Reworked volcaniclastic and epiclastic deposits

*Reworked* volcaniclastic deposits are composed of particles sourced from primary volcaniclastic deposits that have been redeposited by surface processes (wind, water, ice, gravity) either concurrent with eruption or after a period of immobility. In reworked volcaniclastic deposits, the volcanic processes that create the particles are not the same as those that transport the particles to their final depositional site. *Epiclastic* (or volcanogenic) sediments are formed from weathering and erosion of volcanic rocks, including previously lithified volcaniclastic rocks. Lithification can occur as part of volcanic emplacement (e.g. welding) or as a secondary process of cementation or compaction.

### 1.3 Spatial and temporal occurrence of mafic volcaniclastic deposits

One of the strengths of mafic volcaniclastic deposits is in the record they preserve of tectono-volcanic facies and province architecture evolution over time. The recognition that pre-volcanic kilometer-scale doming is not an unequivocal feature of large igneous provinces, coupled with recent numerical modeling indicating that large igneous province emplacement can generate substantial and complex patterns of pre- and syn-volcanic subsidence and/or uplift (+/− hundreds to thousands of meters: e.g. Czamanske *et al.*, 1998; Ukstins Peate and Elkins-Tanton, 2009; Elkins-Tanton and Ukstins Peate, 2010; Sobolev *et al.*, 2011), suggests that tectonic evolution may be a significant factor controlling the broad-scale distribution of these deposits. Provinces that contain significant volcaniclastics include the middle-Jurassic Kirkpatrick section of the Ferrar flood basalts in Antarctica, with tuff-breccias up to 400 m thick (Elliot and Flemming, 2008); the Kachchh region in the northwest of the Deccan flood basalts, with lapilli and lithic blocks (Kshirsagar *et al.*, 2011); and the Karoo (McClintock *et al.*, 2008). Here, we briefly describe three additional significant examples: the North Atlantic, the Emeishan and the Siberian.

### *1.3.1 East Greenland, North Atlantic Igneous Province*

Detailed volcanostratigraphic studies in East Greenland illustrate a cyclicity of phreatomagmatism and subsidence during the initial stages of province emplacement, recording three phases of subaqueous to subaerial volcanism, with progressively less hydrovolcanic influence (and inferred downdropping) in each cycle (Figure 1.1; Ukstins Peate *et al.*, 2003). Initiation of volcanism is represented by subaerially deposited phreatomagmatic lapilli-tuffs with accretionary lapilli and abundant quartz and feldspar grains (~ 50%) sourced from underlying upper shoreface sandstones and mid-Paleocene fluvial clastic deposits (Larsen *et al.*, 2003). Overlying these are a series of hyaloclastites and pillow lavas, some forming foreset-bedded units > 300 m thick (Nielsen *et al.*, 1981), suggesting that water depth increased dramatically with the initiation of basaltic volcanism. Hydromagmatic deposits transition to 500 m of compound lava flows, and the entire volcanic succession forms a shield-like structure with a diameter of ~ 40 km (Ukstins Peate *et al.*, 2003). This is, in turn, overlain by a sequence of mafic volcaniclastic deposits

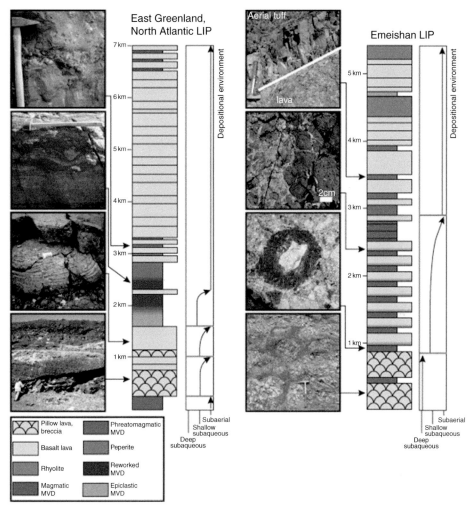

Figure 1.1 Generalized composite stratigraphy of East Greenland, North Atlantic, and Emeishan, China, large igneous provinces (LIPs). While package thicknesses are representative, individual unit thicknesses are schematic, and *x*-axis widths emphasize distinctions in rock types. MVD = mafic volcaniclastic deposit. Intrusive rocks are not shown. East Greenland based on Ukstins Peate *et al.*, (2003), and Emeishan based on Ukstins Peate and Bryan (2008, 2009) and Zhu *et al.* (2014). A black and white version of this figure will appear in some formats. For the colour version, please refer to the plate section.

that preserve a lateral facies change from primary units – including vent sites – in the northeast to reworked volcaniclastic and epiclastic deposits to the southwest.

Primary deposits consist of *c*. 300 m of: fallout tuffs; surge deposits with abundant accretionary and armored lapilli; bomb beds; and scoria deposits with three-dimensional cone morphology (Ukstins Peate *et al.*, 2003). These transition

to 1000 m of reworked and epiclastic deposits of siltstone and sandstone containing up to 80% volcanic material: altered basaltic glass (tachylite, palagonite),
clinopyroxene crystals, basaltic lava clasts, and pyroclastic lithic fragments,
with minor intercalated tuffs. Correlation of reworked and epiclastic deposits
highlight the development of regional syn-volcanic basins with cumulative thicknesses > 3000 m (Larsen *et al.*, 2003; Passey and Bell, 2007).

   Overlying this is the main phase of flood basalt lavas, with a few thin magmatic
tuffs containing Pele's tears and glass shards, concentrated in the lowermost part
of the sequence (Ukstins Peate *et al.*, 2003). A final transition from effusive flood
lava to highly explosive basaltic phreato-Plinian eruptions occurs in the uppermost
sequence, when active lithospheric rifting and subsidence resulted in flooding
of the nascent North Atlantic Rift proto-ocean basin (Larsen *et al.*, 2003; Jolley
and Widdowson, 2005).

### 1.3.2 Emeishan large igneous province

Research on the Emeishan large igneous province highlights the utility of mafic
volcaniclastic deposits in addressing questions of large-scale tectonic evolution
during flood volcanism. A thick and laterally extensive wedge of clastic deposits
(170 m thick, 30 to 80 km wide, 400 km long), emplaced near the base of the
Emeishan lavas, was initially interpreted as an alluvial fan conglomerate, and was
attributed to pre-volcanic, kilometer-scale domal uplift and erosion of underlying
carbonate (He *et al.*, 2003). The ubiquity of dense to poorly vesicular blocky
sideromelane, pyroclastic textures such as accretionary lapilli, volcanic bombs
with bomb sags, and ductile deformation of mafic clasts unequivocally identifies
these rocks as phreatomagmatic lapilli-tuffs and tuff-breccias, and likely represent near-vent deposits (Figure 1.1; Ukstins Peate and Bryan, 2008, 2009). The
abundance of marine limestone lithic fragments – some containing mafic clasts
themselves – and the presence of unbound shelly fossil material, strongly suggests
that active carbonate deposition was contemporaneous with volcanism, and that
these units were emplaced near sea level (Ukstins Peate and Bryan, 2008, 2009).

   Continuing work, focusing on the zone of inferred maximum uplift, has
identified a protracted and extensive record of hyaloclastic and phreatomagmatic
volcanism (Figure 1.1). Microfossil studies show nascent carbonate platform
collapse immediately prior to initiation of volcanism (> 200 m: Sun *et al.*,
2010). The first phase of volcanism is laterally heterogeneous but dominated by
phreatomagmatic and subaqueous volcanism. Eruptions through shallow-water
carbonates generated thin subaqueous hyaloclastites and subaerial tuff deposits
near Daiquo (Ukstins Peate and Bryan, 2008), whereas in the Dali area (the core of
inferred maximum uplift), volcanism initiated with a succession (*c.* 750 m) of

pillow lavas and hyaloclastites with intercalated marine limestones and submarine tuffs (Zhu *et al.*, 2014). Eruptions transitioned to phreatomagmatic lapilli-tuffs and tuff-breccias intercalated with basaltic lava sheet flows displaying peperitic basal zones and carbonates (*c.* 2000 m: Zhu *et al.*, 2014), suggesting a very shallow subaqueous to subaerial depositional environment. This is overlain by > 2500 m of thick a'a and pahoehoe basalts and rhyolite lavas, intercalated with minor, thin (~ 1 m), oxidized basaltic tuffs dominated by glassy vesicular ash shards (Zhu *et al.*, 2014), likely derived from subaerial phreatomagmatic to magmatic pyroclastic eruptions.

### *1.3.3 Siberian flood basalts*

The Siberian flood basalts contain intercalated volcaniclastics to varying extent throughout the most studied sections of that province, for example in Noril'sk (e.g. Fedorenko *et al.*, 1996). A vast literature on the Siberian province exists, but here we focus on the understudied volcaniclastics and present some new results. The most significant volcaniclastics in the Siberian province are the thick, primarily phreatomagmatic deposits underlying the lavas. In the northeast and northwest sections the majority of the basal volcaniclastics are less than 30 m in thickness, and are sometimes absent (Figure 1.2). In the central, eastern and southern regions, however, the volcaniclastics are voluminously present in largely massive, featureless outcrops.

Along almost 200 km of the Angara River north of Ust Ilim'sk, all the river cliffs consist of volcaniclastics, and visible outcrops are as much as 250 m thick, with erosional upper and unexposed basal contacts (Naumov and Ankudimova, 1995). Volcaniclastic units are massive, unbedded and sediment-rich, though near the Kata River there is local bedding and accretionary lapilli. Some outcrops have lithic blocks of underlying sedimentary strata; peperites and sediment dikes indicate an active aquifer and driving force for eruption from depth. Notably absent are pillow basalts and hyaloclastites.

Similar deposits occur along 200 km of the Nizhnaya Tunguska River, stretching east–west past the middle Siberian town of Tura. In Tura, drill cores indicate at least 500 m of tuffs transitioning to overlying effusive lavas (Drenov, 1985). These drill cores demonstrate that voluminous phreatomagmatism immediately preceded the main stage of effusive lava emplacement.

## 1.4 Evidence for volatile loads, temperatures and plume heights

Chemicals released by volcanism will have the greatest effect on global climate, both in terms of destructive chemical reactions and longevity, if they reach the

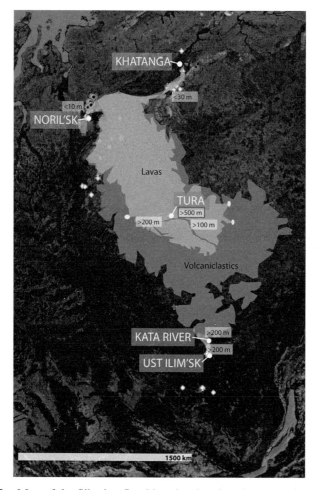

Figure 1.2   Map of the Siberian flood basalts showing the approximate location
of uppermost bedrock lavas and underlying volcaniclastics (after Svensen *et al.*,
2009). Sample locations from Elkins-Tanton's field trips are shown as scattered
dots. Thicknesses indicate outcrops seen on these trips, with the exception of the
Tura dill core (> 500 m), from Drenov (1985). The samples north of Ust Ulim'sk
are on the Angara River, and those east and west of Tura are on the Nizhnaya
Tungusska River.

stratosphere. Material is rapidly washed from the troposphere by rain. Basaltic
magmas are generally less gas-rich (with the possible exception of sulfur) and less
viscous than more silicic eruptions, and are generally less explosive without
interactions with external volatiles. However, basaltic Hawaiian-style fire fountains are capable of injecting material into the stratosphere (Stothers *et al.*, 1986;
Woods, 1993). This is corroborated by the Laki eruption (Iceland 1783–1784,
Thordarson *et al.*, 1996). Laki was largely effusive, but it had significant

fire-fountaining episodes. Witness accounts describe an eruption column more than 9 km high (Thordarson and Self, 2003), and the detection of Laki chemicals in Greenland ice cores confirms that material reached the stratosphere (Fiacco *et al.*, 1994; Wei *et al.*, 2008).

Basaltic eruptions that produce volcaniclastic deposits, however, may have more capacity to implant material into the stratosphere, particularly at northern latitudes with a low-altitude tropopause (~ 9 km near the poles compared to 17 km at the equator, but sensitive to a wide variety of external parameters (Thuburn and Craig, 1997; see also Ross *et al.*, 2008)). Basaltic volcaniclastic deposits are commonly phreatomagmatic, and incorporation of water into an eruptive plume can have a range of effects. If the water vaporizes, the density of the plume decreases with addition of steam, but the temperature also decreases through the consumption of latent heat. The decrease of density means a higher plume may develop with a lower eruption velocity.

Conversely, if added water cannot vaporize, the eruptive fountain will collapse and flow laterally (Koyaguchi and Woods, 1996). As phreatomagmatic basaltic eruptions proceed, they may transition between these states. Koyaguchi and Woods (1996) suggest that accretionary lapilli can form in both the fountains and wet lateral flows, and find that for an initial eruptive velocity of 100 m/s, a temperature of 1000 K, and a volatile content of 3 wt% water, a plume may reach as high as 35 km, at which point its temperature would be below 300 K. In agreement, Walker *et al.* (1984) estimate a vent exit velocity of 250–350 m/s for the basaltic plinian Tarawera eruption of 1886, which generated a column of ash that ascended ~ 30 km.

While individual eruptions may not always reach these conditions, the broad thermal perturbation of a flood basalt province may produce its own weather pattern of thermals in the atmosphere. This concept was first investigated by Emanuel *et al.* (1995), in which they posited that both very large bolide strikes and large-scale volcanic eruptions could produce exceptionally violent storms termed *hypercanes*, capable of injecting large amounts of mass into the stratosphere. More recent work by Kaminski *et al.* (2011) described penetrative convection above large lava flows, where broad temperature perturbations drive large upwellings past the tropopause.

## 1.5 Summary: potential for climate change

Mafic volcaniclastics make up a significant fraction of large igneous province eruptive volume, including in the Siberian, Emeishan, North Atlantic, Karoo, Ferrar, and Columbia River flood basalts. The type and distribution of volcaniclastics can determine the relative impacts of volcanism and tectonism on a region, by filling pre-effusive topography in some cases, and by tracking surface or ground water in others.

Globally, volcaniclastic eruptions are now recognized to have potential for climate-changing atmospheric effects. Their eruptions can inject material into the stratosphere, either from their eruptive plume or with the help of regional weather effects produced by the large igneous province itself.

A significant next step for the scientific community will be estimating the volatile load of these eruptions. Some efforts have been made in Siberia, and demonstrate that Siberian volcaniclastics carried significant sulfur, chlorine and fluorine from the fluids in their eruption-driving aquifer (Black *et al.*, 2012). Further, these eruptions naturally produced halocarbons sufficient to destroy ozone (Black *et al.*, 2013) through reaction in the bedrock and the eruptive plume (Svensen *et al.*, 2009; Black *et al.*, 2014, this volume).

The volcaniclastics in flood basalts, therefore, may be the major missing link between flood basalts and extinctions. The underlying cause may be climate-changing volatiles sourced from continental crustal rocks that chamber the flood basalt magmas, which are missing in ocean basins.

## References

Black, B.A., Elkins-Tanton, L.T., Rowe, M.C. and Ukstins Peate, I. (2012). Magnitude and consequences of volatile release from the Siberian Traps. *Earth and Planetary Science Letters*, **317–318**, 363–373.

Black, B.A., Lamarque, J.-F., Shields, C., Elkins-Tanton, L. and Kiehl, J. (2013). Acid rain and ozone depletion from pulsed Siberian Traps magmatism. *Geology*, **42**, 67–70.

Black, B.A., Lamarque, J.-F., Shields, C., Elkins-Tanton, L., and Kiehl, J. (2015). Environmental effects of large igneous province magmatism: a Siberian perspective, this volume.

Brown, R.J., Blake, S., Thordarson, T. and Self, S. (in press). Pyroclastic edifices record vigorous lava fountains during the emplacement of a flood basalt flow field, Roza Member, Columbia River basalt province, U.S.A. *Geological Society of America Bulletin*, **126**, 265–288.

Bryan, S.E. and Ferrari, L. (2013). Large igneous provinces and silicic large igneous provinces: progress in our understanding over the last 25 years. *Geological Society of America Bulletin*, **125**, 1053–1078.

Busby-Spera, C.J. and White, J.D.L. (1987). Variation in peperite textures associated with differing host-sediment properties. *Bulletin of Volcanology*, **49**, 765–775.

Cashman, K.V., Thornber, C. and Kauahikaua, J.P. (1999). Cooling and crystallization of lavas in open channels, and the transition of pahoehoe lava to a'a. *Bulletin of Volcanology*, **61**, 306–323.

Czamanske, G.K., Gurevitch, A.B., Fedorenko, V. and Simonov, O. (1998). Demise of the Siberian plume: paleogeographic and paleotectonic reconstruction from the pre-volcanic and volcanic record, north-central Siberia. *International Geology Reviews*, **40**, 95–115.

Drenov, N.V., ed. (1985). Federal Geological Map USSR, Tunguska Series, Sheets Q-47 XXXV, XXXVI, compiled by Levitan, M.M. and Zastoina, A.N., Moscow (in Russian).

Elkins-Tanton, L.T. and Ukstins-Peate, I. (2010). On topographic subsidence and magma bursts at initiation of large igneous provinces. Japan Geoscience Union meetings, Tokyo, May 2010.

Elliot, D.H. and Flemming, T.H. (2008). Physical volcanology and geological relationships of the Jurassic Ferrar large igneous province, Antarctica. *Journal of Volcanology and Geothermal Research*, **172**, 20–37.

Emanuel, K.A., Speer, K., Rotunno, R., Srivastava, R. and Molina, M. (1995). Hypercanes: a possible link in global extinction scenarios. *Journal of Geophysical Research*, **100**, 13755–13765.

Ernst, R.E. (2014). Large Igneous Provinces. Cambridge: Fedorenko, V.A., Lightfoot, P.C., Naldrett, A.J. *et al.* (1996). Petrogenesis of the flood-basalt sequence at Noril'sk, north-central Siberia. *International Geology Review*, **38**, 99–135.

Fiacco, R.J., Thordarson, T., Germani, M.S. *et al.* (1994). Atmospheric aerosol loading and transport due to the 1783–84 Laki eruption in Iceland, interpreted from ash particles and acidity in the GISP2 ice core. *Quaternary Research*, **42**, 231–240.

Fisher, R.V. and Smith, G.A. (1991). Volcanism, tectonics and sedimentation. *SEPM Special Publication*, **45**, 1–5.

Guilbaud, M-N., Self, S., Thordarson, T. and Blake, S. (2005). Morphology, surface structures, and emplacement of lavas produced by Laki, A. D. 1783–1784. *Geological Society of America Special Paper*, **396**, 81–102.

Graettinger, A.H., Skilling, I., McGarvie, D. and Hoskuldsson, A. (2013). Subaqueous basaltic magmatic explosions trigger phreatomagmatism: a case study from Askja, Iceland. *Journal of Volcanology and Geothermal Research*, **264**, 17–35.

He, B., Xu, Y-G., Chung, S-L. and Wang, Y. (2003). Sedimentary evidence for a rapid crustal doming before the eruption of the Emeishan flood basalts. *Earth and Planetary Science Letters*, **213**, 389–403.

Jolley, D.W. and Widdowson, M. (2005). Did Paleogene North Atlantic Rift-related eruptions drive early Eocene climate cooling? *Lithos*, **79**, 355–366.

Kaminski, E., Chenet, A.-L., Jaupart, C. and Courtillot, V. (2011). Rise of volcanic plumes to the stratosphere aided by penetrative convection above large lava flows. *Earth and Planetary Science Letters*, **301**, 171–178.

Kokelaar, B.P. (1982). Fluidization of wet sediments during the emplacement and cooling of various igneous bodies. *Journal of the Geological Society of London*, **139**, 21–33.

Kokelaar, B.P. (1986). Magma water interactions in subaqueous and emergent basaltic volcanism. *Bulletin of Volcanology*, **48**, 275–289.

Koyaguchi, T. and Woods, A.W. (1996). On the formation of eruption columns following explosive mixing of magma and surface-water. *Journal of Geophysical Research*, **101**, 5561–5574.

Kshirsagar, P.V., Sheth, H.C. and Shaikh, B. (2011). Mafic alkali magmatism in central Kachchh, India: a monogenetic volcanic field in the northwestern Deccan Traps. *Bulletin of Volcanology*, **73**, 595–612.

Larsen, L.M., Fitton, J.G. and Pedersen, A.K. (2003). Paleogene volcanic ash layers in the Danish Basin: compositions and source areas in the North Atlantic Igneous Province. *Lithos*, **71**, 47–80.

McClintock, M., Marsh, J.S. and White, J.D.L. (2008). Compositionally diverse magmas erupted close together in space and time within a Karoo flood basalt crater complex. *Bulletin of Volcanology*, **70**, 923–946.

Naumov, V.A. and Ankudimova, L.A. (1995). Palynological complexes and age of volcanic sediments of the Angara–Katanga District, Middle Angara Region. *Geology and Geophysics*, **36**, 39–45 (in Russian).

Nielsen, T.D.F., Soper, N.J., Brooks, C.K. *et al.* (1981). The pre-basaltic sediments and the lower basalts at Kangerlussuaq, East Greenland: their stratigraphy, lithology, palaeomagnetism and petrology. *Meddelelser om Grønland, Geosciences*, **6**, 1–25.

Passey, S.R. and Bell, B.R. (2007). Morphologies and emplacement mechanisms of the lava flows of the Faroe Islands Basalt Group, Faroe Islands, NE Atlantic Ocean. *Bulletin of Volcanology*, **70**, 139–156.

Peterson, D.W. and Tilling, R.I. (1980). Transition of basaltic lava from pahoehoe to a'a, Kilauea Volcano, Hawaii: field observations and key factors. *Journal of Volcanology and Geothermal Research*, **7**, 271–293.

Ross, P.-S., Ukstins Peate, I.A., McClintock, M.K. et al. (2005). Mafic volcaniclastic deposits in flood basalt provinces: a review. *Journal of Volcanology and Geothermal Research*, **145**, 281–314.

Ross, P.-S., McClintock, M.K. and White, J.D.L. (2008). Geological evolution of the Coombs-Allan Hills area, Ferrar large igneous province, Antarctica: debris avalanches, mafic pyroclastic density currents, phreatocauldrons. *Journal of Volcanology and Geothermal Research*, **172**, 38–60.

Skilling, I., White, J. and McPhie, J. (2002). Peperite: a review of magma-sediment mingling. *Journal of Volcanology and Geothermal Research*, **114**, 1–17.

Sobolev, S.V., Sobolev, A.V., Kuzmin, D.V. *et al.* (2011). Linking mantle plumes, large igneous provinces and environmental catastrophes. *Nature*, **477**, 312-380.

Stothers, R.B., Wolff, J.A., Self, S. and Rampino, M.R. (1986). Basaltic fissure eruptions, plume heights, and atmospheric aerosols. *Geophysical Research Letters*, **13**, 725–728.

Sun, Y., Lai, X., Wignall, P.B. *et al.* (2010). Dating the onset and nature of the Middle Permian Emeishan large igneous province eruptions in SW China using conodont biostratigraphy and its bearing on mantle plume uplift models. *Lithos*, **119**, 20–33.

Svensen, H., Planke, S., Polozov, A.G. *et al.* (2009). Siberian gas venting and the end-Permian environmental crisis. *Earth and Planetary Science Letters*, **277**, 490–500.

Thordarson, T. and Self, S. (2003). Atmospheric and environmental effects of the 1783–1784 Laki eruption: a review and reassessment. *Journal of Geophysical Research*, **108**, 4011.

Thordarson, T., Self, S., Óskarsson, N. and Hulsebosch, T. (1996). Sulfur, chlorine, and fluorine degassing and atmospheric loading by the 1783–1784 Laki (Skaftár Fires) eruption in Iceland. *Bulletin of Volcanology*, **58**, 205–225.

Thuburn, J. and Craig, G.C. (1997). GCM tests of theories for the height of the tropopause. *Journal of Atmospheric Science*, **54**, 869–882.

Ukstins Peate, I.A. and Bryan, S.E. (2008). Re-evaluating plume-induced uplift in the Emeishan large igneous province. *Nature Geoscience*, **1**, 625–629.

Ukstins Peate, I.A. and Bryan, S.E. (2009). Re-evaluating plume-induced uplift in the Emeishan large igneous province: Reply. *Nature Geoscience*, **2**, 531–532.

Ukstins Peate, I.A. and Elkins-Tanton, L.T. (2009). On topographic subsidence and magma bursts at initiation of magmatic provinces. American Geophysical Union Abstracts, San Francisco, December 2009.

Ukstins Peate, I.A., Larsen, M. and Lesher, C.E. (2003). The transition from sedimentation to flood volcanism in the Kangerlussuaq Basin, East Greenland: basaltic pyroclastic volcanism during initial Palaeogene continental break-up. *Journal of the Geological Society, London*, **160**, 759–772.

Waichel, B.L., de Lima, E.F., Sommer, C.A. and Lubachesky, R. (2007). Peperite formed by lava flows over sediments: an example from the central Paraná continental flood basalts, Brazil. *Journal of Volcanology and Geothermal Research*, **159**, 343–354.

Walker, G.P.L., Self, S. and Wilson, L. (1984). Tarawera 1886, New Zealand – a basaltic plinian fissure eruption. *Journal of Volcanology and Geothermal Research*, **21**, 61–78.

Wei, L.J., Mosley-Thompson, E., Gabrielli, P., Thompson, L.G. and Barbante, C. (2008). Synchronous deposition of volcanic ash and sulfate aerosols over Greenland in 1783 from the Laki eruption (Iceland). *Geophysical Research Letters*, **35**, L16501.

White, J.D.L. (1996). Impure coolants and interaction dynamics of phreatomagmatic eruptions. *Journal of Volcanology and Geothermal Research*, **74**, 155–170.

White, J.D.L. and Houghton, B.F. (2006). Primary volcaniclastic rocks. *Geology*, **34**, 677–680.

White, J.D.L., McPhie, J. and Skilling, I.P. (2000). Peperite: a useful genetic term. *Bulletin of Volcanology*, **62**, 65–66.

White, J.D.L., Bryan, S.E., Ross, P.-S., Self, S. and Thordarson, T. (2009). Physical volcanology of continental large igneous provinces: update and review. *IAVCEI Special Publication*, **2**, 291–321.

Wohletz, K.H. (1983). Mechanisms of hydrovolcanic pyroclast formation: grain size, scanning electron microscopy and experimental results. *Journal of Volcanology and Geothermal Research*, **17**, 31–63.

Wohletz, K.H. (2002). Water/magma interaction: some theory and experiments on peperite formation. *Journal of Volcanology and Geothermal Research*, **114**, 19–35.

Woods, A.W. (1993). A model of the plumes above basaltic fissure eruptions. *Geophysical Research Letters*, **20**, 1115–1118.

Zhu, B., Guo, Z., Liu, R. and Du, W. (2014). No pre-eruptive uplift in Emeishan large igneous province: new evidence from its 'inner zone', Dali area, Southwest China. *Journal of Volcanology and Geothermal Research*, **269**, 57–67.

# 2

# On the nature and consequences of super-eruptions

CLIVE OPPENHEIMER AND AMY DONOVAN

## 2.1 Introduction: what are super-eruptions and super-volcanoes?

Super-eruptions have been described as 'the ultimate geologic hazard' (Self and Blake, 2008). Volcanological use of the term 'super-eruption' can be traced at least back to the title of a 1992 paper on the 74 ka Youngest Toba Tuff eruption in northern Sumatra (Rampino and Self, 1992). However, it is the British Broadcasting Corporation (BBC) that can be credited for the proliferation of the terminology (following a 2000 broadcast), both in the scientific sphere and beyond. While not all volcanologists like the term 'super-eruption' (perhaps considering it is being used more for 'convenience or effect,' Wilson and Charlier, 2009), it is broadly understood (e.g. Sparks *et al.*, 2005) to define a pyroclastic eruption of magnitude, *M*, of 8 or above, where (Pyle, 2000):

$$M = \log_{10}(\text{mass of erupted material in kg}) - 7$$

This threshold thus represents a mass of $10^{15}$ kg. Taking a pumice density of $1000$ kg m$^{-3}$, the bulk volume of a super-eruption deposit would exceed 1000 km$^3$, corresponding to a dense magma volume of approximately 450 km$^3$. While silicic lava (effusive) eruptions can be very large, none approaching this volume has yet been identified. The term 'super-volcano' may be just as contentious as 'super-eruption' but it is also prevalent in the scientific literature, and we consider the definition proposed by Miller and Wark (2008) appropriate: a 'super-volcano' is a volcano associated with one or more super-eruptions.

Notwithstanding the damping effects of the logarithmic term in Equation (2.1), when it comes to reporting eruption magnitudes, it is important to note the considerable uncertainty attached to estimates of eruption volumes or masses. Estimates for super-eruption deposits are typically based on mapping or estimating the thickness

*Volcanism and Global Environmental Change*, eds. Anja Schmidt, Kirsten E. Fristad and Linda T. Elkins-Tanton.
Published by Cambridge University Press. © Cambridge University Press 2015.

of pyroclastic (fall and current) deposits over a sufficiently representative geographic range. Then, estimates of relevant densities of the diverse materials are required to convert to mass (as required in Equation (2.1)). For distal tephra fall deposits, it may be feasible to make direct measurements of mass per unit area. In all cases, an extrapolation or model will be required to account for unmapped material, whether it be the very thin (and potentially unpreserved) ultra-distal ash fall deposits, or proximal ignimbrite with unseen bases resting on unknown pre-eruption topography (e.g. Bonadonna and Costa, 2013; Burden *et al.*, 2013). Many of the published estimates of super-eruption magnitudes are based on limited deposit data and/or 'back of the envelope' calculations of the volumes of intra-caldera deposits, outflow sheets and tephra fallout (see, e.g., Rose and Chesner, 1990; Matthews *et al.*, 2012; Gatti and Oppenheimer, 2013; and Costa *et al.*, 2014 for a digest of estimates of the magnitude of the Youngest Toba Tuff eruption).

### *2.1.1 Collapse calderas*

Super-eruptions are typically caldera-forming events, given the volumes of magma that are expelled from the crust. Thus, the most distinctive feature of a super-volcano is often its sizeable caldera (e.g. Geyer and Marti, 2008). These are generally at least 20 km across; some are closer to 100 km in diameter, though such large structures are often complex resulting from super-imposition of craters formed during more than one episode of collapse (Lipman, 1997). The timing of collapse can be critical in the dynamics of very large eruptions since the ring fractures developed above the evacuating chamber can promote even higher magma discharge rates associated with prodigious discharge of pyroclastic currents at the surface. The depression formed can also end up accommodating a substantial fraction – as much as a third to a half – of the ejecta (Mason *et al.*, 2004). Many large calderas today feature a 'resurgent centre' that typically domes the intra-caldera deposits upwards. The precise origins of such structural uplift are uncertain but probably reflect a combination of magma reservoir recharge and volatile exsolution (Kennedy *et al.*, 2012; Wilcock *et al.*, 2013).

## 2.2 Magma bodies associated with super-volcanoes

Considering the great mass of magma disgorged by a super-eruption, and its exclusively silicic (dacitic through rhyolitic) petrological affinities, it is clear that the magma involved was assembled prior to eruption. A range of geochemical, mineralogical and geochronological studies indicate that the associated timescales of assembly of very large magma chambers may take anything from hundreds of thousands of years to just a few thousand years (e.g. Wilson and Charlier, 2009; Druitt *et al.*,

2012). In the case of long-lived crystal-rich magmas ('mushes') that lack the mobility to erupt, rejuvenation may occur as a result of intrusion by hotter mafic or intermediate magmas (e.g. Bachmann *et al.*, 2002; Wotzlaw *et al.*, 2013), potentially on short timescales (Burgisser and Bergantz, 2011). A related question concerns the particular circumstances that 'keep the lid on' a reservoir of buoyant magma, which may ultimately feed a super-eruption. While super-volcanoes seem consistently to be sited on thick, relatively low-density crust, which promotes accumulation of magmas at depth, super-volcanoes do not only produce super-eruptions – the massive events are interspersed with smaller intrusions, explosions and extrusions (see, e.g., Charlier *et al.*, 2005 on the 'hyperactive' Taupo volcano).

Radioactive-decay series have been widely used to constrain the evolutionary timescales of magma bodies. An important consideration in such studies is to discriminate between the timescales of the chemical processes that generated the silicic magmas responsible for super-eruptions, and the potentially more rapid timescales of assembly and rejuvenation of an eruptible reservoir (Allan *et al.*, 2013). For instance, rubidium–strontium isotope measurements on minerals and glasses from the Bishop Tuff indicate ages at least a million years older than the date of the eruption itself (e.g. Halliday *et al.*, 1989; Christensen and Halliday, 1996) suggesting a very long-lived magma system that was intermittently tapped during eruptions. Uranium–lead and uranium–thorium ages on zircon and allanite crystals isolated from tephra deposits have also revealed timescales of a few hundreds of thousands of years of crystallization prior to eruption. Vazquez and Reid (2004) found evidence for part of the magma body associated with the *M* 8.8 Youngest Toba Tuff being in place up to 150 ka before eruption. A geochemical and geochronological study of zircon crystals from the largest known super-eruption deposit, the 28 Ma, *M* 9 Fish Canyon Tuff (associated with La Garita caldera, Colorado), suggests a 440 ka magmatic history, a significant part of which was taken up by reheating of the pluton that had reached up to 80% crystallinity (Wotzlaw *et al.*, 2013).

### 2.2.1 Eruption triggers

Emerging evidence concerning the thermo-mechanics of magma chambers may give some meaning to the discrimination between super-eruptions and 'ordinary' eruptions. Caricchi *et al.* (2014) have modelled the thermal properties of host rocks and their response to repeated magma intrusions that incrementally inflate the chamber. They argue that *M* 6 eruptions tend to be triggered as a result of the overpressure that develops in smaller chambers in the upper crust as they are repeatedly supplied with magma. This behaviour reflects the elastic behaviour of the relatively cold wall rocks and their propensity to fail under tensile stresses. Such eruptions can be relatively frequent but only partially empty the chamber with each event. Over time, with a net magma supply from depth exceeding the time-averaged outputs, the chamber enlarges (e.g. Reid, 2008). With increasing volume, the wall rocks heat up and deform in a more viscous fashion such that the chamber overpressure reduces. As magma storage increases, rather than the overpressure triggering eruptions, the buoyancy of the magma (whose density is less than that of the host rocks) becomes the driving force for eruption. The critical condition for a super-eruption is then the thickness of the reservoir since it influences the competition between magma supply and cooling and crystallization (the latter process being enhanced in thinner, wider reservoirs).

This theoretical treatment might explain the inflection apparent in the magnitude–frequency relationship for global (terrestrial) volcanism between events of *M* 7 and *M* 8. Taking the statistics at face value, eruptions of size *M* 8 and above are rarer than extrapolation of the data for smaller eruptions would suggest (Figure 2.1). Further support for the hypothesis that buoyancy drives super-eruptions comes from sophisticated measurements of the density of silicic melts at realistic magmatic pressures and temperatures (Malfait *et al.*, 2014).

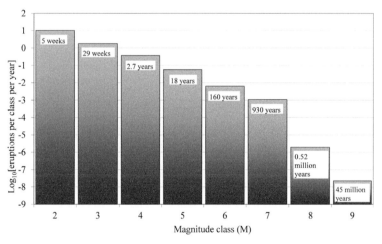

Figure 2.1    Frequency versus magnitude plot for volcanic eruptions based on records for last 300 years for eruption magnitudes *M* 2–6; for the last 2000 years for *M* 6–8; and for all known "super-eruptions" of the past 45 Ma (after Mason *et al.*, 2004). The average return period for each magnitude class is labelled on the plot. The only *M* 9 eruption in this compilation is the 28-Ma-old Fish Canyon Tuff (Colorado, USA). The available data are imperfect and it is probable that not all very great eruptions of the last few million years have yet been identified. Nevertheless, the apparent fall off in the frequency of very great eruptions (*M* 8–9) may be real and reflect a threshold between the contrasting driving forces of eruptions (e.g. Caricchi *et al.*, 2014).

## 2.3 Eruptive style

Are super-eruptions qualitatively different from lesser eruptions? To begin to address this question, we look first at the historical record. The largest eruption of the past century is that of Mount Pinatubo in 1991 (M 6.1). The climactic eruption of 15 June 1991 followed a crescendo in precursory activity over the preceding months. The most intense phase of the Plinian eruption lasted around 3.5 hours (Scott *et al.*, 1996) and was accompanied by the collapse of parts of the ash and gas plume, which generated substantial pyroclastic density currents and a co-ignimbrite ash cloud. A 2.5-km-wide crater was formed as a result of the eruption.

To get close to the *M* 7 class in the modern period for which we have some eyewitness accounts, we turn to the eruption of Tambora (Sumbawa island, Indonesia) in 1815 (*M* 6.9). From what we can infer from the deposits and the contemporary reports, the eruption was of a similar style to that of Pinatubo in 1991, with increasing eruption intensity heralding a collapse of a Plinian eruption column that generated voluminous pyroclastic density currents and associated co-ignimbrite plumes (Sigurdsson and Carey, 1989; Oppenheimer, 2003). Unfortunately, there are no clear indications of the duration of this paroxysmal phase, so it is difficult to say to what extent the greater size of Tambora relative to Pinatubo is due to intensity or duration; but it is probably a combination of both. A comparable *M* 7 eruption, of Rinjani volcano on the Indonesian island of Lombok, has been identified and dated to the mid-thirteenth century (Lavigne *et al.*, 2013).

While intensities and durations of lesser events are somewhat constrained, these parameters can only be inferred for eruptions of *M* 8 and above. In the case of the 0.76-Ma-old *M* 8.3 Bishop Tuff eruption (associated with Long Valley caldera, California) it has been estimated that the duration of the climactic eruption may have been less than a week (Wilson and Hildreth, 1997). Other super-eruptions may have involved intermittent bursts of activity that spanned a few years (Wilson, 2008), or possibly a few centuries (e.g. Ellis *et al.*, 2012; Svensson *et al.*, 2013).

### *2.3.1 Wet versus dry eruptions*

The most recent documented super-eruption is the *M* 8.1 Oruanui eruption of Taupo volcano in New Zealand, which took place *c*. 25.4 ka ago. Triggered and modulated by rift tectonics (Allan *et al.*, 2012), episodes of the eruption involved interaction between magma and lake water (Van Eaton and Wilson, 2013). As a consequence, some of the pyroclastic deposits are particularly fine-grained, reflecting the water-influenced magma fragmentation processes at the vent. The phase changes associated with the interaction of water with magma may impact on the dynamics of explosive eruptions, resulting in alternating phases of Plinian eruption and fountain collapse, and injection of ash into the atmosphere at multiple heights (Van Eaton *et al.*, 2012). Further research in this domain seems warranted since many super-volcanoes are presently occupied, at least partially, by bodies of water.

## 2.4  Where are the super-volcanoes and when did they erupt?

Mason *et al.* (2004) catalogued 42 *M* 8 or larger eruptions spanning the last 36 Ma. However, these events were strongly clustered in both time and space, with clusters between 36 and 27 Ma ago, and from 13.5 Ma ago up to the present. While tectonics may play a role in the triggering of super-eruptions (e.g. Allan *et al.*, 2012), super-volcanoes are found in association with mantle plumes, rifts and subduction zones. However, there are strong biases in the dataset that are readily evident from consideration of the geographic distribution of events – all but four of them are located in the Americas (32 in North America, six in the central Andes). The others are in Sumatra (Toba), New Zealand (Taupo) and the Ethiopian plateau. The more recently compiled 'large magnitude explosive volcanic eruptions' (LaMEVE) database lists over 20 *M* 8 and above eruptions spanning the past 2.2 Ma, suggesting a return period of around 100 ka (Crosweller *et al.*, 2012). A notable difference between the Mason *et al.* (2004) and LaMEVE databases is that the latter identifies six Quaternary super-volcanoes in Japan. However, it is likely that many other super-eruption deposits remain undiscovered or unrecognized (Donovan and Oppenheimer, 2014).

Recently, a case has been made for super-eruptions having occurred on Mars (Michalski and Bleacher, 2013), although the geomorphological evidence presented may not prove compelling for some volcanologists.

## 2.5  Products of super-eruptions

As we have seen, the deposits of super-eruptions can be far flung, yet also concentrated locally at the site of the associated caldera. This section provides an overview of these diverse materials.

### 2.5.1  Tephra fall

One of the astonishing aspects of super-eruption deposits is the areal extent of tephra fall, sourced largely – perhaps exclusively – from the co-ignimbrite plumes (e.g. Matthews *et al.*, 2012). For instance, Youngest Toba Tuff ash has been identified at several terrestrial sites in India; in many deep-sea sediment cores, some as far distant as the Arabian Sea, 4000 km away; and in cores taken from the bed of Lake Malawi (Lane *et al.*, 2013). Studies of the overlying and underlying deposits at such sites offer the potential to investigate the regional environmental impacts of the eruption. The minimum estimate of the mass of the tephra fall-out is $2 \times 10^{15}$ kg (based on a dense magma volume of 800 $km^3$; Rose and Chesner, 1990).

While in many cases a vast proportion of the dust from super-eruptions will have settled in the oceans, in some others, such as past paroxysms of Yellowstone caldera (located mostly in Wyoming, USA), very wide continental areas are likely to have been blanketed in ash (Izett and Wilcox, 1982). Such widespread layers can represent invaluable chronological markers in archaeological and palaeoenvironmental contexts (e.g. Lane *et al.*, 2014).

### 2.5.2 Ignimbrites

The deposits of pyroclastic density currents form some of the most dramatic landscapes around super-volcanoes. They are typically referred to as *ignimbrite*, and are predominantly composed of ash and pumice. The parent flows are characterized by high temperature ($> 550\ °C$), often evident in the deformation and sintering of clasts that form a welded ignimbrite or *sillar*. Ignimbrites can reach hundreds of metres in thickness outside the caldera but as much as a kilometre or two inside. They often form plateaux that become deeply incised, generating complex and spectacular topography with deep, steep-sided canyons.

### 2.5.3 Reworked deposits

The quantities of tephra rapidly deposited on the landscape are clearly very substantial in the case of super-eruptions. In the areas inundated by pyroclastic currents, entire valleys may become choked, resulting in substantial disruption of drainage systems. Distal deposits of tephra fallout, which can cover vast areas, can also have significant hydrological impacts that leave clear signals in the sedimentological record (Figure 2.2; Gatti *et al.*, 2013; Williams *et al.*, 2009).

## 2.6 Impacts of super-eruptions

The global climatic impacts of super-eruptions have been widely debated, especially in the context of the Youngest Toba Tuff eruption (reflecting, in particular, interest in its consequences for extant human populations). However, it has proven

Figure 2.2   Exposed section through the Youngest Toba Tuff ash fallout at Jwalapuram in India. Only the 5 cm or so at the base (at the level of the foot of the ladder) is fallout from the ash cloud. The pale-grey deposits reaching up to ankle-level of the man on the ladder are composed of near-pure ash sourced by reworking of tephra fallout that must have covered the surrounding slopes of the Jurreru valley. Prominent horizontal bands in this reworked ash are 'hardpans' representing periods of aridity and desiccation. This sequence appears to represent five or six monsoon cycles. Above these layers, there is a darker ash-rich deposit, which may be associated with a period of aridity. And above this is an even darker layer of sediments composed of sands and silts. Optically-stimulated luminescence dates suggest that the section spans less than a few thousand years either side of the Youngest Toba Tuff eruption (Petraglia *et al.*, 2007). Strata above and below the ash-rich layers contain abundant Middle Palaeolithic tools.

very challenging to make progress on unravelling the Toba story on account of the eruption's timing during a period of very significant climate oscillations (Dansgaard–Oeschger cycles), and uncertainty in the co-emitted quantity of sulfur reaching the stratosphere, which is critical for evaluating climate forcing (e.g. Oppenheimer *et al.*, 2011). While it remains problematic to match the findings of modelling studies with relevant palaeoclimate data, it is possible to compare the various climate models for extremely large sulfur releases to the atmosphere (e.g. Robock *et al.*, 2009; Timmreck *et al.*, 2012; Segschneider *et al.*, 2013; see also Chapter 13). These studies highlight the significance of dynamical integration of components of Earth system models, of the prevailing climate state and sensitivity at the time of eruption, and of representing adequately the microphysical processes occurring in volcanic aerosol clouds. In the case of the Youngest Toba Tuff eruption, some of the models (e.g. Robock *et al.*, 2009) have assumed a very large sulfur release (200 times greater than that of Mount Pinatubo in 1991). However, petrological considerations point to the Youngest Toba Tuff magma having been relatively low in sulfur content (Scaillet *et al.*, 1998; Chesner and Luhr, 2010).

### 2.6.1 Scenario for a future super-eruption

We can also only speculate when and where the next super-eruption might occur. Toba, Yellowstone or Taupo all remain possible candidates, raising the possibility of future super-eruptions in the tropics or mid-latitudes of both north and south hemispheres. Seismological investigation at Yellowstone revealed a shallow magma reservoir of more than 4300 $km^3$ in total volume, a third of which is molten (Chu *et al.*, 2010). It is also possible that a super-eruption could occur at a volcano that has not experienced one previously.

But how would a super-eruption in the not-too-distant future affect us? Much would depend on the state of preparedness. No rigorous attempts have yet been made to model risk scenarios of such an event, though there have been a number of summaries of generic consequences (e.g. Self, 2006; Oppenheimer, 2011; Donovan and Oppenheimer, 2014) based on inferences from studies of the Youngest Toba Tuff and other eruptions, and the findings of climate and Earth system models. For instance, pyroclastic currents from an *M* 8 or 9 event can be expected to extend up to 100 km radially from the volcano. These would engulf an area of a few tens of thousands of square kilometres in incandescent pumice to a depth of up to 200 m. The case of the eruption of Mont Pelée in 1902, in which approximately 29 000 perished, with almost every last inhabitant of the town of St Pierre killed, demonstrates that the chances of surviving exposure to pyroclastic currents can be vanishingly small. Beyond the fringes of a super-eruption's ignimbrite deposits, there may be some chances of initial survival, though many would subsequently die from exposure, burns and other injuries.

A much wider zone will be affected by thick tephra fallout. Where more than 0.5 m of ash accumulates, substantial building damage can be anticipated, likely claiming many victims. Power lines would be brought down and telecommunications generally compromised by the electromagnetic effects of airborne ash (e.g. Wilson *et al.*, 2012). Air quality and visibility during and for a long time after a future super-eruption would be poor across a vast area, due to windblown ash. This would compound the many challenges of mounting search-and-rescue operations. Meanwhile, many roads and railways would be impassable due to ash fallout, and aviation would be hazardous. Valleys would be inundated by mud flows.

Farming and agriculture would be severely affected where ash has accumulated to depths of more than a few centimetres. As noted above, past eruptions of Yellowstone blanketed much of North America in tephra. In such a scenario, safe food and water resources would become increasingly scarce, and the potential for severe social unrest would likely be very high. Hospitals would be crippled by power cuts and shortages of medicines. In arid regions, water could become scarce. Surface water could be contaminated by fluoride and other chemicals leached from the ash. Pumped water supplies would dwindle due to power shortages. A colossal effort would be required to respond to such a disaster and mitigate loss of life in the most affected regions, especially to stem outbreaks of infectious disease.

For a super-eruption outside the tropics, the human impacts would likely be influenced by the season of eruption – a winter-time scenario might prove deadlier as freezing temperatures would compound the exposure of millions of people. On the other hand a summer eruption would probably result in a deeper hemispheric climate response (Timmreck and Graf, 2006) and immediate impacts on crops and livestock, presenting a greater threat to food security. Agriculture in the zone affected by ash fallout would likely be curtailed for years, and potentially decades, due to deficits in rainfall that can be expected after a super-eruption. This would perturb regional climate for much longer than the 'few years' residence of sulfuric acid aerosol in the stratosphere typical of lesser eruptions. Silicic ash is comparably bright as snow and hence reflects sunlight that would otherwise be absorbed by vegetation and soils. One climate model for the effects of ash cover from an eruption of Yellowstone predicted surface cooling of around 5°C throughout the year for North America (Jones *et al.*, 2007; see also Chapter 17).

The most extreme scenarios for super-eruptions include the demise of technological civilization, and led Mike Rampino to suggest that volcanism might represent a universal constraint on the number of extra-terrestrial civilizations (Rampino, 2002). He argues that the impacts of such a large eruption on worldwide climate would severely reduce global agricultural yields potentially leading to 'widespread starvation, famine, disease, social unrest, financial collapse, and severe damage to the underpinnings of civilization'.

Some idea of the potential political and economic instability that could arise from worldwide reductions in grain supply can be gauged from the impacts of the 2007–2008 and 2010–2011 global food crises (e.g. Rosset, 2008), which some implicate in increased civil unrest such as manifested during the Arab Spring (e.g. Lybbert and Morgan, 2013). While studies linking climate stress to conflict abound (e.g. Hsiang *et al.*, 2013), they are often met with rejections of the implied environmental determinism (e.g. Raleigh *et al.*, 2014). Nevertheless, it is hard, with the present state of knowledge, to rule out the hypothesis that a future super-eruption will provoke an extremely severe international crisis.

## Acknowledgements

We thank the reviewer of the original manuscript for helpful comments and suggestions.

## References

Allan, A. S., Wilson, C. J., Millet, M. A. and Wysoczanski, R. J. (2012). The invisible hand: tectonic triggering and modulation of a rhyolitic supereruption. *Geology*, **40**, 563–566.

Allan, A. S., Morgan, D. J., Wilson, C. J. and Millet, M. A. (2013). From mush to eruption in centuries: assembly of the super-sized Oruanui magma body. *Contributions to Mineralogy and Petrology*, **166**, 1–22.

Bachmann, O., Dungan, M. A. and Lipman, P. W. (2002). The Fish Canyon magma body, San Juan volcanic field, Colorado: rejuvenation and eruption of an upper-crustal batholith. *Journal of Petrology*, **43**, 1469–1503.

Bonadonna, C. and Costa, A. (2013). Plume height, volume, and classification of explosive volcanic eruptions based on the Weibull function. *Bulletin of Volcanology*, **75**, 10–19.

Burden, R. E., Chen, L. and Phillips, J. C. (2013). A statistical method for determining the volume of volcanic fall deposits. *Bulletin of Volcanology*, **75**, 1–10.

Burgisser, A. and Bergantz, G. W. (2011). A rapid mechanism to remobilize and homogenize highly crystalline magma bodies. *Nature*, **471**, 212–215.

Caricchi, L., Annen, C., Blundy, J., Simpson, G. and Pinel, V. (2014). Frequency and magnitude of volcanic eruptions controlled by magma injection and buoyancy. *Nature Geoscience*, **7**, 126–130.

Charlier, B. L. A., Wilson, C. J. N., Lowenstern, J. B. *et al.* (2005). Magma generation at a large, hyperactive silicic volcano (Taupo, New Zealand) revealed by U–Th and U–Pb systematics in zircons. *Journal of Petrology*, **46**, 3–32.

Chesner, C. A. and Luhr, J. F. (2010). A melt inclusion study of the Toba Tuffs, Sumatra, Indonesia. *Journal of Volcanology and Geothermal Research*, **197**, 259–278.

Christensen, J. N. and Halliday, A. N. (1996). Rb–Sr ages and Nd isotopic compositions of melt inclusions from the Bishop Tuff and the generation of silicic magma. *Earth and Planetary Science Letters*, **144**, 547–561.

Chu, R., Helmberger, D. V., Sun, D., Jackson, J. M. and Zhu, L. (2010). Mushy magma beneath Yellowstone. *Geophysical Research Letters*, **37**, L01306.

Costa, A., Smith, V. C., Macedonio, G. and Matthews, M. E. (2014). The magnitude and impact of the Youngest Toba Tuff super-eruption. *Frontiers of Earth Science*, **2**, doi: 10.3389/feart.2014.00016.

Crosweller, H. S., Arora, B., Brown, S. K. *et al.* (2012). Global database on large magnitude explosive volcanic eruptions (LaMEVE). *Journal of Applied Volcanology*, **1**, 1–13.

Donovan, A. R. and Oppenheimer, C. (2014). Extreme volcanism: disaster risks and societal implications. In Ismail-Zadeh, A., Fucugaughi, J., Kijko, A., Takeuchi, K.

and Zaliapin, I. (eds.), *Extreme Natural Hazards, Disaster Risks and Societal Implications*, IUGG Special Publication Series, Cambridge: Cambridge University Press, 29–46.

Druitt, T. H., Costa, F., Deloule, E., Dungan, M. and Scaillet, B. (2012). Decadal to monthly timescales of magma transfer and reservoir growth at a caldera volcano. *Nature*, **482**, 77–80.

Ellis, B. S., Mark, D. F., Pritchard, C. J. and Wolff, J. A. (2012). Temporal dissection of the Huckleberry Ridge Tuff using the $^{40}$Ar/$^{39}$Ar dating technique. *Quaternary Geochronology*, **9**, 34–41.

Gatti, E. and Oppenheimer, C. (2013). Utilization of distal tephra records for understanding climatic and environmental consequences of the Youngest Toba Tuff. In Giosan, L., Fuller, D. Q., Nicoll, K., Flad, R. K. and Clift, P. D. (eds.), *Climates, Landscapes, and Civilizations*. Washington, DC: American Geophysical Union Monograph, 198, 63–74.

Gatti, E., Saidin, M., Talib, K. *et al.* (2013). Depositional processes of reworked tephra from the Late Pleistocene Youngest Toba Tuff deposits in the Lenggong Valley. *Quaternary Research*, **79**, 228–241.

Geyer, A. and Marti, J. (2008). The new worldwide collapse caldera database (CCDB): a tool for studying and understanding caldera processes. *Journal of Volcanology and Geothermal Research*, **175**, 334–354.

Halliday, A. N., Mahood, G. A., Holden, P. *et al.* (1989). Evidence for long residence times of rhyolitic magma in the Long Valley magmatic system: the isotopic record in precaldera lavas of Glass Mountain. *Earth and Planetary Science Letters*, **94**, 274–290.

Hsiang, S. M., Burke, M. and Miguel, E. (2013). Quantifying the influence of climate on human conflict. *Science*, **341**, 1235367.

Izett, G. A. and Wilcox, R. E. (1982). Map showing localities and inferred distribution of the Huckleberry Ridge, Mesa Falls, and Lava Creek ash beds (Pearlette family ash beds) of Pliocene and Pleistocene age in the western United States and southern Canada: US Geological Survey Miscellaneous Investigations Series Map I-1325, scale 1:4000000.

Jones, M. T., Sparks, R. S. J. and Valdes, P. J. (2007). The climatic impact of supervolcanic ash blankets. *Climate Dynamics*, **29**, 553–564.

Kennedy, B., Wilcock, J. and Stix, J. (2012). Caldera resurgence during magma replenishment and rejuvenation at Valles and Lake City calderas. *Bulletin of Volcanology*, **74**, 1833–1847.

Lane, C. S., Chorn, B. T. and Johnson, T. C. (2013). Ash from the Toba supereruption in Lake Malawi shows no volcanic winter in East Africa at 75 ka. *Proceedings of the National Academy of Sciences*, **110**, 8025–8029.

Lane, C. S., Cullen, V. L., White, D., Bramham-Law, C. W. F. and Smith, V. C. (2014). Cryptotephra as a dating and correlation tool in archaeology. *Journal of Archaeological Science*, **42**, 42–50.

Lavigne, F., Degeai, J.-P., Komorowski, J.-C. *et al.* (2013). Source of the great A.D. 1257 mystery eruption unveiled, Samalas Volcano, Rinjani Volcanic Complex, Indonesia, *Proceedings of the National Academy of Science*, doi:10.1073/pnas.1307520110.

Lipman, P. W. (1997). Subsidence of ash-flow calderas: relation to caldera size and magma-chamber geometry. *Bulletin of Volcanology*, **59**, 198–218.

Lybbert, T. J. and Morgan, H. (2013). Lessons from the Arab Spring: food security and stability in the Middle East and North Africa. In Barrett, C. B. (ed.), *Food Security and Sociopolitical Stability*, Oxford: Oxford University Press, pp. 357–380.

Malfait, W. J., Seifert, R., Petitgirard, S. *et al.* (2014). Supervolcano eruptions driven by melt buoyancy in large silicic magma chambers. *Nature Geoscience*, **7**, 122–125.

Mason, B. G., Pyle, D. M. and Oppenheimer, C. (2004). The size and frequency of the largest explosive eruptions on Earth. *Bulletin of Volcanology*, **66**, 735–748.

Matthews, N. E., Smith, V. C., Costa, A. *et al.* (2012). Ultra-distal tephra deposits from super-eruptions: examples from Toba, Indonesia and Taupo volcanic zone, New Zealand. *Quaternary International*, **258**, 54–79.

Michalski, J. R. and Bleacher, J. E. (2013). Supervolcanoes within an ancient volcanic province in Arabia Terra, Mars. *Nature*, **502**, 47–52.

Miller, C. F. and Wark, D. A. (2008). Supervolcanoes and their explosive supereruptions. *Elements*, **4**, 11–16.

Oppenheimer, C. (2003). Climatic, environmental and human consequences of the largest known historic eruption: Tambora volcano (Indonesia) 1815. *Progress in Physical Geography*, **27**, 230–259

Oppenheimer, C. (2011). *Eruptions that Shook the World*. Cambridge: Cambridge University Press.

Oppenheimer, C., Fischer, T. and Scaillet, B. (2014) Volcanic degassing: process and impact. In H. D. Holland and K. K. Turekian (eds.), Treatise on Geochemistry, 2$^{nd}$ edn, Oxford: Elsevier, pp. 111–179.

Petraglia, M. P., Korisettar, R., Boivin, N. *et al.* (2007). Middle Paleolithic assemblages from the Indian subcontinent before and after the Toba super-eruption. *Science*, **317**, 114–116.

Pyle, D. M. (2000). Sizes of volcanic eruptions. In Sigurdsson, H., Houghton, B., McNutt, S. R., Rymer, H. and Stix, J. (eds.), *Encyclopedia of Volcanoes*. New York: Academic Press, pp. 263–269.

Raleigh, C., Linke, A. and O'Loughlin, J. (2014). Extreme temperatures and violence. *Nature Climate Change*, **4**, 76–77.

Rampino, M. R. and Self, S. (1992). Volcanic winter and accelerated glaciation following the Toba super-eruption, *Nature*, **359**, 50–53.

Rampino, M. R. (2002). Supereruptions as a threat to civilizations on Earth-like planets. *Icarus*, **156**, 562–569.

Reid, M. R. (2008). How long does it take to supersize an eruption? *Elements*, **4**(1), 23–28.

Robock, A., Ammann, C. M., Oman, L. *et al.* (2009). Did the Toba volcanic eruption of ~74 ka B.P. produce widespread glaciation? *Journal of Geophysical Research*, **114**, D10107.

Rose, W. I. and Chesner, C. A. (1990). Worldwide dispersal of ash and gases from Earth's largest known eruption: Toba, Sumatra, 75 ka. *Global and Planetary Change*, **3**(3), 269–275.

Rose, W. I. and Chesner, C. A. (1990). Worldwide dispersal of ash and gases from earth's largest known eruption: Toba, Sumatra, 75 ka. *Palaeogeography, Palaeoclimatology, Palaeoecology*, **89**(3), 269–275.

Rosset, P. (2008). Food sovereignty and the contemporary food crisis. *Development*, **51**, 460–463.

Scaillet, B., Clemente, B., Evans, B. W., Pichavant, M. (1998). Redox control of sulfur degassing in silicic magmas. *Journal of Geophysical Research*, **103**, B10, 23937–23949.

Scott, W. E., Hoblitt, R. P., Torres, R. C. *et al.* (1996). Pyroclastic flow deposits from the 15 June 1991 eruption of Mount Pinatubo. In C. G. Newhall and R. S. Punongbayan (eds.), *Fire and Mud: Eruptions and Lahars of Mount Pinatubo, Philippines*. Quezon City; Seattle: Philippine Institute of Volcanology and Seismology; University of Washington Press, pp. 545–570.

Segschneider, J., Beitsch, A., Timmreck, C. *et al.* (2013). Impact of an extremely large magnitude volcanic eruption on the global climate and carbon cycle estimated from ensemble Earth System Model simulations. *Biogeosciences*, **10**, 669–687.

Self, S. (2006). The effects and consequences of very large explosive volcanic eruptions. *Philosophical Transactions of the Royal Society A: Mathematical, Physical and Engineering Sciences*, **364**, 2073–2097.

Self, S. and Blake, S. (2008). Consequences of explosive supereruptions. *Elements*, **4**, 41–46.

Sigurdsson, H. and Carey, S. (1989). Plinian and co-ignimbrite tephra fall from the 1815 eruption of Tambora volcano. *Bulletin of Volcanology*, **51**, 243–270.

Sparks, R. S. J., Self, S., Grattan, J. P. *et al.* (2005). Super-eruptions: global effects and future threats. Report of a Geological Society of London Working Group, London: The Geological Society. See: http://www.geolsoc.org.uk/Education-and-Careers/Resources/Papers-and-Reports/Super-eruptions.

Svensson, A., Bigler, M., Blunier, T. *et al.* (2013). Direct linking of Greenland and Antarctic ice cores at the Toba eruption (74 ka BP). *Climate of the Past*, **9**, 749–766.

Timmreck, C. and Graf, H. F. (2006). The initial dispersal and radiative forcing of a northern hemisphere mid-latitude super volcano: a model study. *Atmospheric Chemistry and Physics*, **6**, 35–49.

Timmreck, C., Graf, H. F., Zanchettin, D. *et al.* (2012). Climate response to the Toba super-eruption: regional changes. *Quaternary International*, **258**, 30–44.

Van Eaton, A. R. and Wilson, C. J. (2013). The nature, origins and distribution of ash aggregates in a large-scale wet eruption deposit: Oruanui, New Zealand. *Journal of Volcanology and Geothermal Research*, **250**, 129–154.

Van Eaton, A. R., Herzog, M., Wilson, C. J. N. and McGregor, J. (2012). Ascent dynamics of large phreatomagmatic eruption clouds: the role of microphysics. *Journal of Geophysical Research*, **117**, B03203, doi:10.1029/2011JB008892.

Wilcock, J., Goff, F., Minarik, W. G. and Stix, J. (2013). Magmatic recharge during the formation and resurgence of the Valles caldera, New Mexico, USA: evidence from quartz compositional zoning and geothermometry. *Journal of Petrology*, **54**, 635–664.

Williams, M. A., Ambrose, S. H., van der Kaars, S. *et al.* (2009). Environmental impact of the 73 ka Toba super-eruption in South Asia. *Palaeogeography, Palaeoclimatology, Palaeoecology*, **284**, 295–314.

Wilson, C. J. (2008). Supereruptions and supervolcanoes: processes and products. *Elements*, **4**, 29–34.

Wilson, C. J. N. and Charlier, B. L. A. (2009). Rapid rates of magma generation at contemporaneous magma systems, Taupo volcano, New Zealand: insights from U–Th model-age spectra in zircons. *Journal of Petrology*, **50**, 875–907.

Wilson, C. J. and Hildreth, W. (1997). The Bishop Tuff: new insights from eruptive stratigraphy. *The Journal of Geology*, **105**, 407–440.

Wilson, T. M., Stewart, C., Sword-Daniels, V. *et al.* (2012). Volcanic ash impacts on critical infrastructure. *Physics and Chemistry of the Earth, Parts A/B/C*, **45**, 5–23.

Wotzlaw, J. F., Schaltegger, U., Frick, D. A. *et al.* (2013). Tracking the evolution of large-volume silicic magma reservoirs from assembly to supereruption. *Geology*, **41**, 867–870.

# 3

# Large igneous province locations and their connections with the core–mantle boundary

TROND H. TORSVIK AND KEVIN BURKE

## 3.1 Introduction

In 2004, Burke and Torsvik demonstrated for the first time that reconstructed large igneous provinces (LIPs) of the past 200 Myr overlie the edges of two equatorial, antipodal large low shear-wave velocity provinces (LLSVPs) in the lowermost mantle beneath Africa (Tuzo, named for Tuzo Wilson) and the Pacific Ocean (Jason, named for Jason Morgan). In that work, palaeomagnetic data were used to reconstruct the continents, and their contained LIPs were compared with the seismic tomography of the lowermost mantle (at 2800 km) using the S20RTS (Ritsema *et al.*, 1999) and SMEAN models (Becker and Boschi, 2002). Because the pattern of LIPs correlating with present-day core–mantle boundary (CMB) tomography was recognizable for times back to 200 Ma, we concluded that LIPs originated above plumes from the edges on CMB of the two LLSVPs, and also that the LLSVPs have been stable for the past 200 Myr.

This first documented stability extending back to 200 Ma was ignored or received with scepticism by the geodynamic modelling community who considered that there was no physical basis for stationary structures in the lowermost mantle because they could not model this stability nor produce plumes from the margins of such stable features. Nowadays, many modellers accept stability back to about 200 Ma but not before that time. Some argue that the formation of the African LLSVP (Tuzo) is intimately linked to the Pangea supercontinent formation and destruction (e.g. Zhang *et al.*, 2010).

Here we first review LIPs of the past 300 Myr, defining them from their distinguishing characters and emphasizing current ideas about their provenance including their origin at depth, and distinguish oceanic plateau LIPs from intra-continental LIPs. Next, we outline the reconstruction methods that we have used to rotate LIP outcrops from their present locations to the locations in which they

*Volcanism and Global Environmental Change*, eds. Anja Schmidt, Kirsten E. Fristad and Linda T. Elkins-Tanton. Published by Cambridge University Press. © Cambridge University Press 2015.

were originally erupted. We follow that by defining plume generation zones (PGZs) on the CMB and describe the role of deep mantle seismological data sets in refining PGZ locations. The results of various reconstruction methods are summarized and compared with several of the seismological data sets used to define the edges of the LLSVPs and smaller LLSVPs (LSVPs) on the CMB. Finally, we outline the evidence for the long-term stability of Tuzo and Jason.

## 3.2 Overview of LIPs of the past 300 Myr

LIPs (Figure 3.1) are areas of dominantly basaltic igneous rock mainly in flows and sills, and very large in volume (often $1 \times 10^6$ km$^3$ or more). They erupted in a very short time (about 1 Myr or no more than a few million years at most) over an extensive area (outcropping over about one million or more square kilometres) but not over areas representing large fractions of the Earth's surface such as those that are occupied by the rocks making up the floors of the oceans. The criterion of eruption duration being limited to a short interval is the strongest in identifying LIPs, although its application appears to require the availability of modern high-resolution U–Pb age determination (see Bowring *et al.*, this volume).

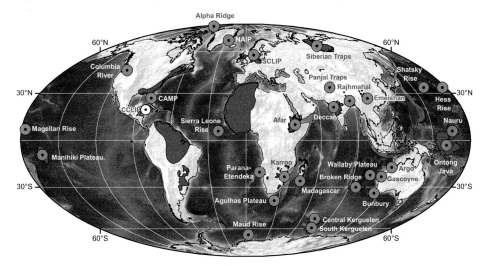

Figure 3.1    Thirty-one large igneous provinces (LIPs), 15 to 297 Myr old, and their estimated eruption centres. One LIP is not reconstructed in this paper (CCLIP, Caribbean Colombian LIP) because of reconstruction uncertainties and therefore not listed in Table 3.1. The Panjal Traps are allochthonous and thus associated with some reconstruction uncertainties in Figure 3.2. The Emeishan LIP has been moved in longitude so that it falls above the western edge of Jason in Figure 3.2. The areal extent of the Central Atlantic Magmatic Province (CAMP) is very liberal in the inclusion of *c.* 201 Myr basalts, sills and dykes. Mollweide projection (updated from Torsvik *et al.*, 2006). A black and white version of this figure will appear in some formats. For the colour version, please refer to the plate section.

The 'great volume' and 'great areal extent of outcrop' criteria are in some cases difficult to apply because of limited surface outcrop and the effects of extensive erosion. LIPs, because of their size, have been recognized as a distinct class of igneous rocks for many decades (e.g. Eldholm & Coffin, 2000). Specialist study groups and comprehensive reviews (e.g. Ernst *et al.*, 2005) have included a strong emphasis on LIPs of Precambrian times. Giant dike swarms are particularly abundant in areas of Precambrian outcrop because rocks erupted at shallower depths, such as flows and sills, have commonly been eroded away. Studies have in some cases attempted to identify LIPs as being characterized by distinctive minor-element and isotopic composition; however, Herzberg's (2011) initiative in characterizing LIP basalts as having been erupted at relatively high temperatures seems more promising as an identification criterion. It has been suggested that noble-gas signatures reflecting diffusion from the hot basal interiors of long-lived LLSVPs and LSVPs are also distinctive of LIPs (Mukhopaday, 2012; Burke *et al.*, 2008).

LIPs have been suggested to be of deep-seated hot source origin because their formation requires partial melting of a large volume of mantle rock in a short time. Because such melting makes basalt it must take place dominantly at shallow depths (typically $< c.$ 100 km). The implication is that LIP origin is likely to require the rise of bodies of hot ultramafic rock from depth. Bodies of that kind have, for the past 50 years, been postulated to exist and have been called 'mantle plumes' or 'deep mantle plumes' (see, e.g., Wilson, 1963; Morgan, 1972).

We have found that LIPs have been erupted above deep mantle plumes generated on the CMB. We are currently testing the hypothesis that LIPs, hotspot volcanoes and rocks derived from sources within the continental mantle lithosphere (e.g. kimberlites) erupt near the Earth's surface close to vertically above plume generation zones (PGZs) on the edges of LLSVPs (Burke *et al.*, 2008; Torsvik *et al.*, 2006, 2010). We find it useful to distinguish oceanic plateaus (OPs), which are LIPs erupted within the world ocean, from LIPs called continental large igneous provinces (CLIPs), which are erupted within the continents (Eldholm & Coffin, 2000).

### 3.2.1 Oceanic plateaus (OPs)

We recognize nearly 20 OPs (73–155 Ma) within the oceans of the world (18 listed in Table 3.1 and shown in Figure 3.1 that we can confidently reconstruct to their locations at the time of their eruption). That is about an average of one preserved for every 10 Myr since the formation age of the world's oldest ocean floor; the youngest recognized OP, the Sierra Leone Rise is about 73 Myr old. We discern no periodicity of eruption within OPs but emphasize that the population is small and the length of time represented by the life of the world's current ocean floor

Table 3.1 Location of LIP centres of the past 300 Myr (today and at eruption time). OP = oceanic plateau, CLIP = continental LIP (at least nine erupted in existing rifts listed in the last column). LIPs are reconstructed according to a hybrid reference frame (moving hotspot frame back to 120 Ma and a TPW-corrected palaeomagnetic frame before that time; Doubrovine et al., 2012; Torsvik et al., 2012).

| | LIP | Abbreviation | Age (Ma) | Type | Centre today | | At eruption time | | Pre-existing rifts (age in Ma) |
|---|---|---|---|---|---|---|---|---|---|
| | | | | | Latitude | Longitude | Latitude | Longitude | |
| 1 | Columbia River Basalts[a] | CRB | 15 | CLIP | 46.0 | 241.0 | 47.7 | −116.0 | |
| 2 | Ethiopia[a] | ET | 31 | CLIP | 10.0 | 39.5 | 5.5 | 36.1 | Najd (610) |
| 3 | North Atlantic Igneous Province[a] | NAIP | 62 | CLIP | 69.9 | 332.8 | 63.7 | −14.3 | Northeast Atlantic (65[k]) |
| 4 | Deccan Traps[a] | D | 65 | CLIP | 21.0 | 73.0 | −14.8 | 53.6 | Narmada (320) |
| 5 | Sierra Leone Rise[a] | SL | 73 | OP | 6.0 | 338.0 | 5.8 | −32.4 | |
| 6 | Madagascar[a] | M | 87 | CLIP/OP | −26.0 | 46.0 | −42.7 | 31.2 | Morondava (310) |
| 7 | Broken Ridge[a] | BR | 95 | OP | −30.0 | 96.0 | −49.5 | 62.7 | |
| 8 | Wallaby Plateau[a] | W | 96 | OP | −22.0 | 104.0 | −46.2 | 77.9 | |
| 9 | Hess Rise[a] | HR | 99 | OP | 34.0 | 177.0 | 5.4 | −140.0 | |
| 10 | Central Kerguelen[a] | CK | 100 | OP | −52.0 | 74.0 | −48.9 | 59.2 | |
| 11 | Agulhas Plateau[e] | AP | 100 | OP | −39.0 | 26.0 | −53.2 | −1.0 | |
| 12 | Nauru[a] | N | 111 | OP | 6.0 | 166.0 | −23.3 | −142.0 | |
| 13 | Southern Kerguelen[a] | SK | 114 | OP | −59.0 | 79.0 | −48.6 | 57.6 | |
| 14 | Rajmahal Traps[a] | R | 118 | OP | 25.0 | 88.0 | −37.9 | 60.7 | |
| 15 | Ontong Java[d] | OJ | 121 | OP | −3.0 | 161.0 | −34.5 | −141.2 | |
| 16 | Manihiki Plateau[a] | MP | 123 | OP | −9.0 | 196.0 | −31.6 | −97.2 | |
| 17 | High Arctic LIP (Alfa Ridge)[f] | HALIP | 124 | OP/CLIP | 85.7 | 225.0 | 77.9 | −6.4 | |
| 18 | Maud Rise[c] | MR | 125 | OP | −65.0 | 3.0 | −52.2 | 2.2 | |
| 19 | Bunbury Basalts[a] | BB | 132 | CLIP | −34.0 | 115.0 | −54.7 | 72.5 | |
| 20 | Paraná—Etendeka[a] | PR | 134 | CLIP | −20.0 | 11.0 | −31.8 | −15.0 | South Atlantic (150) |
| 21 | Gascoyne[b] | G | 136 | OP | −23.0 | 114.0 | −46.3 | 81.2 | |

Table 3.1 (cont.)

| LIP | Abbreviation | Age (Ma) | Type | Centre today | | At eruption time | | Pre-existing rifts (age in Ma) |
|---|---|---|---|---|---|---|---|---|
| | | | | Latitude | Longitude | Latitude | Longitude | |
| 22 Magellan Rise [a] | MR | 145 | **OP** | 7.0 | 183.0 | -1.1 | -108.0 | |
| 23 Shatsky Rise [a] | SR | 147 | **OP** | 34.0 | 160.0 | 7.3 | -108.0 | |
| 24 Argo Margin [b] | AM | 155 | **OP** | -17.0 | 120.0 | -43.4 | 82.9 | |
| 25 Karroo [a] | K | 182 | CLIP | -23.0 | 32.0 | -37.0 | -5.2 | Dwyka (320) |
| 26 Central Atlantic Magmatic Province [a] | CAMP | 201 | CLIP | 27.0 | 279.0 | 16.8 | -25.4 | Newark (230) |
| 27 Siberian Traps [g] | SBT | 251 | CLIP | 65.0 | 97.0 | 62.5 | 44.2 | Pudom (270) |
| 28 Emeishan LIP [h] | E | 258 | CLIP | 26.6 | 104.0 | -4.0 | 134.2 | |
| 29 Panjal Traps [i] | PT | 285 | CLIP | 34.0 | 75.0 | -42.9 | 59.5 | |
| 30 Skagerrak-centred LIP [j] | SCLIP | 297 | CLIP | 57.5 | 9.0 | 9.8 | -2.0 | Oslo (310) |

[a] Originally used by Burke and Torsvik (2004) and mostly based on Eldholm & Coffin (2000).
[b] Not used in subsequent paper because of age uncertainties but shown here.
[c] Originaly listed as 73 Ma (revised in Torsvik et al., 2006).
[d] Originally listed as Ontong Java 1 and 2.
[e] First reconstructed in Torsvik et al. (2009).
[f] First reconstructed here using U–Pb ages from Svalbard (Corfu et al., 2013) but ages of 90 Ma reported from Alfa Ridge.
[g] First reconstructed in Torsvik et al. (2006).
[h] First reconstructed in Torsvik et al. (2008a).
[i] First reconstructed here.
[j] First defined as a LIP and reconstructed in Torsvik et al. (2008b).
[k] Youngest rift phase (Late Cretaceous—Early Tertiary).

is short. The Caribbean Colombian LIP (CCLIP) could be the OP that was parent to the active Galapagos hotspot. The CCLIP was erupted at about 90 Ma and collided with the Great Arc of the Caribbean at about 80 Ma (Burke, 1988). However, because of reconstruction uncertainties we have not included the CCLIP in our analysis.

### 3.2.2 LIPs of the continents (CLIPs)

Fourteen CLIPs have erupted into and onto continental crust for the past 300 Myr. Two of those CLIPs also lie partly on oceanic crust and a majority of the remainder lie at the time of, or soon after, their eruption within no more than a few hundred kilometres of a continental margin. More generally, locations of CLIPs have shown a close association with intra-continental rifts (Table 3.1); in trying to apply the ideas of plate tectonics to past continental geology it was long ago suggested that eruption of intra-continental LIPs and hotspots above mantle plumes led to the formation of topographic domes crested by newly formed radial intra-continental rifts. Horizontally linked, dog-legged patterned sets of those new rifts, thousands of kilometres in total length, with hotspot locations at nodes, were suggested to evolve into the kinds of continuous rifts that mark the initiation of Atlantic-type continental margins (e.g. Burke and Whiteman, 1973; Burke and Dewey, 1973). More recently, temporally better resolved isotopic, stratigraphic and magnetic anomaly information about the relative times of formation of CLIPs, rifts and newly formed Atlantic-type ocean floor has become available and those ideas have been shown to have been wrong (see, e.g., Sengor, 2001; Burke *et al.*, 2003). It has now become clear that CLIPs and hotspots have erupted into already existing intra-continental rifts. As in the case of the Paraná–Etendeka rift (Table 3.1) some of those rifts were no more than about 20 Myr old at the time of CLIP eruption; however, in other cases, such as the Deccan CLIP, the rift into which the CLIP erupted was already hundreds of millions of years old. CLIPs that formed above underlying plumes from the CMB intrude into existing intra-continental rifts; they have not initiated rifts. Sleep's (1997) idea of 'upside-down drainage' provides a good explanation of how CLIPs can come to be erupted into pre-existing rifts. Sengor and Natalin (2001) catalogued and plotted on a world map over 400 such structures. If the rifts were uniformly distributed within the $170 \times 10^6$ km$^2$ area of the continents then there would be one rift in every $0.5 \times 10^6$ km$^2$ area. The only significance of this estimate is that it shows that the upside-down drainage of the LIP magma is unlikely to have to travel for more than a few hundred kilometres along the base of continental lithosphere to find a rift into which to erupt.

## 3.3 Reconstruction methods

Until recently, Earth scientists have had no way of calculating the longitudes of continents before the Cretaceous, leaving palaeomagnetism, which cannot determine longitude, as the only quantitative means of positioning continents on the globe before that time. However, by choosing a reference continent that has moved the least longitudinally (i.e. Africa), longitudinal uncertainty can be minimized. The analytical trick is to rotate all palaeomagnetic poles to Africa and calculate a global apparent polar wander path (GAPWaP) in African coordinates, which serves as the basis for subsequent global reconstructions. This method, dubbed the 'zero-longitudinal motion' approximation for Africa (Burke and Torsvik, 2004), has also allowed us to confidently estimate true polar wander (TPW) since Pangea formation at around 320 Ma (Steinberger & Torsvik, 2008; Torsvik *et al.*, 2012).

The past decade has also seen the dawn of so-called hybrid reference frames, which combine different frames for different time periods. The first of these (Torsvik *et al.*, 2008a) was based on a mantle reference (moving hotspot) frame for the past 100 Myr, and, before that, a reference frame derived from the African GAPWaP, making the assumption that Africa has not moved in longitude. The longitude offset for Africa at 100 Ma (5°) was applied to all the older reconstructions. Here we use a new moving hotspot reference frame (Dubrovine *et al.*, 2012) for the past 125 Myr and – before that – the updated African GAPWaP of Torsvik *et al.* (2012) with TPW correction.

Choosing Africa as a reference plate that has remained quasi-stationary with respect to longitude only works back to Pangea's assembly. This is because most relative plate circuits are reasonably well-known from that time, except for blocks or continents not belonging to Pangea, such as North and South China. The latter blocks did not join Eurasia before the early Cretaceous closure of the Mongol–Okhotsk Ocean, and thus well after the main Pangea break-up. Reconstruction of the *c.* 258 Ma Emeishan LIP in South China therefore requires a different approach. Using the surface-to-CMB correlation that we have identified (i.e. LIPs from the edges of Tuzo or Jason) we can relocate South China in longitude so that Emeishan falls above the western margin of Jason (Torsvik *et al.*, 2008b).

## 3.4 Seismic tomography and refinement of the plume generation zones

Tuzo and Jason (Figure 3.2) are prominent features in all shear-wave tomographic models. LIPs have erupted above their margins, dubbed the plume generation zones (PGZs) by Burke *et al.* (2008). The exact definition of the PGZs depends

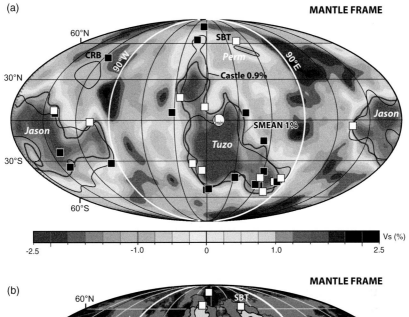

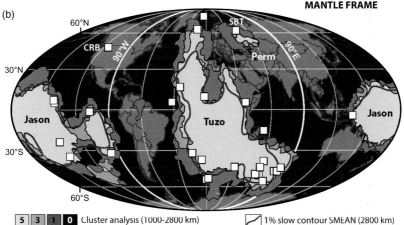

Figure 3.2 (a) Reconstructed LIPs (Figure 3.1) superimposed on the SMEAN tomographic model (Becker & Boschi, 2002) at 2800 km depth ($\delta V_s$ is S-wave anomaly). LIPs are reconstructed using a moving hotspot frame (black squares) or a TPW-corrected palaeomagnetic frame (white squares) before 125 Ma (see text). We also show the 1% slow contour in this model (SMEAN 1%) that we have extensively used as the proxy for the plume generation zones (Torsvik *et al.*, 2006) as well as the 0.9% slow contour in the S-SKS Castle *et al.* (2000) model (Burke *et al.*, 2008). CRB, Columbia River Basalts; SBT, Siberian Traps. The African and Pacific LLSVPs are dubbed the Tuzo and Jason provinces. Columbia River Basalts (CRB) and the Siberian Traps (SBT) are not directly correlated with the edges of Tuzo and Jason but to smaller anomalies in the lowermost mantle beneath regions such as Perm (Lekic *et al.*, 2012). The mean axis (0° N, 11° E) for Mesozoic TPW is marked by a white circle ($I_{min}$) and approximates the longitude of minimum moment of inertia associated with Tuzo and Jason. (b) Reconstructed LIPs as in Figure 3.2a but shown together with seismic voting-map contours (Lekic *et al.*, 2012). Contours 5–1 define the LLSVPs and 0 (blue) denotes faster regions in the lower mantle. The 1% slow SMEAN contour is shown for comparison. A black and white version of this figure will appear in some formats. For the colour version, please refer to the plate section.

on the tomography model used, and in most of our papers the PGZs have been approximated by the 1% slow SMEAN tomographic contour near the CMB (Torsvik *et al.*, 2006). The SMEAN model is an average of three shear-wave velocity anomaly models but we have also examined numerous other models (e.g. Torsvik *et al.*, 2008b). They all provide broadly similar characteristics near the CMB; thus, the choice of tomographic model is not critical, although different models lead to slightly different statistical correlations. We have also compared whole-mantle shear-wave velocity models with S-SKS models (Burke *et al.*, 2008, Figs. 3 and 4), globally at the CMB, and have found that the 1% slow (negative) contour of the SMEAN model compares with the 0.96% slow contour in the Castle *et al.* (2000) model (Figure 3.2a), and the 0.77% slow contour in the Kuo *et al.* (2000) model.

The PGZs were similarly located using shear-wave velocity cluster analysis by Lekic *et al.* (2010) who produced a 'voting' map, which described footprints on the CMB for five global tomographic models. They determined whether a geographical data point was located in a seismically slower-than-average region. The data were contoured in such a way that a contour score of 5 stated that all the models agreed about the definition of a data point. The voting map (Figure 3.2b) confirms that the Siberian Traps originate from the PGZ of an LSVP (Burke *et al.*, 2008), dubbed Perm by Lekic *et al.* (2010). This smaller anomaly is also discernible in the SMEAN model (–0.5% slow in Figure 3.2a).

Below, we reconstruct 30 LIPs (Table 3.1) in different ways and compare our results with SMEAN, S-SKS models and the seismic voting-map contours.


### 3.4.1 Mantle reference frame

Using a hybrid reference frame, i.e. a moving hotspot frame back to 125 Ma and a TPW-corrected palaeomagnetic frame before that time, we find that the reconstructed LIPs plot on average 7.9° from the 1% slow SMEAN contour (Table 3.2, Figure 3.3a); however, there are notably large deviations (Figure 3.3b) for the Columbia River Basalt (*c.* 39°), North Atlantic Igneous Province (NAIP, *c.* 19°), the Alpha Ridge (or High Arctic LIP, *c.* 34°) and the Siberian Traps (*c.* 36°). The voting-map contours (C1–C5 in Figure 3.3) reduce this discrepancy considerably because they extend the African LLSVP further northwards as well as identifying the Perm anomaly. Seismic contour 5 provides a very good fit. LIPs plot on average 5.4° from that contour (Table 3.2, Figure 3.3a). The best fit is yielded by the Castle contour (4.4°) because it also includes a small anomaly (LSVP) at the CMB below the Columbia River Basalts (Figure 3.2a) that has not yet been identified in global shear-wave tomographic models. The overall similarity of the several seismologically defined PGZs tested is remarkable as is made clear in Figure 3.3a and Table 3.2.

Table 3.2 *Great circle distance (in degrees) from reconstructed LIP location (Table 3.1) to the 1% slow SMEAN contour (Becker & Bosci, 2002), the seismic voting-map contours C1 to C5 (Lekic et al., 2012), the 0.96% slow contour in the Castle et al. (2000) model and the 0.77% slow contour in the Kuo et al. (2000) model (see Burke et al., 2008). We also list the mean and the standard deviation.*

| LIP | SMEAN | C1 | C2 | C3 | C4 | C5 | Kuo | Castle |
|---|---|---|---|---|---|---|---|---|
| Columbia River Basalts | 39.2 | 36.3 | 38.3 | 40.4 | 41.4 | 45.1 | 5.9 | 4.9 |
| Ethiopia | 3.6 | 4.8 | 0.5 | 1.9 | 2.1 | 2.4 | 14.5 | 7.5 |
| North Atlantic Igneous Province | 19.5 | 4.2 | 3.8 | 0.4 | 2.0 | 5.8 | 3.3 | 2.5 |
| Deccan Traps | 4.1 | 7.0 | 7.7 | 7.7 | 7.7 | 10.6 | 7.2 | 9.9 |
| Sierra Leone Rise | 11.0 | 5.2 | 7.0 | 7.0 | 7.0 | 7.0 | 9.4 | 13.0 |
| Madagascar | 1.6 | 12.0 | 6.4 | 3.6 | 3.8 | 0.3 | 5.7 | 4.4 |
| Broken Ridge | 6.8 | 9.9 | 8.0 | 7.5 | 5.3 | 1.8 | 2.7 | 2.8 |
| Wallaby Plateau | 2.5 | 10.2 | 8.6 | 5.8 | 5.7 | 3.1 | 3.2 | 5.8 |
| Hess Rise | 4.1 | 7.1 | 6.3 | 0.8 | 1.1 | 1.3 | 9.2 | 1.1 |
| Central Kerguelen | 6.6 | 9.2 | 7.5 | 7.2 | 4.2 | 0.3 | 3.7 | 2.6 |
| Agulhas Plateau | 4.2 | 0.5 | 3.2 | 5.2 | 5.2 | 5.2 | 3.8 | 2.7 |
| Nauru | 10.9 | 14.5 | 10.2 | 9.7 | 9.3 | 9.0 | 3.9 | 6.6 |
| Southern Kerguelen | 6.5 | 9.2 | 7.6 | 7.4 | 4.1 | 0.8 | 4.2 | 2.5 |
| Rajmahal Traps | 7.0 | 16.0 | 13.9 | 12.1 | 9.4 | 8.9 | 7.1 | 7.2 |
| Ontong Java | 1.7 | 4.5 | 4.4 | 4.1 | 3.9 | 1.6 | 2.6 | 2.0 |
| Manihiki Plateau | 3.9 | 4.7 | 2.2 | 1.0 | 0.7 | 0.7 | 4.8 | 0.1 |
| High Arctic LIP (Alfa Ridge) | 33.7 | 5.1 | 9.0 | 10.3 | 15.2 | 19.7 | 15.0 | 9.8 |
| Maud Rise | 2.9 | 2.2 | 1.3 | 4.5 | 4.5 | 4.5 | 3.1 | 1.3 |
| Banbury Basalts | 1.8 | 8.0 | 6.1 | 6.1 | 3.3 | 7.2 | 4.0 | 1.8 |
| Paraná–Etendeka | 2.5 | 11.9 | 1.1 | 1.4 | 1.4 | 1.4 | 0.9 | 3.5 |
| Gascoyne | 0.2 | 8.0 | 7.1 | 3.8 | 3.5 | 3.3 | 2.5 | 3.8 |
| Magellan Rise | 2.2 | 5.7 | 0.8 | 0.9 | 1.8 | 1.8 | 9.3 | 2.4 |
| Shatsky Rise | 3.8 | 5.6 | 5.5 | 0.2 | 1.3 | 1.4 | 9.4 | 2.4 |
| Argo Margin | 1.5 | 7.8 | 7.9 | 5.6 | 4.4 | 1.5 | 4.8 | 3.8 |
| Karroo | 6.4 | 11.9 | 7.1 | 4.5 | 4.2 | 4.0 | 7.5 | 4.8 |
| Central Atlantic Magmatic Province | 0.6 | 4.0 | 3.6 | 3.4 | 1.8 | 0.8 | 0.3 | 0.9 |
| Siberian Traps | 36.2 | 2.5 | 0.3 | 3.1 | 3.3 | 3.8 | 4.4 | 1.1 |
| Emeishan LIP | 0.7 | 0.9 | 1.3 | 1.3 | 1.3 | 1.3 | 19.9 | 4.8 |
| Panjal Traps | 11.4 | 14.8 | 13.0 | 13.0 | 9.8 | 5.0 | 9.6 | 8.2 |
| Skagerrak-centred LIP | 0.6 | 21.0 | 15.0 | 1.6 | 0.9 | 3.0 | 4.0 | 7.3 |
| **Mean** | **7.9** | **8.8** | **7.1** | **6.0** | **5.6** | **5.4** | **6.2** | **4.4** |
| **Standard deviation** | **10.5** | **7.0** | **7.1** | **7.4** | **7.5** | **8.5** | **4.4** | **3.1** |

### 3.4.2 Palaeomagnetic reference frame

TPW is the motion of the geographic pole in a reference frame representative of the entire solid Earth, caused by rotation of the planet relative to the spin axis. Both Tuzo and Jason in the lower mantle and the crust are affected by TPW but these are kept fixed in the mantle in our correlative exercises. The motion of the continents

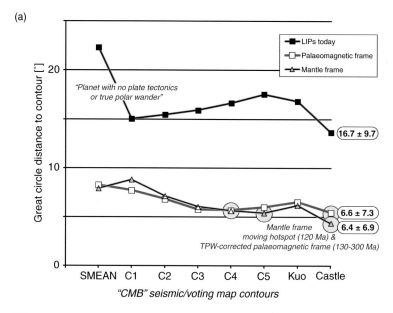

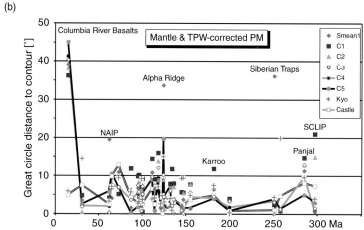

Figure 3.3 (a) Average great circle distance from present-day LIP eruption
centres ('Planet with no plate tectonics or true polar wander") and reconstructed
LIP locations (two different reference frames: mantle frame and Palaeomagnetic
frame) to the 1% slow SMEAN contour (Becker & Boschi, 2002), seismic
voting-map contours C1 to C5 (Lekic *et al.*, 2012), the 0.96% slow contour in
the Castle *et al.* (2000) model and the 0.77% slow contour in the Kuo *et al.*
(2000) model. Mean values and the standard deviation for the mantle frame
are listed in Table 3.2. (b) Great circle distance as in (a) but here we show
the distance for individual LIPs reconstructed in the mantle frame (see text
for details).

must therefore be corrected for TPW. It is critical to test how large the effect of the TPW correction is. Steinberger and Torsvik (2008) and Torsvik *et al.* (2012) found the overall effect for the past 300 Myr to be very small because the TPW rotation axis ($I_{min}$) is located close to the centres of mass of Tuzo and Jason (Figure 3.2a), and the TPW shows an oscillating pattern of slow ($\leq 0.8°$/Myr) clockwise or counter-clockwise rotations about that axis. Mean values for the different seismic contours are quite similar (Figure 3.3) for the past 300 Myr.

## 3.5 The long-term stability of Tuzo and Jason

Originally we argued that Tuzo and Jason had been stable for at least 200 Myr (Burke & Torsvik, 2004). This was subsequently extended to 300 Myr (Torsvik *et al.*, 2008c), perhaps for the entire Phanerozoic (Torsvik *et al.*, 2010), and possibly since Hadean times (Burke *et al.*, 2012). This long-term stability dominated by a degree-2 mode-mantle is not supported by many modellers. As an example, based on modelling, Zhang *et al.* (2010) argued that planet Earth was in a degree-1 mode during Pangea formation with only one LLSVP-like object, which they called a 'megapile' centred below the Pacific realm. In their model, a changeover from degree-1 to degree-2 mode appears to have been taken place only after 200 Myr although even at 120 Myr their newly generated 'African LLSVP' in the model does not resemble the present-day African LLSVP (Tuzo). Zhang *et al.* (2010) did not favour stability of the LLSVPs, notably before 200 Ma and they stated that: 'between 330 and 200 Ma, there were only three well-documented LIPs globally: Skagerrak, Emeishan, and Siberia Traps'. They further advocated the idea that the Siberian Traps (and Emeishan) were not obviously related to a LLSVP and thus that our model before 200 Ma was based only on a single LIP (the Skagerrak-centered LIP (SCLIP), Torsvik *et al.*, 2008c). We agree that the Siberian Traps (251 Ma) is not a direct fit to Tuzo (although it could be a north-east arm of Tuzo) but can be related to a separate LSVP (Perm in Figure 3.2b, Lekic *et al.*, 2012). Between 100 and 200 Ma, reconstruction of 17 LIPs and 513 kimberlites shows a clear surface pattern that overlies the edges of Tuzo and Jason on the CMB (Figure 3.4a). Before 200 Ma there are fewer LIPs ($N = 5$) and kimberlites ($N = 47$) preserved and identified in our database but the surface-to-CMB correlation (Figure 3.4b) is identical to that of later times (Figure 3.4a). Thus, there is no basis for questioning the stability of Tuzo and Jason back to 300 Ma, and perhaps for much longer.

## 3.6 Conclusions

The main conclusion of this paper (see Figure 3.2 and Table 3.2) is that 30 LIPs of the past 300 Myr (all of those rotable to their locations at eruption) were erupted

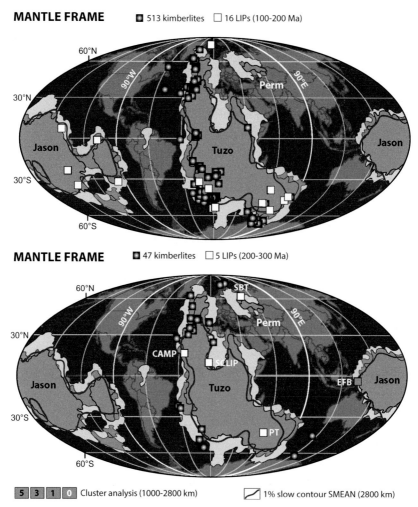

Figure 3.4   Comparison of reconstructed LIPs and kimberlites erupted between 100 and 200 Ma and in the 200–300 Ma range. The pattern is identical, with LIPs and kimberlites sourced by plumes from the edges of Tuzo and Jason, but the total numbers of LIPs and kimberlites differs. The reconstructed LIPs and kimberlites are superimposed on seismic voting-map contours as in Figure 3.2b and the 1% slow SMEAN contour in the lowermost mantle. CAMP, Central Atlantic Magmatic Province (251 Ma); SBT, Siberian Traps (251 Ma); SCLIP, Skagerrak-centred LIP (297 Ma). PT, Panjal Traps (285 Ma); EFB, Emeishan Flood Basalts (258 Ma). Mollweide projection.

close to vertically above the sources of deep mantle plumes. These plumes originated either in the PGZs bounding on the CMB, one or other of the two large LLSVPs of the deep mantle, or above PGZs on the CMB bounding similar but smaller LSVPs.

Secondly, this surface-to-CMB pattern is recognized from both a moving hot-spot frame (black squares in Figure 3.2a), and a TPW-corrected palaeomagnetic frame (white squares in Figure 3.2a) before 125 Ma that uses Africa as a reference plate that has remained quasi-stationary with respect to longitude since Pangea formation.

Thirdly, using the surface-to-CMB correlation for the past 300 Myr we have also relocated one additional LIP – the Emeishan CLIP in South China – above the western margin of the Jason LLSVP.

Fourthly, similar conclusions have previously been reached about the majority of these LIPs (see, e.g., Torsvik *et al.*, 2006; 2010; Burke *et al.*, 2008; and Lekic *et al.*, 2012). What is new about this study is that new TPW results and a new global moving hotspot model have been applied together to yield a consistent and comparable set of results for the locations of the 30 LIPs.

Fifthly, the availability of a new set of restored locations for LIP eruption sites at the surface has enabled us to drop perpendiculars to the CMB from all those 30 eruption sites to find which, if any, of eight different seismological maps of shear-wave velocity variation distribution on the CMB performed best as indicators of the locations of PGZs on the CMB. Figure 3.2 shows that all eight maps gave closely comparable results in both the 'mantle and TPW-corrected frame' and in the 'palaeomagnetic frame'. The SMEAN map performed differently because of different mapping criteria rather than from inadequacy in displaying shear-wave velocity distribution. In summary, all the seismologically prepared and various shear-wave velocity maps of the CMB performed comparably well in displaying the loci of PGZs on the CMB.

Sixthly, the Earth presently contains two anti-podal, equatorial LLSVPs in the deep mantle, i.e. the Earth is of degree-2 mode in that respect. However, that mass distribution has been suggested to date from a fairly recent time in the past (since about 200 Ma) and to have been preceded by degree-1 mode with only a single equatorial-centred mass. Because our newly controlled 30 LIP dataset straddles in age the time during which a transition from degree 1 to degree 2 is suggested to have occurred (e.g. Zhong *et al.*, 2007; Zhang *et al.*, 2010) we took the opportunity, by separately treating the LIP and kimberlite populations of 100 Ma to 200 Ma and 200 Ma to 300 Ma, to test whether there was evidence of change, during the postulated degree-1-to-degree-2 transition in mapable PGZ distribution. Our results are shown in Figure 3.4. No significant differences in PGZ distribution between or within the two intervals are discernible. We conclude that there has not been a change in deep mantle mass distribution from degree 1 to degree 2 during the Pangea assembly and dispersal and that the deep mantle mass distribution has always been of degree 2 since the Late Palaeozoic.

Finally, we emphasize that major global events recorded by the Earth's surface are linked to mantle-core dynamics. LIP interaction with upper crustal rocks may have huge environmental implications and they provide a direct link between plume-generating processes in the deepest mantle and the atmosphere and biosphere. Mantle plumes take us from the deep Earth to sub-lithospheric depths where partial melting occurs, and to the surface where episodic LIP activity through Earth history has led to Earth crises. These mantle plumes are almost entirely sourced from the edges of the two major thermochemical piles beneath Africa and the Pacific. Only the Columbia River Basalts and the Siberian Traps, which probably triggered the largest extinction event ever known, differ from this pattern. The Siberian Traps could potentially be a north-eastern arm of the African LLSVP (Torsvik *et al.*, 2008a) but could very well be related to a separate low shear-velocity province (dubbed Perm) at the core–mantle boundary (Figure 3.2).

## Acknowledgements

The European Research Council under the European Union's Seventh Framework Programme (FP7/2007–2013)/ERC Advanced Grant Agreement Number 267631 (Beyond Plate Tectonics) and the Research Council of Norway through its Centres of Excellence funding scheme (CEED, Project number 223272) are acknowledged for financial support.

## References

Becker, T.W. and Boschi, L. (2002). A comparison of tomographic and geodynamic mantle models. *Geochem. Geophys. Geosyst.*, **3**, doi:10.1029/2001GC000168.

Burke, K. (1988). Tectonic evolution of the Caribbean. *Ann. Rev. Earth Planet. Sci.*, **16**, 210–230.

Burke, K. and Dewey, J.J. (1973). Plume-generated triple junctions: key indicators in applying plate tectonics to old rocks. *J. Geol.*, **81**, 406–433.

Burke, K. and Whiteman, A.J. (1973). Uplift, rifting and break-up of Africa. In Tarling, D.H. and Runcorn, S.K. (eds.), *Implications of Continental Drift to the Earth Sciences*. Academic Press, London, pp. 735–755.

Burke, K., MacGregor, D.S. and Cameron, N.R. (2003). Africa's petroleum systems: four tectonic 'Aces' in the past 600 million years. *J. Geol. Soc. Lond.*, **207**, 21–60.

Burke, K. and Torsvik, T.H. (2004). Derivation of large igneous provinces of the past 200 million years from long-term heterogeneities in the deep mantle. *Earth Planet. Sci. Lett.*, **227**, 531–538.

Burke, K., Steinberger, B., Torsvik, T.H. and Smethurst, M.A. (2008). Plume generation zones at the margins of large low shear velocity provinces on the core–mantle boundary. *Earth Planet. Sci. Lett.*, **265**, 49–60.

Burke, K., Werner, S.C., Steinberger, B. and Torsvik, T.H. (2012). Why is the areoid like the residual geoid? *Geophys. Res. Lett.*, **39**, L17203, doi:10.1029/2012GL052701.

Castle, J.C., Creager, K.C., Winchester, J.P. and van der Hilst, R.D. (2000). Shear wave speeds at the base of the mantle. *J. Geophys. Res.*, **105**, 21543–21558.

Corfu, F., Polteau, S., Planke, S. *et al.* (2013). U–Pb geochronology of Cretaceous magmatism on Svalbard and Franz Josef Land, Barents Sea Large Igneous Province. *Geol. Mag.* 1–9, doi:10.1017/S0016756813000162.

Doubrovine, P.V., Steinberger, B. and Torsvik, T.H. (2012). Absolute plate motions in a reference frame defined by moving hot spots in the Pacific, Atlantic, and Indian oceans. *J.Geophys. Res.*, **117**, B09101, doi:10.1029/2011jb009072.

Ernst, R.E., Buchan, K.L. and Campbell, I.H. (2005). Frontiers in large igneous province research. *Lithos*, **79**, 271–297.

Eldholm, O. and Coffin, M.F. (2000). Large igneous provinces and plate tectonics. In Richards, M.A., Gordon, R.G. and van der Hilst, R.D.(eds.), *The History and Dynamics of Global Plate Motions*. American Geophysical Union, Washington, DC, pp. 309– 326.

Herzberg, C. (2011). Basalts as temperature probes of Earth's mantle. *Geology*, **39**, 1179–1180.

Kuo, B.-Y., Garnero, E.J. and Lay, T. (2000). Tomographic inversion of S-SKS times for shear velocity heterogeneity in D″: degree 12 and hybrid models. *J. Geophys. Res.*, **105**, 28,139–28,157.

Lekic, V., Cottar, S., Dziewonski, A. and Romanowicz, B. (2012). Cluster analysis of global lower mantle tomography: a new class of structure and implications for chemical heterogeneity. *Earth Planet. Sci. Lett.*, **357**, 68–77.

Morgan, W.J. (1972). Plate motions and deep mantle convection. *Geol. Soc. Amer. Memoir*, **132**, 7–22.

Mukhopadhyay, S. (2012). Early differentiation and volatile accretion recorded in deep-mantle neon and xenon. *Nature*, **486**, 101–104.

Ritsema, J., van Heijst, H.J. and Woodhouse, J.H. (1999). Complex shear velocity structure imaged beneath Africa and Iceland. *Science*, **286**, 1925–1928.

Sengor, A.M.C. (2001). Elevation as indicator of mantle plume activity. *Geol. Soc. Amer. Spec. Paper*, **352**, 183–225.

Sengor, A.M.C. and Natalin, B.A. (2001). Rifts of the World. *Geol. Soc. Amer. Spec. Paper*, **352**, 389–482.

Sleep, N.H. (1997). Lateral flow and ponding of starting plume material. *J. Geophys. Res.*, **102**, 10001–10012.

Steinberger, B. and Torsvik, T.H. (2008). Absolute plate motions and true polar wander in the absence of hotspot tracks. *Nature*, **452**, 620–623.

Torsvik, T.H., Smethurst, M.A., Burke, K. and Steinberger, B. (2006). Large igneous provinces generated from the margins of the large low velocity provinces in the deep mantle. *Geophys. J. Int.*, **167**, 1447–1460.

Torsvik, T.H., Steinberger, B., Cocks, L.R.M. and Burke, K. (2008a). Longitude: linking Earth's ancient surface to its deep interior. *Earth Planet. Sci. Lett.*, **276**, 273–282.

Torsvik, T.H., Müller, R.D., Van der Voo, R., Steinberger, B. and Gaina, C. (2008b). Global Plate Motion Frames: toward a unified model. *Rev. Geophys.*, **46**, RG3004, doi:10.1029/2007RG000227.

Torsvik, T.H., Smethurst, M.A., Burke, K. and Steinberger, B. (2008c). Long term stability in deep mantle structure: evidence from the *ca.* 300 Ma Skagerrak-centered Large Igneous Province (the SCLIP). *Earth Planet. Sci. Lett.*, **267**, 444–452.

Torsvik, T.H., Rousse, S., Labails, C. and Smethurst, M.A. (2009). A new scheme for the opening of the South Atlantic Ocean and dissection of an Aptian Salt Basin. *Geophys. J. Int.*, **177**, 1315–1333.

Torsvik, T.H., Burke, K., Steinberger, B., Webb, S.C. and Ashwal, L.D. (2010). Diamonds sourced by plumes from the core mantle boundary. *Nature*, **466**, 352–355.

Torsvik, T.H., Van der Voo, R., Preeden, U. *et al.* (2012). Phanerozoic polar wander, paleogeography and dynamics. *Earth Sci. Rev.*, **114**, 325–368.

Wilson, J.T. (1963). A possible origin of the Hawaiian Islands. *Can. J. Phys.*, **41**, 863–870.

Zhang, N., Zhong, S., Leng, W. and Li, Z.X. (2010). A model for the evolution of the Earth's mantle structure since the Early Paleozoic. *J. Geophys. Res.*, **115**, B06401, doi:10.1029/2009JB006896.

Zhong, S., Zhang, N., Li, Z.-X. and Roberts, J.H. (2007). Supercontinent cycles, true polar wander, and very long-wavelength mantle convection. *Earth Planet. Sci. Lett.*, **261**, 551–564.

# 4

# High-precision U–Pb geochronology of Phanerozoic large igneous provinces

SETH D. BURGESS, TERRENCE J. BLACKBURN AND SAMUEL A. BOWRING

## 4.1 Introduction

Large igneous provinces (LIPs) are characterized by the intrusion and eruption of large volumes ($>1\,\mathrm{km}^3$) of primarily basaltic magmas into and onto Earth's crust and they reflect large-scale melting of the mantle in both continental and oceanic settings (Bryan and Ernst, 2008). The distribution of LIPs in space and time is used to speculate on the role that observed deep mantle structures may play in magmatism (e.g. Burke, 2011) and LIPs featuring large-scale dike swarms are used in continental reconstructions (e.g. Ernst *et al.*, 2013). While their origin is still debated, there has been a focus on the potential link between LIP magmatism, global environmental perturbations such as ocean anoxia, and mass extinctions (e.g. Wignall, 2005). Even when a temporal correlation exists, the details of how magmatism drives environmental stress and extinction remain difficult to evaluate for two major reasons. First, the domains where LIPs are emplaced, erupted and exposed are often geographically disparate from the stratigraphic sections where variations in marine chemistry and biology are documented, requiring high-fidelity stratigraphic proxies in order to correlate between sections. Second, geochronological data from LIPs generally lack the precision needed to evaluate cause and effect at the sub-millennial time-scales over which oceanographic and climate changes occur.

Studies of the major mass extinctions have revealed that in many cases the extinction interval is extremely short-lived (durations on the order of thousands of years) (e.g. Renne *et al.*, 2013; Burgess *et al.*, 2014) but that the recovery interval is protracted (e.g. 0.5–5 Ma; e.g. D'Hondt, 2005; Fraiser and Bottjer, 2007; Ramezani *et al.*, 2007; Chen and Benton, 2012; Renne *et al.*, 2013), and more difficult to constrain. Radioisotopic geochronology and astrochronology are the best tools for establishing the age and duration of biotic crises and recovery, allowing comparison

*Volcanism and Global Environmental Change*, eds. Anja Schmidt, Kirsten E. Fristad and Linda T. Elkins-Tanton. Published by Cambridge University Press. © Cambridge University Press 2015.

with possible drivers such as LIPs, asteroid impacts and global glaciations. In some cases, interpolation between multiple dated horizons within stratigraphic sequences of both LIPs and extinction intervals allow millennial-scale temporal resolution, even in deep time (Burgess *et al.*, 2014).

Although it is commonly assumed that magmatic pulses within LIPs are erupted over time intervals of ~ 1–5 Ma (Bryan and Ernst, 2008), most have not been dated to a level of precision that permits testing of this assumption. Several studies have suggested that massive volatile release in the very earliest stages of LIP emplacement may drive environmental changes that lead to mass extinction (e.g. Wignall, 2005; Svensen *et al.*, 2010; Sobolev *et al.*, 2011; Blackburn *et al.*, 2013), and that the largest volumes are erupted on very short timescales (e.g. Chenet *et al.*, 2009; Barry *et al.*, 2013). Since plagioclase-bearing basalt and gabbro are major rock types within LIPs, the majority of LIP geochronology has utilized the K–Ar and $^{40}$Ar–$^{39}$Ar systems (e.g. Courtillot and Renne, 2003; Jourdan *et al.*, 2008; Reichow *et al.*, 2009; Marzoli *et al.*, 2011; Baski, 2013). Most published $^{40}$Ar–$^{39}$Ar data are characterized by maximum precision on the order of 0.5–1.0% ($2\sigma$ internal), translating to hundreds of thousands to millions of years in uncertainty for Paleozoic events, often preventing a detailed comparison with extinction records and estimates of the duration of magmatism. In many cases U–Pb zircon geochronology offers an order of magnitude improvement in precision but has seen limited application due to an apparent paucity of zircon in dominantly basaltic rocks. However, some lavas and their intrusive equivalents contain uranium-bearing accessory minerals such as zircon, baddeleyite, titanite, perovskite, apatite and zirconolite (e.g. LeCheminant and Heaman, 1989; Heaman, 1997, 2009; Amelin and Naldrett, 1999; Kamo *et al.*, 2003; Rasmussen and Fletcher, 2004; Corfu and Dahlgren, 2008; Wu *et al.*, 2010). While both U–Pb and $^{40}$Ar–$^{39}$Ar geochronology have been applied to the dating of LIPs for decades, renewed interest in the causes of mass extinctions has emphasized that higher levels of precision and accuracy are necessary prerequisites for comparing the timing of magmatism with the timing of environmental change. This comparison necessitates that, where possible, geochronological constraints on LIPs be revisited with the latest techniques to improve both accuracy and precision. In the following contribution we discuss the occurrence and behavior of zircon in mafic LIP rocks, discuss the sources of major improvements to the precision and accuracy of zircon dates, and look forward to the future of high-precision geochronology on LIP rocks.

## 4.2 U-Pb zircon geochronology of mafic rocks

Zircon ($ZrSiO_4$) is the premier high-precision chronometer for most of Earth history, primarily because most zircons have no initial common-Pb. High-precision analysis of small amounts of radiogenic Pb in single zircons is routine (e.g. Schoene *et al.*, 2013), and Pb-loss can in most cases be eliminated. Rocks of nominally basaltic composition do not typically yield single zircon crystals that are large enough (>20 µm) to be dated by high-precision methods. Many LIPs, however, do contain thick flows with relatively coarse-grained interiors as well as pegmatitic segregations, both of which can contain zircon and baddeleyite (Kouvo, 1977). In our experience, zircons from these rocks have distinctive morphologies, high U contents, and easily fracture, requiring specialized separation and handling. Below we describe occurrences and guidelines for finding primary zircon in mafic rocks from LIPs based on our observations from the Siberian, Karoo, Ferrar, and CAMP (Central Atlantic Magmatic Province) LIPs, outline the petrographic setting of zircons, describe their behavior during the chemical abrasion process, and discuss the importance of accurate correction for initial Th exclusion.

### 4.2.1 Zircon crystallization in mafic rocks

The vast majority of mafic LIP rocks, both intrusive and extrusive rocks, are fine-grained with limited textural variation and contain rare zircon and baddeleyite crystals in excess of 10 µm. As a general rule, zircons are most common in the interior, coarsest-grained parts of a flow or intrusion or where fractional crystal-lization and physical segregation work to concentrate melts with high volatile and incompatible element concentrations. Much of this remaining melt buoyantly rises through either conduits or diapirs (Puffer and Horter, 1993; Philpotts *et al.*, 1996) to concentrate below the upper quench surface of a flow or sill. Other enriched segregations occur as irregularly bounded pods of coarse-grained plagioclase, pyroxene and oxides surrounded by a matrix of finer-grained material. These pods are seemingly quenched during their upward migration (Kontak, 2008; Figure 4.1). Consistent with this model are crystal size data for the Palisades sill, a ~ 300-m-thick intrusive basaltic member of the CAMP, which shows that crystal size is finest at both the top and bottom boundaries, and increases towards the sill center. Maximum crystal size is ~ 40 m below the sill top, where we presume that volatile-rich fluids have buoyantly risen and collected beneath the upper quench surface. Mimicking the observed trend in crystal size is the concentration of incompatible elements, such as Zr, which is shown to increase in concentration within the center of the Palisades sill, with the very highest concentrations corresponding to the coarse-grain horizon (Figure 4.2). A seminal increase in incompatible element concentration is seen within coarse-grained segregations in the CAMP basalt flows (Puffer and Horter, 1993; Philpotts *et al.*, 1996). In the Pallisades sill, large zircon crystals were found in multiple samples collected at the stratigraphic height of ~300 m, and yielded a weighted mean $^{206}Pb/^{238}U$ date of 201.520 ± 0.034 Ma (Blackburn *et al.*, 2013).

(a)                                    (b)

(c)

Figure 4.1   Photographs of a doloritic intrusion from the Siberian LIP. (a) View
of sill thickness and columnar jointing. (b) View of coarse-grained segregation
and finer-grained matrix. (c) View of plagioclase and pyroxene-rich coarse-
grained segregation from which zircon crystals were isolated. Photo credit Dougal
Jerram.  A black and white version of this figure will appear in some formats. For
the colour version, please refer to the plate section.

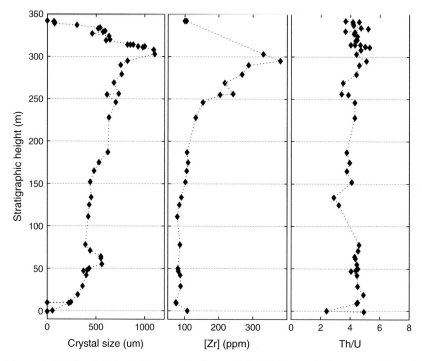

Figure 4.2   Crystal size, Zr concentration and Th/U for vertical transect through
Palisades sill, New Jersey, USA. Data from Shirley, 1987. Sample NB08–13,
which contained zircon crystals, was dated by Blackburn *et al.*, 2013, and was
sampled at a stratigraphic height of ∼ 300 m.

Below we summarize evidence linking zircons that form in basaltic rocks with a residual liquid or mesostatis, an observation that implies that zircons crystallize from a highly enriched residual liquid whose Zr concentrations likely far exceed the whole-rock values.

### 4.2.2 Zircon morphology in mafic rocks

Zircon crystals isolated from diabase or dolorite sills, basalts and gabbros commonly have distinct morphologies. Rather than being clear, faceted, doubly terminated crystals as are common in felsic rocks, these grains are more often blocky, brittle, and are easily fragmented during mineral separation. When not fragmented, grains commonly exhibit aspect ratios of 7:1 or greater (e.g. Figure 4.3a-2). In plane light, these grains are translucent to brownish and if unaltered (e.g. Figure 4.3a-1,2,3) can be transparent and faceted. In the more common case (e.g. Figure 4.3a-4), grains are fragmented and translucent brown. Grains often have crystal surfaces that are tightly corrugated, a morphology that is likely evidence of having crystallized in contact with a preexisting plagioclase crystal (e.g. Corfu *et al.*, 2003). Our experience suggests that grains which depart from this seemingly characteristic morphology are often xenocrysts.

Melt inclusions are common within LIP zircons and are often found radiating from a central point near the middle of the long axis of the crystal (e.g. Figure 4.3a-2,3), and also as a single, dark band in a similar orientation (e.g. Figure 4.3a-4). In rare cases where the entire zircon crystal is recovered, these opaque melt inclusions appear to narrow in the middle and flare towards the end of the crystal. Devitrified melt-inclusion compositions fall into two distinct categories: (1) feldspars that are unmixed into (cathodoluminescent) CL-bright and -darker zones – with the precision afforded by wavelength dispersive spectrometry, these zones are chemically indistinguishable (Figure 4.3d-2,3); and (2), an Fe-rich composition that is also separated into CL-light and -dark domains with no measurable compositional difference. Melts rich in silica, alkalis, and Fe are expected in the late stages of crystallization (e.g. Walker, 1969; Philpotts, 1979; Charlier and Grove, 2012).

Mineral inclusions are found primarily along the long axis of crystals, although the boundary between inclusion and host crystal is sharp rather than undulatory, as is the case with melt inclusions. Wavelength dispersive spectrometry suggests that these mineral inclusions have bulk compositions similar to the devitrified melt, either an Fe-rich phase or a feldspar, such as albite. Partial removal of inclusion material during polishing (e.g. Figure 4.3d-1) exposes evidence that suggests that the zircon nucleated on a pre-existing mineral, such as feldspar, and subsequently overgrew it, corroborating late zircon crystallization. Both mineral and melt inclusions are often surrounded by a halo of CL-bright zircon, which is then surrounded by CL-dark zircon. In grains without mineral and/or melt inclusions,

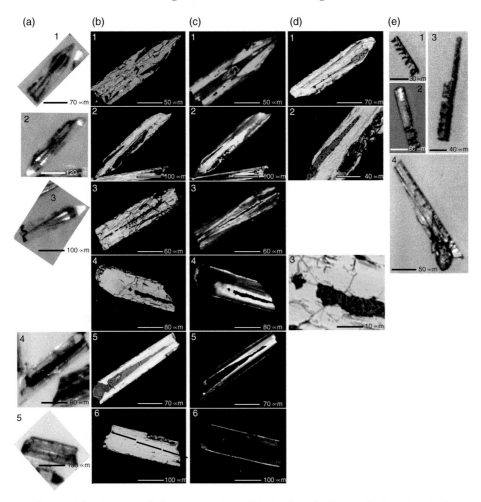

Figure 4.3  Images of zircon crystals and inclusions before and after chemical abrasion. (a) Plane-light images of thermally annealed zircon crystals from a Ferrar LIP diabase/dolorite, showing the range of inclusion morphology. (b) Backscatter electron microprobe images of zircon crystals. (c) Cathodoluminescence images of zircon crystals. (d) Backscatter images of inclusions within zircon crystals. (e) Plane-light images of zircon crystals after the chemical abrasion process.

CL images reveal a homogenous, relatively CL-dark grain interior surrounded by a very thin rim of CL-bright material (Figure 4.3c-6).

### *4.2.3 Application of the chemical abrasion technique*

Most high-precision U–Pb geochronology routinely utilizes the chemical abrasion method (Mattinson, 2005) to eliminate Pb-loss. Zircon crystals isolated from many basaltic rocks commonly dissolve completely during this process. However, if a slightly lower temperature is used for the partial dissolution step (12 hours at 180 °C vs. 210 °C), some zircon survives, although often as skeletal remnants, with clear evidence of having been partially dissolved (e.g. Figure 4.3e-1,2,3). In cases where the grains contain elongate inclusions, chemical abrasion completely dissolves the inclusions, leaving thin, blade-like fragments of zircon, that presumably had overgrown the inclusion. These shards are then completely dissolved and usually result in high-precision concordant analyses. In cases where chemical abrasion of a single grain yields a single large fragment and many smaller fragments, dates on both populations are identical within uncertainty.

### 4.2.4 Correction for initial Th exclusion

While the U–Pb zircon system may yield the most precise results for LIP eruptions, the accuracy of these measured dates is in part controlled by the necessary correction for the initial preferential exclusion of $^{230}$Th (and corresponding $^{206}$Pb deficiency) when zircon crystallizes. The magnitude of the correction can be significant when the highest levels of precision are achieved. Accurate correction requires knowledge of the relative partitioning of Th and U for a zircon and the magma from which it crystallized. The bulk partition coefficient for Th/U is determined using the Th/U of both the zircon and magma it crystallized from (Schärer, 1984). The zircon Th/U value is often determined from the measured $^{208}$Pb–$^{206}$Pb, assuming concordance between the $^{232}$Th–$^{208}$Pb and $^{238}$U–$^{206}$Pb systems. Zircons from flood basalts often have high Th/U values compared with those from silicic rocks (e.g. Palisades sill ~ 1.2). A Th/U for the magma may be measured directly from melt inclusions or inferred from whole-rock composition. It is unlikely that whole-rock data reliably represent the liquid from which a zircon in a flood basalt crystallized, as the magmatic processes leading to the concentration of incompatible elements, such as Zr, could very likely fractionate Th from U. Efforts to map the Th and U concentrations within sills reveals that although the concentrations of U and Th mimic that of Zr, the Th/U value remains nearly constant (~ 4.0). This was first observed by Gottfried *et al.* (1968) in diabase sills from CAMP and Tasmania, and is shown here, using data from Shirley (1987) for the Palisades sill (Figure 4.2).

   The magnitude of the $^{238}$U–$^{206}$Pb Th-correction is correlated with the difference between Th/U of the zircon crystal and that of the magma from which it crystallizes. There exists, however, a threshold at which this correction will reach a maximum, resulting in a near constant $^{238}$U–$^{206}$Pb date for some plausible Th/U magma compositions. For the Palisades sill, a series of Th-corrected weighted mean $^{238}$U–$^{206}$Pb dates, calculated over a range of assumed Th/U

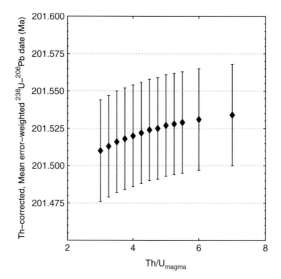

Figure 4.4   Model Th-corrected $^{238}$U–$^{206}$Pb error-weighted mean dates for the Palisades sill for a range of Th/U magma compositions. This correction yields results that are within uncertainty of one another, even over this broad range of Th/U compositions. The best estimate for the age of the Palisades sill comes from using a Th/U value observed for coarse-grained samples (~ 300 m, Th/U ~ 4, Figure 4.2).

values, remain within analytical uncertainty of one another. For the Palisades sill, the Th-correction is near maximized at Th/U magma compositions of ~ 3.5 (Figure 4.4). This range of assumed Th/U for the parental magma composition well exceeds the variation observed within the Palisades sill (best estimate Th/U = 4) (Figure 4.2), indicating the sill's emplacement date is both accurate and precise.

## 4.3  Why accuracy and precision matter

The ability to (1) determine the emplacement/eruption rates of LIP magmas and the spatial/temporal patterns in lava/sill chemistry, all of which are crucial for understanding their origin; and (2) critically evaluate the synchroneity and thus potential casual role of magmatism as a trigger for environmental forcing of biological crises requires having high-precision, accurate age models for both. For example, the age and duration of the end-Permian extinction is constrained by dates that are precise to better than 0.02% (Burgess *et al.*, 2014). Current temporal constraints on the timing and duration of the majority of LIP magmatism are consistent with the coincidence of magmatism and mass extinction (e.g. Courtillot and Renne, 2003), but prevent detailed evaluation of the synchrony and relative timing of magmatism and environmental perturbation at the timescale over which each is likely to have occurred. To illustrate the disparity in precision between the most precise published dates on the timing of biotic crises and the majority of dates characterizing LIP eruption/emplacement, Figure 4.5 compares the analytical uncertainty (at the $2\sigma$ percent level) on the weighted mean $^{206}Pb-^{238}U$ date defining cessation of the end-Permian mass extinction with recently-published U–Pb weighted mean $^{206}Pb-^{238}U$ dates and $^{40}Ar-^{39}Ar$ plateau dates for four Phanerozoic LIPs that have been temporally associated with, and implicated as triggers for, mass extinction: The Siberian Traps, CAMP, Karoo–Ferrar, and Deccan Traps. It is clear that the precision on dates constraining the age and duration of mass extinction events, e.g. the end-Permian biotic crisis (Reichow *et al.*, 2009; Burgess *et al.*, 2014) using both zircon and sanidine are more than one order of magnitude more precise than the majority of dates available on mafic LIP rocks.

The accuracy of U–Pb and $^{40}Ar-^{39}Ar$ dates has improved over the past decade, illustrated in Figure 4.5 using dates for the Permian–Triassic boundary. Published dates have evolved over the past 16 years as analytical capabilities and adoption of protocols including chemical abrasion have led to improvements in accuracy and precision (Mattinson, 2005; Schoene *et al.*, 2006; Kuiper *et al.*, 2008; Renne *et al.*, 2010; Bowring *et al.*, 2011; McLean *et al.*, 2011; Hiess *et al.*, 2012; Burgess *et al.*, 2014). A similar trend is expected with application of these advances to LIP dating.

### 4.3.1 Case study

#### Central Atlantic Magmatic Province (CAMP)

The CAMP LIP and the end-Triassic extinction serve as an example of the complex relationship between eruption, environmental effects and extinction. Estimates on the total duration of CAMP magmatism provided by astrochronology (Whiteside *et al.*, 2007) and $^{40}Ar-^{39}Ar$ dates (e.g. Nomade *et al.*, 2007; Hames and Renne, 2000; Beutel *et al.*, 2005; Marzoli *et al.*, 2011; Knight *et al.*, 2004) differed by an order of magnitude. Most of the available $^{40}Ar-^{39}Ar$ geochronologic data for CAMP in the eastern USA and Morocco permit a causal relationship by overlapping within uncertainty with both the end-Triassic extinction (201.56 ± 0.015 Ma) and the subsequent biologic recovery (Triassic–Jurassic boundary at 201.46 ± 0.040 Ma (Blackburn *et al.*, 2013; Figure 4.6a). Problematically, this includes geochronologic data from rocks that lie stratigraphically *above* the extinction horizon. Though these lower-precision $^{40}Ar-^{39}Ar$ dates permit a causal relationship with CAMP and the extinction, they prevent a detailed comparison of the eruption, extinction and recovery histories and permit an emplacement interval of up to several million years – durations that far exceed the effusion rates required to explain the inferred increase in the partial pressure of $CO_2$ ($pCO_2$) in the atmosphere (Rampino and Caldeira, 2011).

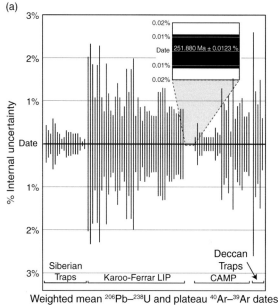

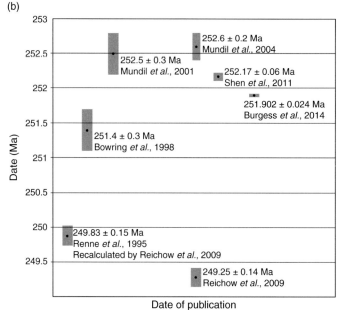

Figure 4.5   (a) $2\sigma$ percent internal uncertainty on U–Pb weighted-mean
$^{206}$Pb–$^{238}$U dates and $^{40}$Ar–$^{39}$Ar plateau dates published on the Siberian Traps
LIP, the Deccan Traps LIP, the CAMP, and the Karoo–Ferrar LIP. Each vertical
line represents a single published plateau age or weighted-mean date, the length
of which is equal to the $2\sigma$ percent uncertainty on that date. The black line through
the entire dataset, which is shown zoomed in the inset, is the $2\sigma$ percent internal
uncertainty on the $^{206}$Pb–$^{238}$U weighted-mean date defining cessation of the

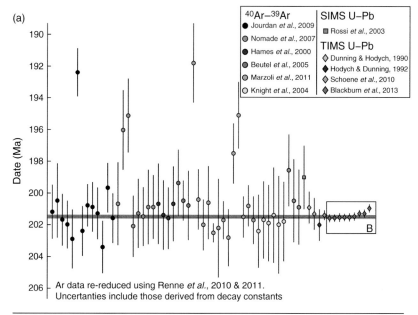

(a)

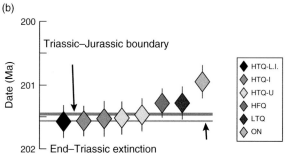

(b)

Figure 4.6 (a) Comparison between $^{40}$Ar–$^{39}$Ar and U–Pb zircon data for CAMP-related basalt flows and intrusions from the eastern USA and Morocco. The relatively large uncertainties associated with $^{40}$Ar–$^{39}$Ar or SIMS U–Pb data provides insufficient age resolution for clarifying the relationship between CAMP, the end-Triassic extinction and the biologic recovery or Triassic–Jurassic boundary. (b) High-precision zircon U–Pb dates place the earliest CAMP eruptions within uncertainty of the end-Triassic extinction. A black and white version of this figure will appear in some formats. For the colour version, please refer to the plate section.

Figure 4.5 (*cont.*) end-Permian mass extinction (Burgess *et al.*, 2014). Data from (Duncan *et al.*, 1997; Kamo *et al.*, 2003; Jourdan *et al.*, 2008; Reichow *et al.*, 2009; Svensen *et al.*, 2010; Marzoli *et al.*, 2011; Baksi, 2014). (b) Evolution of the age of the Permian–Triassic boundary since 1995. $^{40}$Ar–$^{39}$Ar plateau dates from (Renne *et al.*, 1995; Reichow *et al.*, 2009) and weighted mean $^{206}$Pb–$^{238}$U dates from (Bowring *et al.*, 1998; Mundil *et al.*, 2001; 2004; Shen *et al.*, 2011; Burgess *et al.*, 2014).

Previous syntheses of stratigraphic, paleontological and astrochronologic data in North America placed the end-Triassic extinction – recorded as a turnover in vertebrae trace fossils, palynology and a fern spike – to pre-date North America's earliest CAMP basalts by 20 kyr (summarized in Blackburn *et al.*, 2013). Across the Atlantic Basin in Morocco, this same terrestrial extinction horizon is not observed, where instead, Triassic sporomorphs are present up to the base of the lowermost CAMP basalts. In Blackburn *et al.* (2013), eight basalt flows and sills from each stratigraphically defined geochemical member were dated using the zircon U–Pb method (Figure 4.6b). These high-precision dates, combined with cyclo- and magneto-stratigraphic data, provide the ability to correlate the timing and relative position of the extinction event observed within the North American stratigraphy to Morocco (where the extinction is not observed), where the very oldest CAMP flood basalts can be placed directly on top of or coincident with the extinction event (Blackburn *et al.*, 2013). This coincidence between extinction and the eruption of the very earliest lavas provides a key data point for future climate models aiming to understand precisely the mechanism(s) that led to global biologic collapse. These data further provide a measure of: (1) LIP duration, resolving four discrete pulses of magma within a ~ 620 kyr eruptive interval, a duration that is consistent with astrochronologic estimates (Whiteside *et al.*, 2007), and (2) the rate of CAMP's geochemical evolution – from the most primitive, voluminous early units to the increasingly more mid-ocean-ridge basalt (MORB)-like, smaller bodies that are emplaced in three discreet pulses over the next 620 kyr.

## 4.6 Future of high-precision LIP geochronology

The past decade has seen significant advancements in U–Pb and $^{40}$Ar–$^{39}$Ar geochronology and most importantly the recognition, quantification and minimization of different sources of uncertainty in the calculation of a radiometric date, including inter-technique and inter-laboratory biases. This has in turn led to major advances in technique development for both geochronometers and in many cases agreement at the 0.1% level between laboratories and techniques. This progress allows more focused and sophisticated questions to be asked of the geologic record and allows for robust integration of different geochronometers, which in turn drives further advance in geochronology. An excellent example of this iterative approach is detailed study of the possible links between LIPs and mass extinctions. Although the approximate coincidence has been suggested for decades, high-precision geochronology allows us to better understand important aspects of possible links including the timing, duration and tempo, of eruptions. These datasets can also elucidate the temporal relationships between intrusive and extrusive magmatism, high-precision records of reversal of the paleomagnetic field in the lava stratigraphy, and the relationships between specific pulses of LIP eruption/emplacement and mass extinctions/biotic crises.

We propose that the active field of U–Pb geochronology of mafic rocks in LIPs for tectonic applications (e.g. Ernst *et al.*, 2013) can, in some cases be taken to higher levels of precision and accuracy by: (1) utilizing a variety of U-bearing accessory minerals in addition to zircon, including perovskite, baddelyite and zirconolite to better understand the age, duration, and tempo of LIP magmatism; and (2) integrating the U–Pb and $^{40}Ar$–$^{39}Ar$ geochronometers. Additionally, it is clear that to fully understand the origin and significance of LIPs in Earth history as well as the links between LIPs and global environmental perturbation, multiple, stratigraphically well-constrained dates on mafic rocks are essential. This will allow detailed testing of the environmental effects of large eruptions with the emerging highly resolved records of extinction intervals. The next level of inquiry must involve testing model predictions for collapse of ecosystems based on the relative timing between eruption, the relative timing of increased $CO_2$ levels in the oceans and atmosphere, variations in pH, and anoxia, for example. Ideally this will involve integration of paleomagnetic data (including secular variations), astrochronology, and $^{40}Ar$–$^{39}Ar$ and U–Pb geochronology. Doing so will put age models with millennial-scale resolution for thc last 542 Myr within our grasp.

## References

Amelin, Y., Li, C. and Naldrett, A. J. (1999) Geochronology of the Voisey's Bay intrusion, Labrador, Canada, by precise U–Pb dating of coexisting baddeleyite, zircon, and apatite, *Lithos*, **47**, pp. 33–51.

Baksi, A. K. (2014) The Deccan Trap Cretaceous–Paleogene boundary connection; new $^{40}Ar/^{39}Ar$ ages and critical assessment of existing argon data pertinent to this hypothesis, *Journal of Asian Earth Sciences*, **84**, pp. 9–23.

Barry, T. L., Kelley, S. P., Reidel, S. P. *et al.* (2013) Eruption chronology of the Columbia River Basalt Group, *Geological Society of America Special Papers*, **497**, pp. 45–66.

Baski, A. K. (2013) $^{40}Ar/^{39}Ar$ ages of flood basalt provinces in Russia and China and their possible link to global faunal extinction events: a cautionary tale regarding alteration and loss of $^{40}Ar$, *Journal of Asian Earth Sciences*, pp. 1–13.

Beutel, E. K., Nomade, S. and Fronabarger, A. K. (2005) Pangea's complex breakup: a new rapidly changing stress field model, *Earth and Planetary Science Letters*, **236**, pp. 471–485.

Blackburn, T. J., Olsen, P. E., Bowring, S. A. *et al.* (2013) Zircon U–Pb geochronology links the end-Triassic extinction with the Central Atlantic Magmatic Province, *Science*, pp. 941–945.

Bowring, J. F., McLean, N. M. and Bowring, S. A. (2011) Engineering cyber infrastructure for U–Pb geochronology: Tripoli and U–Pb_Redux, *Geochemistry Geophysics Geosystems*, **12**, pp. 1–19.

Bowring, S. A., Erwin, D. H., Jin, Y. G. *et al.* (1998) U/Pb zircon geochronology and tempo of the end-Permian mass extinction, *Science*, **280**, pp. 1039–1045.

Bryan, S. E. and Ernst, R. E. (2008) Revised definition of large igneous provinces (LIPs), *Earth-Science Reviews*, **86**, pp. 175–202.

Burgess, S. D., Bowring, S. A. and Shen, S.-Z. (2014) A new high-precision timeline for Earth's most severe extinction, *Proceedings of the National Academy of Sciences*, **111**, pp. 1–37.

Burke, K. (2011) Plate tectonics, the Wilson cycle, and mantle plumes: geodynamics from the top, *Annual Review of Earth and Planetary Sciences*, **39**, pp. 1–29.

Charlier, B. and Grove, T. L. (2012) Experiments on liquid immiscibility along tholeiitic liquid lines of descent, *Contributions to Mineralogy and Petrology*, **164**, pp. 27–44.

Chen, Z. Q. and Benton, M. J. (2012) The timing and pattern of biotic recovery following the end-Permian mass extinction, *Nature Geoscience*, **5**, pp. 375–383.

Chenet, A.-L., Courtillot, V., Fluteau, F. *et al.* (2009) Determination of rapid Deccan eruptions across the Cretaceous-Tertiary boundary using paleomagnetic secular variation: 2. Constraints from analysis of eight new sections and synthesis for a 3500-m-thick composite section, *Journal of Geophysical Research*, **114**, p. B06103.

Corfu, F. and Dahlgren, S. (2008) Perovskite U–Pb ages and the Pb isotopic composition of alkaline volcanism initiating the Permo-Carboniferous Oslo Rift, *Earth and Planetary Science Letters*, **265**, pp. 256–259.

Corfu, F., Hanchar, J. M., Hoskin, P. W. O. *et al.* (2003) Atlas of zircon textures, *Reviews in Mineralogy and Geochemistry. Zircon*, pp. 469–500.

Courtillot, V. E. and Renne, P. R. (2003) On the ages of flood basalt events, *Comptes Rendus Geoscience*, **335**, pp. 113–140.

D'Hondt, S. (2005) Consequences of the Cretaceous/Paleogene mass extinction for marine ecosystems, *Annual Review of Ecology, Evolution, and Systematics*, **36**, pp. 295–317.

Duncan, R. A., Hooper, P. R. and Rehacek, J. (1997) The timing and duration of the Karoo igneous event, southern Gondwana, *Journal of Geophysical Research*, **102**, pp. 18127–18138.

Ernst, R. E., Bleeker, W., Söderlund, U. *et al.* (2013) Large igneous provinces and supercontinents: toward completing the plate tectonic revolution, *Lithos*, **174**, pp. 1–14.

Fraiser, M. L. and Bottjer, D. J. (2007) Elevated atmospheric $CO_2$ and the delayed biotic recovery from the end-Permian mass extinction, *Palaeogeography, Palaeoclimatology, Palaeoecology*, **252**, pp. 164–175.

Gottfried, D., Greenland, L. P. and Campbell, E. Y. (1968) Variation of Nb–Ta, Zr–Hf, Th–U and K–Cs in two diabase-granophyre suites, *Geochimica et Cosmochimica Acta*, **32**, pp. 925–947.

Hames, W. and Renne, P. (2000) New evidence for geologically instantaneous emplacement of earliest Jurassic Central Atlantic Magmatic Province basalts on the North American margin, *Geology*, **28**, pp. 859–862.

Heaman, L. M. (1997) Global mafic magmatism at 2.45 Ga: remnants of an ancient large igneous province?, *Geologic Society of America Special Paper*, **25**(4), p. 299.

Heaman, L. M. (2009) The application of U–Pb geochronology to mafic, ultramafic and alkaline rocks: an evaluation of three mineral standards, *Chemical Geology*, **261**, pp. 43–52.

Hiess, J., Condon, D. J., McLean, N. *et al.* (2012) $^{238}U/^{235}U$ systematics in terrestrial uranium-bearing minerals, *Science*, **335**, pp. 1610–1614.

Jourdan, F., Féraud, G., Bertrand, H. *et al.* (2008) The $^{40}Ar/^{39}Ar$ ages of the sill complex of the Karoo large igneous province: implications for the Pliensbachian–Toarcian climate change, *Geochemistry Geophysics Geosystems*, **9**, pp. 1–20.

Kamo, S. L., Czamanske, G. K., Amelin, Y. *et al.* (2003) Rapid eruption of Siberian flood-volcanic rocks and evidence for coincidence with the Permian–Triassic boundary and mass extinction at 251 Ma, *Earth and Planetary Science Letters*, **214**, pp. 75–91.

Knight, K. B., Nomade, S., Renne, P. R. *et al.* (2004) The Central Atlantic Magmatic Province at the Triassic–Jurassic boundary: paleomagnetic and $^{40}$Ar/$^{39}$Ar evidence from Morocco for brief, episodic volcanism, *Earth and Planetary Science Letters*, **228**, pp. 143–160.

Kontak, D. J. (2008) On the edge of CAMP: geology and volcanology of the Jurassic North Mountain Basalt, Nova Scotia, *Lithos*, **101**, pp. 74–101.

Kouvo, O. (1977) The use of mafic pegmatoids in geochronometry, *Fifth European Colloquium of Geochronology, Cosmochronology and Isotope Geology*, Pisa, Abstracts.

Kuiper, K. F., Deino, A., Hilgen, F. J. *et al.* (2008) Synchronizing rock clocks of Earth history, *Science*, **320**, pp. 500–504.

LeCheminant, A. N. and Heaman, L. M. (1989) Mackenzie igneous events, Canada: Middle Proterozoic hotspot magmatism associated with ocean opening, *Earth and Planetary Science Letters*, **96**, pp. 38–48.

Marzoli, A., Jourdan, F., Puffer, J. H. *et al.* (2011) Timing and duration of the Central Atlantic Magmatic Province in the Newark and Culpeper basins, eastern U.S.A., *Lithos*, **122**, pp. 175–188.

Mattinson, J. M. (2005) Zircon U–Pb chemical abrasion ("CA-TIMS") method: combined annealing and multi-step partial dissolution analysis for improved precision and accuracy of zircon ages, *Chemical Geology*, **220**, pp. 47–66.

McLean, N., Bowring, J. and Bowring, S. (2011) An algorithm for U–Pb isotope dilution data reduction and uncertainty propagation, *Geochemistry Geophysics Geosystems*, **12**, pp. 1–26.

Mundil, R., Ludwig, K. R., Metcalfe, I. *et al.* (2004) Age and timing of the Permian mass extinctions: U/Pb dating of closed-system zircons, *Science*, **305**, pp. 1760–1763.

Mundil, R., Metcalfe, I., Ludwig, K. R. *et al.* (2001) Timing of the Permian–Triassic biotic crisis: implications from new zircon U/Pb age data (and their limitations), *Earth and Planetary Science Letters*, **187**, pp. 131–145.

Nomade, S., Knight, K. B., Beutel, E. *et al.* (2007) Chronology of the Central Atlantic Magmatic Province: implications for the Central Atlantic rifting processes and the Triassic–Jurassic biotic crisis, *Palaeogeography, Palaeoclimatology, Palaeoecology*, **244**, pp. 326–344.

Philpotts, A. R. (1979) Silicate liquid immiscibility in tholeiitic basalts, *Journal of Petrology*, **20**(1), pp. 99–118.

Philpotts, A. R., Carroll, M. and Hill, J. M. (1996) Crystal-mush compaction and the origin of pegmatitic segregation sheets in a thick flood-basalt flow in the Mesozoic Hartford Basin, Connecticut, *Journal of Petrology*, **37**, pp. 811–836.

Puffer, J. H. and Horter, D. L. (1993) Origin of pegmatitic segregation veins within flood basalts, *Geological Society of America Bulletin*, **105**, pp. 738–748.

Ramezani, J., Bowring, S. A., Martin, M. W. *et al.* (2007) Timing of recovery from the end-Permian extinction: geochronologic and biostratigraphic constraints from south China: COMMENT AND REPLY: REPLY, *Geology*, **35**, pp. e137–e138.

Rampino, M. R. and Caldeira, K. (2011) Comment on "Atmospheric *p*CO$_2$ perturbations associated with the Central Atlantic Magmatic Province", *Science*, **334**, pp. 594–594.

Rasmussen, B. and Fletcher, I. R. (2004) Zirconolite: a new U–Pb chronometer for mafic igneous rocks, *Geologic Society of America Special Paper*, **32**, p. 785.

Reichow, M. K., Pringle, M. S., Al'Mukhamedov, A. I. *et al.* (2009) The timing and extent of the eruption of the Siberian Traps large igneous province: implications for the end-Permian environmental crisis, *Earth and Planetary Science Letters*, **277**, pp. 9–20.

Renne, P. R., Black, M. T., Zichao, Z. *et al.* (1995) Synchrony and causal relations between Permian–Triassic boundary crises and Siberian flood volcanism, *Science*, **269**, pp. 1413–1416.

Renne, P. R., Deino, A. L., Hilgen, F. J. *et al.* (2013) Time scales of critical events around the Cretaceous–Paleogene boundary, *Science*, **339**, pp. 684–687.

Renne, P. R., Balco, G., Ludwig, K. R. *et al.* (2011) Response to the comment by W.H. Schwarz *et al.* on "Joint determination of $^{40}$K decay constants and $^{40}$Ar*/$^{40}$K for the Fish Canyon sanidine standard, and improved accuracy for $^{40}$Ar/$^{39}$Ar geochronology" by P.R. Renne *et al.* (2010), *Geochemica et Cosmochemica Acta.*, **75**, pp. 5097–5100.

Renne, P. R., Mundil, R., Balco, G. *et al.* (2010) Joint determination of $^{40}$K decay constants and $^{40}$Ar–$^{40}$K for the Fish Canyon sanidine standard, and improved accuracy for $^{40}$Ar/$^{39}$Ar geochronology, *Geochimica et Cosmochimica Acta*, **74**, pp. 5349–5367.

Rossi, P., Cocherie, A., Fanning, C. M. *et al.* (2003) Datation U-Pb sur zircons des dolérites tholéiitiques pyrénéennes (ophites) à la limite Trias–Jurassique et relations avec les tufs volcaniques dits « infra-liasiques » nord-pyrénéens, *Comptes Rendus Geoscience*, **335**, pp. 1071–1080.

Schärer, U. (1984) The effect of initial $^{230}$Th disequilibrium on young U–Pb ages: the Makalu case, Himalaya, *Earth and Planetary Science Letters*, **67**, pp. 191–204.

Schoene, B., Condon, D. J., Morgan, L. *et al.* (2013) Precision and accuracy in geochronology, *Elements*, **9**, pp. 19–24.

Schoene, B., Crowley, J. L., Condon, D. J. *et al.* (2006) Reassessing the uranium decay constants for geochronology using ID-TIMS U–Pb data, *Geochimica et Cosmochimica Acta*, **70**, pp. 426–445.

Schoene, B., Guex, J., Bartolini, A. *et al.* (2010) Correlating the end-Triassic mass extinction and flood basalt volcanism at the 100 ka level, *Geology*, **38**, pp. 387–390.

Shen, S. Z., Crowley, J. L., Wang, Y. *et al.* (2011) Calibrating the end-Permian mass extinction, *Science*, **334**, pp. 1367–1372.

Shirley, D. N. (1987) Differentiation and compaction in the Palisades sill, New Jersey, *Journal of Petrology*, **28**, pp. 835–865.

Sobolev, S. V., Sobolev, A. V., Kuzmin, D. V. *et al.* (2011) Linking mantle plumes, large igneous provinces and environmental catastrophes, *Nature*, **477**, pp. 312–316.

Svensen, H., Corfu, F., Polteau, S. *et al.* (2010) Rapid magma emplacement in the Karoo Large Igneous Province, *Earth and Planetary Science Letters*, **325–326**, pp. 1–9.

Walker, K. R. (1969) *The Palisades Sill, New Jersey: A Reinvestigation*. Boulder, CO: Geological Society of America, Special Paper, vol. 111.

Whiteside, J. H., Olsen, P. E., Kent, D. V. *et al.* (2007) Synchrony between the Central Atlantic Magmatic Province and the Triassic–Jurassic mass-extinction event?, *Palaeogeography, Palaeoclimatology, Palaeoecology*, **244**, pp. 345–367.

Wignall, P. (2005) The link between large igneous province eruptions and mass extinctions, *Elements*, **1**, pp. 293–297.

Wu, F. Y., Yang, Y. H., Mitchell, R. H. *et al.* (2010) In situ U–Pb age determination and Nd isotopic analysis of perovskites from kimberlites in southern Africa and Somerset Island, Canada, *Lithos*, **115**, pp. 205–222.

# 5

# Volcanic pulses in the Siberian Traps as inferred from Permo-Triassic geomagnetic secular variations

VLADIMIR PAVLOV, FRÉDÉRIC FLUTEAU, ROMAN VESELOVSKIY, ANNA
FETISOVA, ANTON LATYSHEV, LINDA T. ELKINS-TANTON, ALEXANDER V.
SOBOLEV AND NADEZHDA A. KRIVOLUTSKAYA

## 5.1 Introduction

There is little reason to believe that eruption of the Siberian Traps occurred gradually. On the contrary, trap emplacement likely occurred in the form of brief but voluminous volcanic pulses, as for example demonstrated for the Deccan Traps (Chenet *et al.*, 2008, 2009) in India and for the Karoo Traps (Moulin *et al.*, 2011, 2012) in South Africa.

To search for evidence of such pulses during formation of the Siberian Traps, and to obtain time constraints on eruptive activity we have undertaken a paleomagnetic study of several important trap volcanic sections from the Norilsk and Maymecha–Kotuy areas (Figure 5.1). To complement our work, we include paleomagnetic directions published by Heunemann *et al.* (2004) from the Norilsk area. A brief description of the sampling areas and the paleomagnetic results are presented below.

## 5.2 Sampling

Detailed paleomagnetic sampling was carried out on the Medvejia and Truba sections in the Maymecha–Kotuy region (Figure 5.1a, b) during field seasons in 2008 and 2009, and on the Sunduk and Ergalakh sections in the Norilsk region in 2010 (Figure 5.1c). Between 8 and 20 oriented hand samples were collected from individual lava flows or tuff units. In total, about 1400 samples, representing 133 lava flows, were collected and studied.

The Medvejia and Truba sections (~600 m thick) are on the Kotuy River and include lava flows of the Arydzhangsky and Onkuchaksky Formations. The sections collectively constitute the most complete (among those exposed) sequence of volcanic rocks of these formations in the region (Figure 5.2). The Medvejia section contains in total 37 units of the Arydzhangsky suite, which consists of

*Volcanism and Global Environmental Change*, eds. Anja Schmidt, Kirsten E. Fristad and Linda T. Elkins-Tanton.
Published by Cambridge University Press. © Cambridge University Press 2015.

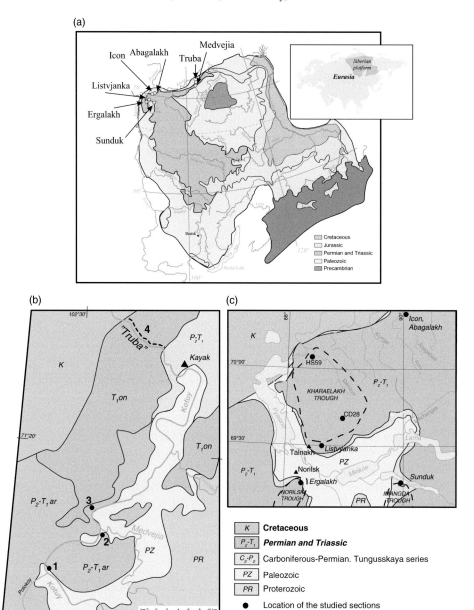

Figure 5.1 (a) Sketch geological map of the Siberian platform with location of the studied sections; (b) studied volcanic sections of the Kotuy River valley: 1–3, Medvejia; 4, Truba (ar = Arydzhangsky Formation, on = Onkuchaksky Fm (Kogotoksky series)); (c) studied volcanic sections of Norilsk region (CD28 and HS59 = boreholes). A black and white version of this figure will appear in some formats. For the colour version, please refer to the plate section.

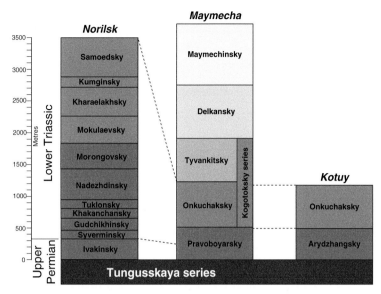

Figure 5.2    Stratigraphy of volcanic sequences of Norilsk, Kotuy and Maymecha regions. Correlation scheme after (Kamo *et al.*, 2003).

melanephelinite, augitite and other alkaline-ultramafic lavas interlayered in the lower part of the section with alkaline-ultramafic tuffs. The Truba section contains 42 basaltic flows of the Onkuchaksky Formation.

The Ergalakh section (~200 m thick) is located 15 km south of Norilsk (Figure 5.1c). The lowermost part of the Norilsk lava section, including the Ivakinsky, Syverminsky and Gudchikhinsky formations (Figure 5.2), is exposed there. Samples were collected from 12 basaltic flows and from the Norilsk-2 intrusion, which cuts the volcanic sequence. The Sunduk section is located about 90 km to the east of Norilsk (Figure 5.1c) and contains the lower half of the Norilsk volcanic sequence from the Ivakinsky to Nadezhdinsky formations, with a total thickness of about 750 m. The section consists of 42 tholeitic and picritic lava flows and mafic tuffs.

The rocks composing the studied sections are virtually flat-lying except in the Truba section, where flows are slightly tilted with dip to the northwest at 3–5°. Sedimentary rocks of nearby outcropping Tungusskaya sedimentary series are flat-lying, so we propose that the dip of the Truba flows represents local paleo-topography due to erosion.

Additionally, we have used Heunemann *et al.* (2004) from four sections (Listvjanka, Icon, Abagalakh – ~2200 m – and Talnakh – ~180 m) located in the Norilsk  area (Figure 5.1c). The Listvjanka section contains the three lowermost units from the Ivakinsky to Gudchikhinsky formations, the Icon section contains three units from the Tuklonsky to Morongovsky formations, the Abagalakh section contains four units from the Morongovsky to Kumginsky formations, and the

Talnakh section contains the two lowermost units (Ivakinsky and Syverminsky formations).

## 5.3 Paleomagnetism

All samples were thermally demagnetized up to 580–680 °C with an average of 12 steps to isolate the components of the natural remanent magnetization. The measurements were performed in the paleomagnetic laboratories of the Institut de Physique du Globe de Paris (France), the Institute of Physics of the Earth and Lomonosov Moscow State University (Moscow, Russia).

The characteristic remanent magnetizations were determined after inspection of orthogonal projections (Zijderveld, 1967) using principal component analysis (Kirschvink, 1980). Site mean directions based on four to ten sample results were calculated using Fisher statistics (Fisher, 1953). The paleomagnetic analysis was performed using the PaleoMac software (Cogné, 2003) or R. J. Enkin's (Enkin, 1994) paleomagnetic software packages.

The quality of the paleomagnetic record differs from flow to flow. In most samples, from all studied sections two components of magnetization can be isolated. A low-temperature component, destroyed by heating at 200–250 °C, has a direction close to the modern geomagnetic field. We therefore consider this component as a recent viscous overprint and do not discuss it further. The second, high-temperature component decays to the origin of an orthogonal vector diagram between 200–600 °C (Figure 5.3a, d, g, j).

We sampled 20 lithic clasts from a tuff layer from the lower part of the Arydzhangsky Formation. In 19 of them we isolated the high-temperature component of magnetization, which decays to the origin of the Zijderveld diagram between 350 and 560 °C. The distribution of vectors of this component is shown in Figure 5.3c. The length of the resultant vector, $R = 2.33$, is much less than the critical (at the 95% confidence level) value $R_o = 6.98$ (Watson, 1956) for $N = 19$. Therefore, the result of the conglomerate test is positive, and indicates a primary origin of high-temperature component isolated in the Arydzhangsky Formation.

The lowermost Arydzhangsky Formation (sometimes referred to as the Khardakhsky Formation (Egorov, 1995)) in the lower part of the composite Kotuy section has a reversed polarity (Table 5.1 and Figure 5.3f). Above this, the larger part of the Arydzhangsky and the lowermost flows of the Onkuchaksky formations yield a normal polarity, while the overlying middle and upper parts of the Onkuchaksky Formation are again reversely magnetized.

A reversal test (McFadden and McElhinny, 1990) applied to the whole section indicates some minor non-antipodality (see Figure 5.3b) that could result either

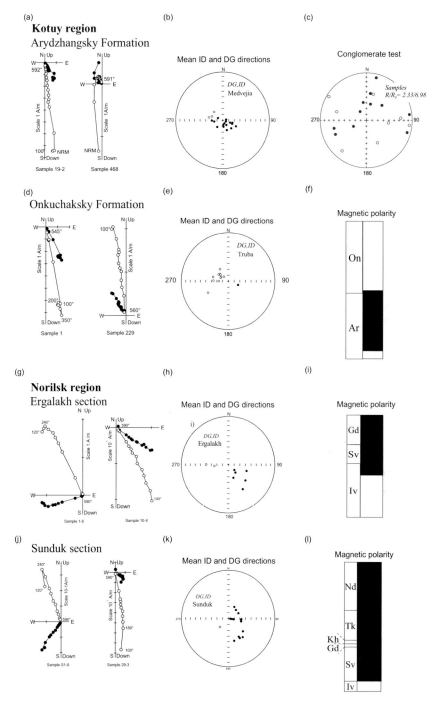

Figure 5.3  Paleomagnetism of trap formations of the Kotuy River valley and the Norilsk region. (a), (d) Orthogonal component plots for the samples from the Arydzhangsky (a) and Onkuchaksky (d) formations. (g), (j) Orthogonal component

from insufficient averaging of the secular variation, or from weak contamination by the low-temperature component.

Several of the lowest flows in the Ivakinsky Formation in both the Sunduk and Ergalakh sections yield a reverse polarity. All upper-lying flows in these sections are magnetized in normal polarity.

The primary origin of the high-temperature component isolated in these rocks relies on:

(1) the close proximity of these mean directions to the expected Permo-Triassic directions for this region (Pavlov *et al.*, 2007), and their distinction from younger Siberian directions (Pavlov, 2012);
(2) a positive conglomerate test for the Arydzhangsky Formation;
(3) the fact that normal and reversed polarity directions are virtually antiparallel;
(4) a positive fold test for the Morongovsky Formation (Pavlov *et al.*, 2013); and
(5) the existence of primary magmatic magnetic minerals (Latyshev *et al.*, 2012).

A primary origin of the isolated magnetization is also supported by the statistically different mean directions of the directional groups (see Section 5.4).

## 5.4 Discussion

In total we have studied 79 flows from the Kotuy region (Medvejia and Truba sections) and 54 flows from the Norilsk region (Sunduk and Ergalakh sections). In 116 out of 133 sampled sites, we have isolated the high-temperature component, which we interpret to be primary.

To estimate the time interval during which the studied trap sections were formed, we applied the procedure described in Chenet *et al.* (2008, 2009). This procedure, based on the analysis of the secular variations recorded in lava flows, has permitted the identification of 17 directional groups (DGs) and 13 individual directions (IDs) in the composite Kotuy section (Table 5.1 and Figure 5.4), four

---

Figure 5.3 (*cont.*) plots for samples from the Ergalakh (g) and Sunduk (j) sections. Closed (open) circles represent projection onto the horizontal (vertical) plane. (b), (e) Equal area projections showing mean ID and DG directions for the Arydzhangsky (b) and Onkuchaksky (e) formations. (h), (k) Equal area projections showing mean flow directions for the Ergalakh (h) and Sunduk (k) sections. (c) Equal area projection showing the result of the conglomerate test for the Arydzhangsky Formation. Closed (open) circles on the equal area net indicate down (up) direction. (f) Magnetic polarity of volcanic traps of the Kotuy River valley. (Ar = Arydzhangsky Formation, On = Onkuchaksky Formation.). (i), (l) Magnetic polarity of Ergalakh (i) and Sunduk (l) sections (Iv = Ivakinsky, Sv = Syverminsky, Gd = Gudchikhinsky, Kh = Khakanchansky, Tk = Tuklonsky, Nd = Nadezhdinsky formations). Black (white) = normal (reversed) polarity.

Table 5.1. *Mean paleomagnetic directions for IDs and DGs of studied sections of the Maymecha–Kotuy region. Arydzhangsky Formation – Medvejia section, 71.2° N, 102.7° E; flows: hard1–flow26. Onkuchaksky Formation – Truba section, 71.6°N, 102.9°E; flows fl1–fl42.*

| nn | Flows[a] | DG (ID) | $N_{DG}$ ($N_{ID}$) | D (°) | I (°) | K | α95 (°) |
|---|---|---|---|---|---|---|---|
| | | DG and ID statistics | | | | | |
| 1 | hard1 | ID1 | (5) | 281.3 | −65.3 | 15.1 | 16.1 |
| 2 | hard2–hard5 | DG1 | 4 | 312.5 | −69.4 | 255.1 | 4.4 |
| 3 | hard6 | ID2 | (3) | 289.0 | −70.0 | 10.8 | 24.6 |
| 4 | medv1–medv2 | DG2 | 2 | 132.5 | 70.0 | 66.1 | 12.2 |
| 5 | medv3–medv4 | DG3 | 2 | 78.4 | 84.1 | 73.7 | 11.5 |
| 6 | flow0–flow3 | DG4 | 4 | 152.3 | 84.1 | 231.0 | 4.6 |
| 7 | flow4–flow6b | DG5 | 4 | 107.1 | 76.7 | 200.5 | 4.9 |
| 8 | flow7–flow10 | DG6 | 4 | 173.0 | 79.1 | 252.0 | 4.4 |
| 9 | flow3 | ID3 | (10) | 262.1 | 75.0 | 21.3 | 9.6 |
| 10 | flow2 | ID4 | (15) | 294.0 | 80.9 | 136.0 | 3.1 |
| 11 | flow1 | ID5 | (7) | 268.3 | 71.0 | 51.1 | 7.4 |
| 12 | flow13 | ID6 | (7) | 109.8 | 84.3 | 23.6 | 10.9 |
| 13 | flow14–flow16 | DG7 | 3 | 104.3 | 72.7 | 122.4 | 7.3 |
| 14 | flow1718–flow19 | DG8 | 2 | 174.1 | 82.3 | 81.9 | 10.9 |
| 15 | flow20 | ID7 | (13) | 220.7 | 75.2 | 55.8 | 5.2 |
| 16 | flow21–flow22 | DG9 | 2 | 141.5 | 79.9 | 52.0 | 13.7 |
| 17 | flow23 | ID8 | (6) | 113.9 | 64.8 | 33.3 | 9.9 |
| 18 | flow 25 | ID9 | (8) | 156.6 | 72.0 | 78.1 | 5.6 |
| 19 | flow26–fl3 | DG10 | 4 | 110.0 | 72.6 | 247.4 | 4.5 |
| 20 | fl4 | ID10 | (15) | 292.6 | −75.4 | 52.3 | 5.0 |
| 21 | fl5 | ID11 | (7) | 285.1 | −62.2 | 24.5 | 10.7 |
| 22 | fl12–fl13 | DG11 | 2 | 308.1 | −72.4 | 46.0 | 14.6 |
| 23 | fl14–fl17 | DG12 | 5 | 272.1 | −63.7 | 207.4 | 4.3 |
| 24 | fl18–fl21 | DG13 | 4 | 270.7 | −73.3 | 252.6 | 4.4 |
| 25 | fl21u | ID12 | (10) | 313.4 | −79.1 | 85.1 | 4.8 |
| 26 | fl22–fl26 | DG14 | 5 | 299.2 | −75.5 | 246.0 | 4.0 |
| 27 | fl27–fl32 | DG15 | 6 | 309.5 | −71.0 | 202.9 | 4.0 |
| 28 | fl33 | ID13 | (9) | 324.1 | −65.0 | 47.1 | 6.8 |
| 29 | fl34–fl40 | DG16 | 7 | 303.2 | −69.7 | 732.6 | 2.0 |
| 30 | fl41–fl42 | DG17 | 2 | 241.0 | −48.8 | 85.3 | 10.7 |
| | **Mean**[b] | | **26** | **122.4** | **75.1** | **62.0** | **3.5** |

[a] Flows in each section are sorted from the bottom (first row) to the top (last row) of stratigraphic sequence. DG = directional group, ID = individual direction (see text); nn = sequence; DG (ID) = name of DG or ID correspondingly; $N_{DG}$ = number of flows within the DG; $N_{ID}$ = number of samples within the ID; D/I = declination/inclination in geographic coordinates; K and α95 = parameters of Fisher statistics.
[b] DG and ID mean paleomagnetic direction for the Kotuy composite section, calculated without anomalous ID3, ID4, ID5 and DG17 (see discussion in Pavlov et al., 2011). Corresponding paleomagnetic pole is: $S_{lat}$ = 71.3°, $S_{long}$ = 102.5°, $P_{lat}$ = 49.1°, $P_{long}$ = 139.8°, dp/dm = 5.8°/6.4° ($S_{lat}$, $S_{long}$ = latitude and longitude of the mean sampling point; $P_{lat}$, $P_{long}$ = latitude and longitude of the paleomagnetic pole; dp/dm = semi-axes of the ellipse of confidence about this paleomagnetic pole.)

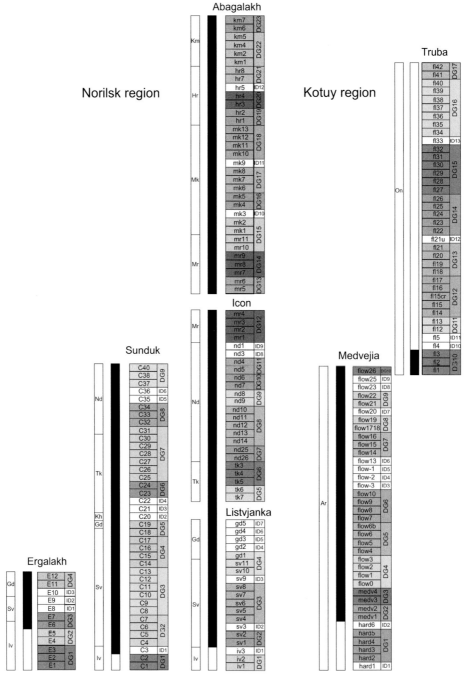

Figure 5.4 Magnetic stratigraphy of the Kotuy and Norilsk regions sections. DG = directional groups (pulses), ID = individual directions (individual eruptions). Formations: Iv = Ivakinsky, Sv = Syverminsky, Gd = Gudchikhinsky, Kh = Khakanchansky, Tk = Tuklonsky, Nd = Nadezhdinsky, Mr = Morongovsky, Mk = Mokulaevsky, Kh = Kharaelakhsky, Km = Kumginsky, Sm = Samoedsky. A black and white version of this figure will appear in some formats. For the colour version, please refer to the plate section.

DGs and three IDs in the Ergalakh section, and nine DGs and six IDs in the Sunduk section (Table 5.2 and Figure 5.4).

The same approach has been applied to the composite section containing the Listvjanka, Icon and Abagalakh sections studied earlier by Heunemann *et al.* (2004) in the Norilsk region. In this composite section, which includes 76 flows and represents the majority of the trap volcanic sequence in the Norilsk region, we have isolated 23 DGs and 12 IDs (Figure 5.4). In our analysis, we did not consider the Talnakh section because it is located in immediate proximity to the Listvjanka section and virtually repeats its lower part.

Correlating directional groups (individual directions) with volcanic pulses (individual eruptions) and taking into account the time constraints discussed in Chenet *et al.* (2008, 2009), we conclude that the duration of active volcanism that produced the composite Kotuy and Norilsk sections does not exceed 9000 and 11 000 years respectively.

These estimates do not include, of course, periods of quiescence separating volcanic pulses and individual eruptions. The absence of sedimentary layers and developed weathering crusts between the flows, however, is a clear indication that such quiescent periods were brief (nevertheless, we must mention that quiescent periods can also be marked by an absence of any evidence).

This conclusion is also supported by the rather high values of the non-random-ordering factor (Biggin *et al.*, 2008), calculated for the composite Kotuy and Norilsk sections (0.9999 and 0.8750 respectively). The non-random-ordering factor was suggested by Biggin *et al.* (2008) as a measure of serial correlation of the successive directions in lava sections. The calculated values point to high correlation of ID and DG directions in the studied sections and, therefore, indicate brief time gaps within them.

Heunemann *et al.* (2004) and Gurevitch *et al.* (2004) subdivided the composite Norilsk section (Listvjanka, Icon and Abagalakh) into consecutive R, T, E and N intervals, each recording reversed, transitional, excursional and normal states, respectively, of the geomagnetic field. They note also that between the transitional and excursional intervals (flows gd5, tk7, tk6), the geomagnetic field briefly reaches normal polarity. The authors further suggest that the excursional interval may correspond to a post-transitional rebound effect (Merrill *et al.*, 1996; Valet *et al.*, 2012).

We observe the same features in the Ergalakh and Sunduk sections. While in the Ergalakh section only the reversed and transitional intervals can be identified (levels DG1, DG2 and DG3, ID1, ID2, ID3, DG4, respectively), the Sunduk section contains almost all (Figure 5.5) of the intervals isolated in the composite Listvjanka–Icon–Abagalakh section by Heunemann *et al.* (2004).

Table 5.2 *Mean paleomagnetic directions for IDs and DGs of studied sections of the Norilsk region[a]*

| nn | Flow[a] | DG (ID) | $N_{DG}$ ($N_{ID}$) | D (°) | I (°) | K | $\alpha95$ (°) |
|----|---------|---------|---------|-------|-------|---|--------|
| | | DG and ID statistics | | | | | |
| Sunduk section (69.3°N, 90.1°E) | | | | | | | |
| 1 | C1–C2 | DG1 | 2 | 226.0 | −69.3 | 85.8 | 10.7 |
| 2 | C3 | ID1 | (10) | 165.1 | 50.2 | 78.4 | 5.0 |
| 3 | C4–C7 | DG2 | 4 | 149.7 | 53.9 | 170.9 | 5.4 |
| 4 | C8–C13 | DG3 | 6 | 141.8 | 49.1 | 222.3 | 3.8 |
| 5 | C14–C17 | DG4 | 4 | 149.2 | 50.7 | 142.2 | 5.9 |
| 6 | C18–C19 | DG5 | 2 | 130.5 | 56.4 | 168.0 | 7.6 |
| 7 | C20 | ID2 | (9) | 98.5 | 83.7 | 34.0 | 8.0 |
| 8 | C21 | ID3 | (10) | 95.8 | 67.9 | 33.9 | 7.6 |
| 9 | C22 | ID4 | (10) | 96.3 | 86.7 | 21.7 | 9.5 |
| 10 | C23–C24 | DG6 | 2 | 102.5 | 67.3 | 129.3 | 8.7 |
| 11 | C25–C31 | DG7 | 7 | 98.2 | 80.5 | 243.4 | 3.4 |
| 12 | C32–C34 | DG8 | 3 | 42.6 | 70.6 | 359.5 | 4.3 |
| 13 | C35 | ID5 | (9) | 92.8 | 67.4 | 31.6 | 8.3 |
| 14 | C36 | ID6 | (5) | 27.9 | 68.0 | 37.7 | 10.2 |
| 15 | C37–C40 | DG9 | 3 | 63.2 | 70.5 | 175.6 | 6.1 |
| Ergalakh section (69.3°N, 88.3°E) | | | | | | | |
| 1 | E1–E3 | DG1 | 3 | 265.5 | −65.4 | 421.1 | 3.9 |
| 2 | E4–E5 | DG2 | 2 | 271.1 | −50.7 | 84.5 | 10.8 |
| 3 | E6–E7 | DG3 | 2 | 136.7 | 72.0 | 135.8 | 8.5 |
| 4 | E8 | ID1 | (9) | 157.9 | 66.3 | 29.4 | 8.6 |
| 5 | E9 | ID2 | (5) | 148.5 | 40.6 | 33.0 | 10.9 |
| 6 | E10 | ID3 | (6) | 131.6 | 49.7 | 120.8 | 5.2 |
| 7 | E11–E12 | DG4 | 2 | 108.2 | 56.4 | 74.5 | 11.5 |
| **Norilsk composite section (Heunemann *et al.*, 2004)** | | | | | | | |
| Listvjanka section (69.5°N, 88.7°E) | | | | | | | |
| 1 | iv1–iv2 | DG1 | 2 | 247.8 | −64.0 | 299.5 | 5.7 |
| 2 | iv3 | ID1 | (5) | 258.0 | −70.0 | 202.5 | 4.4 |
| 3 | sv1–sv2 | DG2 | 2 | 150.1 | 56.5 | 139.5 | 8.4 |
| 4 | sv3 | ID2 | (7) | 148.0 | 48.0 | 92.6 | 5.5 |
| 5 | sv4–sv8 | DG3 | 5 | 151.4 | 55.0 | 1190.0 | 1.8 |
| 6 | sv9 | ID3 | (4) | 161.0 | 50.0 | 84.8 | 7.6 |
| 7 | sv10–gd1 | DG4 | 3 | 149.6 | 55.4 | 237.6 | 5.2 |
| 8 | gd2 | ID4 | (7) | 127.0 | 69.0 | 175.0 | 4.0 |
| 9 | gd3 | ID5 | (7) | 114.0 | 58.0 | 132.3 | 4.6 |
| 10 | gd4 | ID6 | (5) | 82.0 | 51.0 | 71.6 | 7.4 |
| 11 | gd5 | ID7 | (5) | 94.0 | 73.0 | 360.0 | 3.3 |
| Icon section (70.4° N, 90.1° E) | | | | | | | |
| 12 | tk7–tk6 | DG5 | 2 | 71.2 | 76.0 | 428.0 | 4.8 |
| 13 | tk5–tk3 | DG6 | 3 | 28.4 | 71.4 | 197.7 | 5.7 |
| 14 | nd26–nd25 | DG7 | 2 | 34.6 | 63.0 | 106.6 | 9.6 |
| 15 | nd14–nd10 | DG8 | 5 | 13.8 | 67.1 | 546.2 | 2.7 |

Table 5.2 (*cont.*)

| nn | Flow[a] | DG (ID) | $N_{DG}$ ($N_{ID}$) | $D$ (°) | $I$ (°) | $K$ | $\alpha95$ (°) |
|----|---------|---------|---------------------|---------|---------|-----|----------------|
|    |         | DG and ID statistics | | | | | |
| 16 | nd9–nd8 | DG9 | 2 | 31.3 | 68.0 | 624.9 | 4.0 |
| 17 | nd7–nd6 | DG10 | 2 | 9.4 | 70.6 | 1012.6 | 3.1 |
| 18 | nd5–nd4 | DG11 | 2 | 62.2 | 73.7 | 433.6 | 4.8 |
| 19 | nd3 | ID8 | 7 | 40.0 | 81.0 | 634.9 | 2.1 |
| 20 | nd1 | ID9 | 8 | 100.0 | 73.0 | 1088.9 | 1.5 |
| 21 | mr1–mr4 | DG12 | 4 | 89.0 | 78.1 | 692.5 | 2.7 |
| Abagalakh section (70.4° N, 90.1° E) | | | | | | | |
| 22 | mr5–mr6 | DG13 | 2 | 159.9 | 82.3 | 789.6 | 3.5 |
| 23 | mr7–mr9 | DG14 | 3 | 86.4 | 75.9 | 261.3 | 5.0 |
| 24 | mr10–mr12 | DG15 | 4 | 120.0 | 77.3 | 340.2 | 3.8 |
| 25 | mk3 | ID10 | 9 | 161.0 | 75.0 | 277.8 | 2.8 |
| 26 | mk4–mk5 | DG16 | 2 | 75.5 | 73.2 | 657.7 | 3.9 |
| 27 | mk6–mk8 | DG17 | 3 | 105.8 | 79.5 | 225.7 | 5.4 |
| 28 | mk9 | ID11 | 9 | 68.0 | 67.0 | 188.4 | 3.4 |
| 29 | mk10–mk13 | DG18 | 4 | 93.5 | 70.8 | 191.9 | 5.1 |
| 30 | hr1–hr2 | DG19 | 2 | 89.0 | 57.1 | 440.4 | 4.7 |
| 31 | hr3–hr4 | DG20 | 2 | 107.1 | 65.0 | 260.4 | 6.1 |
| 32 | hr5 | ID12 | 8 | 96.0 | 73.0 | 56.2 | 6.6 |
| 33 | hr7–hr8 | DG21 | 2 | 53.4 | 78.0 | 264.8 | 6.1 |
| 34 | km1–km5 | DG22 | 4 | 71.7 | 72.3 | 2423.2 | 1.4 |
| 35 | km6–km7 | DG23 | 2 | 42.5 | 75.6 | 173.9 | 7.5 |
| Talnakh section (69.5° N, 88.7° E) | | | | | | | |
| 1 | ta8 | ID | 8 | 146.0 | 58.0 | 84.0 | 5.4 |
| 2 | ta7 | ID | 8 | 152.0 | 49.0 | 56.2 | 6.6 |
| 3 | ta6 | ID | 6 | 166.0 | 56.0 | 17.2 | 13.8 |
| 4 | ta5 | ID | 4 | 195.0 | −61.0 | 3402.8 | 1.2 |
| 5 | ta4 | ID | 8 | 280.0 | −78.0 | 43.6 | 7.5 |
| 6 | ta3 | ID | 7 | 236.0 | −66.0 | 72.8 | 6.2 |
| | **Mean**[b] | | **18** | **88.1** | **75.6** | **68** | **4.2** |

[a] See Table 5.1 for further details and abbreviations.
[b] DG and ID mean paleomagnetic direction for the composite Norilsk section, calculated without reversal, transitional and excursional directions (for details see text and the discussion in Pavlov *et al.*, 2011: paleomagnetic directions from DG11–DG23 stratigraphic interval only have been used. Corresponding paleomagnetic pole is: $S_{lat} = 70.0°$, $S_{long} = 89.3°$, $Pl_{at} = 57.2°$, $P_{long} = 146.9°$, dp/dm = 7.1°/7.8°.

The reversed interval of the Sunduk section includes the two lowermost reversely magnetized flows that form the volcanic pulse DG1. The next 17 flows (ID1, DG2, DG3, DG4, DG5) make up the transitional interval. Excursional directions (identified as it was done by Heunemann *et al.* (2004) and Gurevitch *et al.* (2004)) are recorded by flows in DG8 and ID6. Similarly to the section studied by Heunemann *et al.* (2004), in the Sunduk section between the transitional

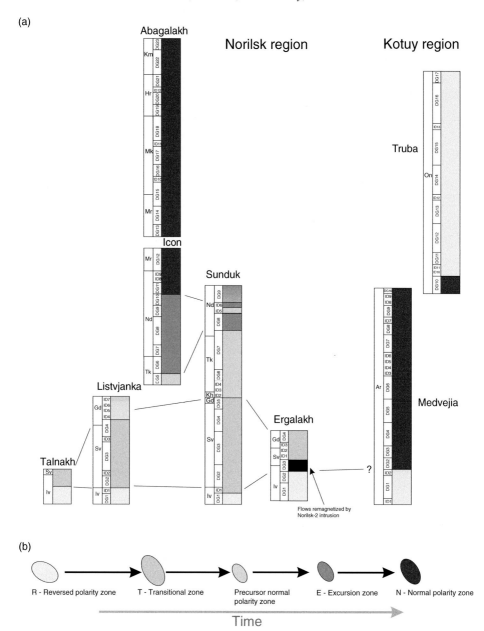

Figure 5.5    Correlation of sections of Kotuy and Norilsk regions (a) and a color key (b). A black and white version of this figure will appear in some formats. For the colour version, please refer to the plate section.

and excursional intervals we observe several flows (C20–C31) that mark the first arrival of the geomagnetic field to a "precursor" normal state, succeeded by a possible post-transitional excursion (DG8, ID6, DG9(?)). The uppermost three flows (DG9) in the Sunduk section possibly record the final arrival of the field to a fully normal state.

Thus, peculiarities of the geomagnetic field recorded in the sections allow clear correlation among these sections (Figure 5.5). The correlation indicates that at least one-quarter of the Norilsk region volcanic sequence was likely formed during a relatively short time interval at the time of a reversal of the geomagnetic field. Studies converge on estimates of the duration of a geomagnetic field reversal of between 5000 and 10 000 years, though durations as long as 20–30 thousand years have been suggested (Merrill *et al.*, 1996). Recently, Valet *et al.* (2012) analyzed similarities of several well-recorded reversals of different ages. Three phases can be observed during a reversal: a precursor, the transit, and a rebound. In our case, only the transit and rebound are present, as in the Karoo Traps (Moulin *et al.*, 2011). Valet *et al.* (2012) have estimated that the transit and the rebound may have not exceeded 1 kyr and 2.5 kyr, respectively. Thus, our data indicate that a significant part (up to 1200 m thick) of the volcanic sequence of the Norilsk region may have been formed over periods as short as several thousand years and even shorter.

It must be noted, however, that at this stage of the study, we cannot completely exclude the possibility that the observed transitional and excursional intervals are linked with independent geomagnetic events that do not constitute parts of the same reversal. If so, the time estimation above should be considered very conservative.

It is interesting to note that the transitional interval (referred to as Group 2 by Gurevitch *et al.* (2004)), corresponding to the Gudchikhinsky and Syverminsky formations, is characterized by directions with, on average, smaller inclinations than those of the overlying intervals. We can therefore try to make correlations between our measurements in this interval and those from magnetostratigraphic borehole studies (Gurevitch *et al.*, 2004; Mikhal'tsov *et al.*, 2012). Indeed, in two of three paleomagnetically studied boreholes crossing the Gudchikhinsky and Syverminsky formations (CD28 and HS59, see Figure 5.1), mean inclinations calculated for these formations are close to those obtained for the transitional interval (Heunemann *et al.*, 2004; Gurevitch *et al.*, 2004) and, at a 95% level of confidence, are smaller than the mean inclinations calculated for overlying forma-tions. Therefore, the data suggest that a magmatic event corresponding to the transitional and, most probably, excursional intervals occurred over a large region, including at least the Kharaelakh, Norilsk and Imangda troughs. Using the total area of these troughs (more than 18 000 km$^2$) and the thickness (~ 1100 m) of the transitional and excursional intervals in the studied sections, we obtain a minimum volume estimate for the lava of 20 000 km$^3$ (the greater thickness in the centers of the troughs is not considered, and thus this is a minimum estimate). If the subdivision of our studied sections is correct, all this volume was erupted during only a few volcanic pulses and individual eruptions.

## 5.5 Conclusions

(1) Paleomagnetic studies of several key sections of the Permo-Triassic traps of the northern Siberian platform indicate that the majority of volcanism in the Norilsk and Maymecha–Kotuy regions was formed during a limited number of volcanic pulses and individual eruptions. Specifically, for the composite Kotuy section our calculations reveal 17 pulses and 13 individual eruptions, and for the composite Norilsk section, 23 pulses and 12 individual eruptions. This conservatively implies that the duration of active volcanism in the Kotuy and Norilsk sections did not exceed a time interval on the order of 10 000 years. This estimate does not include the quiescence periods that separate the volcanic pulses and individual eruptions. Nevertheless, the general absence of sedimentary layers and weathering crusts between the flows in the studied areas is a clear indication that the quiescence periods were brief.

(2) Our study confirms the occurrence of thick transitional and excursional intervals in sections of the Norilsk region, as suggested by Heunemann *et al.* (2004) and Gurevitch *et al.* (2004). These intervals can be traced across the entire Norilsk region. This observation indicates that at least one-quarter of the Norilsk region volcanic sequence may have been formed during a relatively short time interval at the time of a reversal of the geomagnetic field, i.e. within several tens of thousands years or even faster. However, we stress that available data are not sufficient to affirm that observed transitional and excursional intervals are parts of the same geomagnetic reversal. Therefore, the time period between observed transitional and excursional intervals may have been significantly larger, as was suggested in recent petrological and geodynamical studies (Sobolev *et al.*, 2009, 2011). Tracing the transitional interval through the Kharaelakh, Norilsk and Imangda troughs as well as the subdivision of sections suggested in this study, implies that more than 20 000 km$^3$ of lava may have been erupted in just a few volcanic pulses and individual eruptions.

## Acknowledgements

This work was supported by the US National Science Foundation "The Siberian flood basalts and the end-Permian Extinction"(EAR-0807585); Russian Foundation for Basic Research, projects #12-05-00403, #13–05-12030 and #10-05-00557; and The Ministry of Education and Science of the Russian Federation (grant 14. Z50.31.0017).

## References

Biggin, A. J., van Hinsbergen, D. J. J., Langereis, C. G., Straathof, G. B. and Deenen, M. H. L. (2008). Geomagnetic secular variation in the Cretaceous Normal Superchron and in the Jurassic. *Physics of the Earth and Planetary Interiors,* **169**, 3–19.

Chenet, A. L., Courtillot, V., Fluteau, F. *et al.* (2009). Determination of rapid Deccan eruptions across the Cretaceous–Tertiary boundary using paleomagnetic secular variation: 2. Constraints from analysis of eight new sections and synthesis for a 3500-m-thick composite section. *Journal of Geophysical Research*, **114**(B06103), doi:10.1029/2008JB005644.

Chenet, A. L., Fluteau, F., Courtillot, V. *et al.* (2008). Determination of rapid Deccan eruptions across the Cretaceous–Tertiary boundary using paleomagnetic secular variation: results from a 1200-m-thick section in the Mahabaleshwar escarpment. *Journal of Geophysical Research*. **113**(B04101), doi:10.1029/2006JB004635.

Cogné, J. P. (2003). PaleoMac: a Macintosh TM application for paleomagnetic data and making plate reconstructions. *Geochemistry, Geophysics, Geosystems*, **4**, 1007.

Egorov, V. N. (1995). Dismemberment and correlation of the Triassic volcanic rocks of the Maymecha–Kotuy province. *Nedra Taymyra. Collected Articles*, **1**, VSEGEI, pp. 141–154 (in Russian).

Enkin, R.J. (1994). A computer program package for analysis and presentation of paleomagnetic data. *Pacific Geoscience Centre, Geological Survey of Canada*, 16.

Fisher, R. (1953). Dispersion on a sphere. *Proceedings of the Royal Society of London, Series A, Mathematical and Physical Sciences*, **217**, 295–305.

Gurevitch, E. L., Heunemann, C., Rad'ko, V. *et al.* (2004). Palaeomagnetism and magnetostratigraphy of the Permian–Triassic northwest central Siberian Trap Basalts. *Tectonophysics*, **379**, 211–226.

Heunemann, C., Krasa, D., Soffel, H. *et al.* (2004). Directions and intensities of the Earth's magnetic field during a reversal: results from the Permo-Triassic Siberian Trap basalts, Russia. *Earth and Planetary Science Letters*, **218**, 197–213.

Kamo, S. L., Czamanske, G. K., Amelin, Yu. *et al.* (2003). Rapid eruption of Siberian flood-volcanic rocks and evidence for coincidence with the Permian-Triassic boundary and mass extinction at 251 Ma. *Earth and Planetary Science Letters*, **214**, 75–91.

Kirschvink, J. L. (1980). The least-square line and plane and the analysis of paleomagnetic data. *Geophysical Journal of the Royal Astronomical Society*, **62**, 699–718.

Latyshev, A. V., Tselmovich, V. A. and Markov, G. P. (2012). Magnetic mineralogy of the volcanic rocks of the Noril'sk and the Maymecha–Kotuy regions (Siberian Trap province). *Transactions of All-Russia Conference of Young Scientists "Minerals: Composition, Properties, Methods of Investigations"*. Ekaterinbourg, pp. 212 (in Russian).

McFadden, P. L. and McElhinny, M. W. (1990). Classification of the reversal test in palaeomagnetism. *Geophysical Journal International*, **103**, 725–729.

Merrill, R. T., McElhinny, M. W. and McFadden, P. L. (1996). The Magnetic Field of the Earth. Academic Press, San Diego.

Mikhal'tsov, N., Kazansky, A., Ryabov, V. *et al.* (2012). Paleomagnetism of trap basalts in the northwestern Siberian craton, from core data. *Russian Geology and Geophysics*, **11**, 1228–1242.

Moulin, M., Fluteau, F., Courtillot, V. *et al.* (2011). An attempt to constrain the age, duration, and eruptive history of the Karoo flood basalt: Naude's Nek section (South Africa). *Journal of Geophysical Research*, **116**(B07403), doi:10.1029/2011JB008210.

Moulin, M., Fluteau, F., Courtillot, V. and Valet, J. P. (2012). The "van Zijl" Jurassic geomagnetic reversal revisited. *Geochemistry, Geophysics, Geosystems*, **13**, doi:10.1029/2011GC003910.

Pavlov, V. E., Courtillot, V., Bazhenov, M. L. and Veselovskiy, R. V. (2007). Paleomagnetism of the Siberian Traps: new data and a new overall 250 Ma pole for Siberia. *Tectonophysics*, **443**, 72–92.

Pavlov, V. E., Fluteau, F., Veselovskiy, R. V., Fetisova, A. M. and Latyshev, A. V. (2011). Secular geomagnetic variations and volcanic pulses in the Permian–Triassic Traps of the Norilsk and Maimecha–Kotui provinces. *Izvestiya, Physics of the Solid Earth*, **47**, 402–417.

Pavlov, V. (2012). Siberian paleomagnetic data and the problem of rigidity of the northern Eurasian continent in the post-Paleozoic. *Izvestiya, Physics of the Solid Earth*, **48**, 721–737.

Pavlov, V., Latyshev, A., Veselovskiy, R., Fetisova, A. and Fluteau, F. (2013). Permo-Triassic paleosecular variations and their bearing on estimation of duration of the Siberian Trap emplacement. Transaction of the International Conference on Paleomagnetism and Petromagnetism, Kazan (Volga region) Federal University (in Russian).

Sobolev, S. V., Sobolev, A. V., Kuzmin, D. V. *et al.* (2011). Linking mantle plumes, large igneous provinces and environmental catastrophes. *Nature*, **477**, 312–316.

Sobolev, A. V., Krivolutskaya, N. A. and Kuzmin, D. V. (2009). Petrology of the parental melts and mantle sources of Siberian trap magmatism. *Petrology*, **17**, 253–286.

Valet, J-P., Fournier, A., Courtillot, V. and Herrero-Bervera, E. (2012). Dynamical similarity of geomagnetic field reversals. *Nature*, **490**, 89–94.

Zijderveld, J. D. A. (1967). *A.C. Demagnetization of Rocks: Analysis of Results. Methods in Paleomagnetism,* Amsterdam, Elsevier, pp. 254–286.

# Part Two

Assessing gas and tephra release in the
present day and palaeo-record

# 6

# Volcanic-gas monitoring

ALESSANDRO AIUPPA

## 6.1 Introduction

The environmental impact of volcanoes is closely related to the rates, style and chemistry of their gas emissions (Delmelle, 2003). Monitoring the composition and mass flux of volcanic gases is therefore central to understanding how volcanism impacts our planet, on both global and local scales.

There are two main modes of volcanic-gas release on Earth (Chapter 14): (i) the impulsive emission of large quantities of gases during episodic, large-scale volcanic eruptions, and (ii) the far more sluggish, but persistent, passive gas release from quiescent or mildly erupting volcanoes. Characterising the chemical composition of impulsive emissions has remained a challenge, and direct measurements have remained limited to satellite-based $SO_2$ mass estimates (Carn *et al.*, 2003). These measurements have been combined with indirect estimates of gas budgets for other gases based on petrological models (e.g. Gerlach *et al.*, 1996). However, it is the persistent type of emissions that dominates the global volcanic-gas budget over long-term (> decadal) timescales (Chapter 14), and this chapter focuses on the chemical composition of such gas emissions.

It has been known for more than a century that the analysis and interpretation of compositions and fluxes of passive volcanic-gas emissions can provide profound insights into how active volcanoes work. Such insight can contribute to understanding and possibly even forecasting the transition from quiescence to an eruption. The basic concept is that magmatic volatiles have finite solubilities in silicate melts, and hence they inexorably degas to a vapour phase as magmas are decompressed and cooled as they ascend to the surface (Giggenbach, 1996). Since gases are more mobile than magma itself, volcano geochemists aim to capture a precursor signal of magma ascent by tracking changes in the

*Volcanism and Global Environmental Change*, eds. Anja Schmidt, Kirsten E. Fristad and Linda T. Elkins-Tanton. Published by Cambridge University Press. © Cambridge University Press 2015.

chemical composition and flux of gases emitted at the surface. Such monitoring efforts, now run by many volcano observatories worldwide, have greatly improved our knowledge of volcanic-gas chemical compositions and fluxes. This chapter summarises the current state-of-the-art of volcanic-gas monitoring, focusing on recent progress in instrumentation techniques for measuring major volcanic volatiles.

## 6.2  Chemical composition of volcanic-gas emissions: clues from direct sampling techniques

Most information on volcanic-gas chemical composition comes from conventional laboratory analyses of gas samples that are directly collected from fumaroles using evacuated bottles and caustic solutions (Giggenbach, 1975; Figure 6.1a). Direct sampling makes it possible to characterise undiluted and uncooled gas samples, and therefore detect a large number of chemicals down to the parts-per-billion level (Symonds *et al.*, 1987) in fluids that are most representative of quenched high-temperature magma-gas equilibria (Symonds *et al.*, 1994). Direct sampling also allows the isotope composition of fluids to be characterised, and therefore to fully constrain their origin (Hilton *et al.*, 2002).

Direct sampling has demonstrated that the major components of volcanic gases comprise molecular combinations of H, O, C, S and halogens, with $H_2O$, $CO_2$ and either $SO_2$ or $H_2S$ being the dominant molecular species (Symonds *et al.*, 1994). The compositions of a representative selection of high-temperature, directly sampled magmatic gases are illustrated in Figure 6.2. The figure illustrates the considerable intervolcano variability of gas composition. Two dominant source processes, the mantle-inherited hydrous composition of source magmas (Wallace, 2005) and the extensive hydrothermal interactions (Giggenbach, 1996), contribute to the volcanic gases released by arc volcanoes being richer in $H_2O$ than their counterparts in within-plate/divergent margin settings (Figure 6.2). Scrubbing of water-soluble magmatic gases at depth (Symonds *et al.*, 2001), combined with the addition of meteoric fluids, result in a higher $H_2O$ content of low-temperature hydrothermal gases (Figure 6.2).

Interpretation of fumarolic gas data has provided invaluable insights into the time-dependent mixing relations between magmatic and hydrothermal fluids, which vary with the restlessness of the volcano (Chiodini *et al.*, 1993, 2012). Hydrothermal interactions also control geothermal gas equilibria. These equilibria, pioneered by Giggenbach (1987) and Chiodini and Marini (1998), have allowed the *P–T*–compositional properties of hydrothermal systems to be characterised, which is key to interpreting gas time series collected at restless volcanoes. For instance, both models and observations indicate the $CO_2/CH_4$ ratio as a

(a)

(b)

(c)

Figure 6.1 Techniques for monitoring volcanic gases: (a) direct sampling (credit, G. Chiodini); (b) deployment of an open-path FTIR during the Eyjafjal-lajökull eruption on May 8, 2010 (credit, P. Allard); and (c) *in-situ* Multi-GAS observations in the plume of Mount Etna (credit, M. Liuzzo).

key parameter for detecting the influx of magmatic fluids during volcanic unrest (Chiodini, 2009). Moreover, improved models of the solubility of volatiles (e.g. noble gases) can be used to infer the ascent dynamics and storage depths of feeding magmas (Paonita *et al.*, 2013).

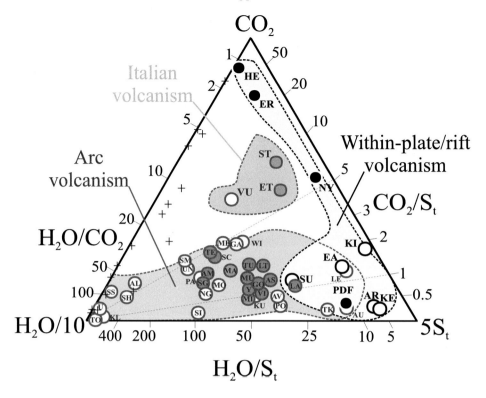

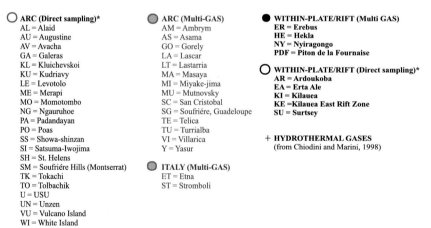

Figure 6.2   $H_2O$–$CO_2$–$S_t$ (i.e. $SO_2$ + $H_2S$) gas compositions for several volcanoes. Direct sampling data (white circles) are from Symonds *et al.* (1994) and Fischer (2008), expect for the Soufrière Hills volcano at Montserrat (Hammouya *et al.*, 1998). Multi-GAS data (grey circles) are from the UniPa–INGV database, except for Villarica (Shinohara and Witter, 2005); each symbol represents the mean $H_2O$–$CO_2$–($SO_2$ + $H_2S$) gas composition, obtained by averaging Multi-GAS records over measurement periods from hours up to several years (in the case of Etna and Stromboli). A black and white version of this figure will appear in some formats. For the colour version, please refer to the plate section.

## 6.3 Instrument-based gas monitoring techniques

The last 15 years have seen major advances in the ability to make high-rate, near-real-time volcanic-gas observations using compact, fully automated instruments. The routine gas measurements now made with permanent monitoring networks enable characterisation of magmatic degassing processes with unprecedented temporal detail and resolution.

### 6.3.1 Fourier-transform infrared spectroscopy

The introduction of open-path Fourier-transform infrared (FTIR) spectrometers in the mid 1990s (Mori *et al.*, 1993; Francis *et al.*, 1998) represented a major breakthrough in volcanic-gas research that has promoted studies of the chemistry of volcanic plumes. In particular, until the development of FTIR techniques, open-vent (mafic) volcanoes were relatively unstudied compared with more silicic arc volcanoes. A major advance occurred when incandescent rocks/magma were used as a source of radiation, which allowed first rapid (1-Hz), real-time observations (Figures 6.1b, 6.3) of the compositions of gases emitted during lava-fountaining episodes (Allard *et al.*, 2005), gas piston and lava spattering events (Edmonds and Gerlach, 2007), Strombolian explosions (Burton *et al.*, 2007) and lava lake degassing (Oppenheimer *et al.*, 2009). FTIR measurements have demonstrated, for instance, a more $CO_2$-rich chemistry of fluids released during mild (Hawaiian to Strombolian) explosive activity, relative to quiescent gas emissions (Figure 6.3). This observation has important consequences for understanding the mechanisms of gas generation, segregation and loss that drive basaltic explosions (Allard *et al.*, 2005; Burton *et al.*, 2007). FTIR has also contributed to volcano monitoring (Notsu and Mori, 2010), with the first prototype fully autonomous FTIR systems now being tested (La Spina *et al.*, 2013).

### 6.3.2 Multicomponent gas analyser system

A considerable amount of information on the chemical composition of volcanic plumes has come from the Multi-GAS (multicomponent gas analyser system) technique, which was first applied to volcanoes in the mid 2000s (Aiuppa *et al.*, 2005; Shinohara, 2005). The Multi-GAS is based on assembling commercially available infrared and electrochemical gas sensors into a single sensor kit (Figures 6.1c, 6.4). The reasonable cost, light weight and compact configuration ($< 2$ kg; $40 \times 20 \times 15$ cm), robustness, and acquisition frequency up to 0.5 Hz has made the Multi-GAS ideal for analysing volcanic gas. Figure 6.2

(a)

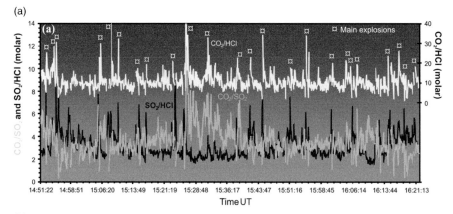

(b)

Figure 6.3    (a) Open-path FTIR sensing of the gas composition at Yasur volcano (Vanuatu) during (b) ash-rich explosive activity (courtesy of P. Allard; data from Allard *et al.*, 2012). A black and white version of this figure will appear in some formats. For the colour version, please refer to the plate section.

illustrates the Multi-GAS results obtained during field surveys at several arc, hot-spot and rift-related volcanoes. Multi-GAS-based volcanic arc gas compositions appear to span a far more restricted range than direct sampling results would suggest (Figure 6.2), with an apparent cluster of gas samples at $CO_2/S$ ratios of $\sim 2.0 \pm 0.5$ (mean $\pm$ standard deviation) and $H_2O/CO_2$ ratios of $\sim 35 \pm 10$. The implications of this observation are further discussed in Section 6.4.2.

The main contribution of the Multi-GAS to volcanology is the demonstration that fully automated systems can be deployed to make continuous, unattended, near-real-time field measurements of the compositions of gas plumes (Aiuppa *et al.*, 2007, 2009, 2010; Shinohara, 2013). Permanent Multi-GAS observations have targeted measurements of $CO_2/SO_2$ ratios of volcanic-gas plumes. Numerical simulations of volcanic degassing suggest that high $CO_2/SO_2$ ratios can be used to track pre-eruptive degassing of more primitive (gas-rich) magma (Aiuppa *et al.*, 2007). Continuing improvements in the temporal resolution of Multi-GAS instrumentation networks have allowed precursor cyclic changes in the volcanic-gas $CO_2/SO_2$ ratio to be systematically detected prior to basaltic eruptions (Figure 6.5) (Aiuppa *et al.*, 2009, 2010).

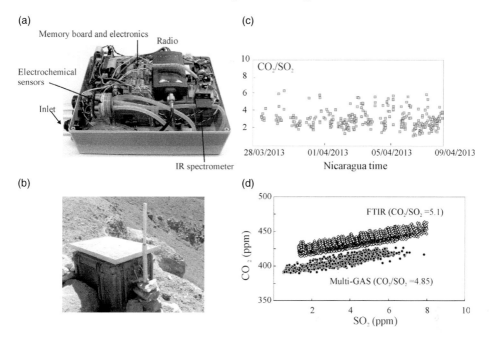

Figure 6.4 (a) Interior of an INGV-type Multi-GAS unit; (b) a Multi-GAS permanent installation (Santorini, Greece); (c) an example Multi-GAS-derived time series of $CO_2/SO_2$ ratios (Telica volcano, Nicaragua; courtesy of P. Robidoux, modified from Conde *et al.*, in press); and (d) scatter plot of $CO_2$ vs. $SO_2$ concentrations in the plume of Telica volcano (Nicaragua) on 23 March 2013, demonstrating an excellent match between FTIR and Multi-GAS $CO_2/SO_2$ ratios (courtesy of V. Conde; see Conde *et al.*, 2014). Different intercepts of regression lines (e.g. $CO_2$ levels at zero $SO_2$) for the two data sets reflect the different measurement setups of FTIR and Multi-GAS (e.g. open-path vs. punctual measurements).

## 6.4 Volcanic-gas fluxes

The environmental consequences of volcanic degassing are to a large extent determined by the mass of gas released per unit time (gas flux). However, only the $SO_2$ fluxes can be measured directly, which means that the exact magnitudes of the volcanic budgets of environmentally significant volatiles such as $CO_2$ and halogens remain unclear.

### *6.4.1 SO₂*

The abundance of volcanic $SO_2$ is relatively easy to quantify because it is spectroscopically active and present in air at a much lower concentration than in a volcanic gas/plume. Systematic efforts to quantify volcanic $SO_2$ began in the late 1970s with observations made using the correlation spectrometer (COSPEC)

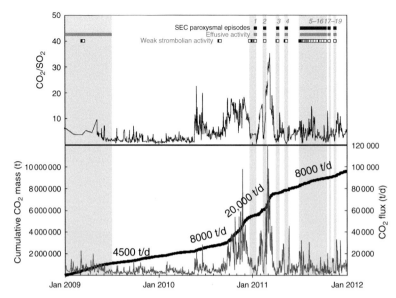

Figure 6.5   Temporal evolution of $CO_2/SO_2$ ratios (upper panel) and $CO_2$ fluxes (lower panel, solid thin line, right scale) at Etna during 2009–2011 (modified from Patanè *et al.*, 2013). Resumption of lava-fountaining activity (vertical grey bars) at the new south-east crater in early 2011 was preceded by several months of increasing $CO_2/SO_2$ ratios and $CO_2$ fluxes, with peaks of > 20 and > 20 000 t/day, respectively. Four main distinct $CO_2$ degassing regimes are identified by changes in the gradient of the cumulative $CO_2$ mass curve (lower panel, solid thick curve, left scale), with the highest fluxes being observed in the preparatory phase of the eruptions (September to December 2010).

(Stoiber *et al.*, 1983). The COSPEC was central to volcanology for more than 20 years, before it was replaced with ultraviolet (UV) spectrometers based on charge-coupled device (CCD) detectors (Galle *et al.*, 2003). A major advantage of UV spectrometers is a high degree of automation, which opened the way for the first networks of fully autonomous volcanic $SO_2$-flux scanning systems (Edmonds *et al.*, 2003; Galle *et al.*, 2010). The higher time resolution of differential optical absorption spectroscopy (DOAS) scanning networks represented a major advance in the geochemical surveillance of active volcanoes (e.g. Burton *et al.*, 2009). Ultraviolet cameras with temporal resolutions up to ~ 1 Hz (Mori and Burton, 2006) are now used to study rapid $SO_2$ flux variations associated with passive degassing (Tamburello *et al.*, 2013) and transient volcanic phenomena such as Strombolian (Tamburello *et al.*, 2012) and Vulcanian (Holland *et al.*, 2011) explosions.

   $SO_2$-flux observations made over the last 40 years have greatly advanced several central issues in volcanology. They have served as the basis for calculations of gas and magma-degassing budgets for individual volcanic systems (e.g. Allard *et al.*,

1994), and for verifying magma convection and passive gas-melt separation in conduits as likely causes of excess sulfur degassing (Shinohara, 2008). As observations increased in number, the first calculations of the global volcanic sulfur output were proposed, and there is now general consensus that present-day annual $SO_2$ emissions from volcanoes amount to 13–18 Tg annually (Mather and Pyle, this volume).

There is now considerable evidence of precursor variations in $SO_2$ fluxes being detectable prior to the eruption of intermediate to silicic volcanoes. Silicic volcanoes, which frequently undergo periods of dormancy lasting centuries or even millennia, generally comprise low-temperature ($H_2S$-dominated) fumarolic systems, which constitute the surface expression of subsurface hydrothermal systems (Giggenbach, 1996). Pre-eruptive degassing of fresh magma commonly leads to hydrothermal fluids being replaced by hotter, more oxidised magmatic fluids that are rich in $SO_2$. This was evident in the evolution of the gas chemical composition at Soufrière Hills volcano on Montserrat prior to and during the emplacement of the 1995 dome (Hammouya *et al.*, 1998) and at the re-awaking of Turrialba in Costa Rica (Vaselli *et al.*, 2010). This process has also been detected in the $SO_2$ flux record, such as at Pinatubo in 1991, where the cataclysmal plinian eruption was preceded by the $SO_2$ flux increasing by an order of magnitude over a period of $\sim 2$ weeks, indicating the intrusion of pre-eruptive magma at shallow depth (Daag *et al.*, 1996). Improved networks for measuring $SO_2$ fluxes are now being used to increase our understanding of the transition from quiescence to eruption. There have been recent reports of long-term and large changes in $SO_2$ degassing prior to small phreatic eruptions, such as at Santa Ana (El Salvador) and Turrialba (Costa Rica) (Olmos *et al.*, 2007; Conde *et al.*, 2013). During long-lived dome eruptions, degassing trends have shown particularly large $SO_2$ fluctuations, reflecting variable rates of mafic magma recharge and degassing (Christopher *et al.*, 2010), and/or changes in conduit/dome permeability (Stix *et al.*, 1993; Fischer *et al.*, 1994; Edmonds *et al.*, 2003). At mafic volcanoes, such as Etna (Caltabiano *et al.*, 1994) and Kilauea (Sutton *et al.*, 2001), $SO_2$ fluxes are strongly correlated with the rates of shallow magma supply and degassing (Allard *et al.*, 1994), and ultimately with the magma extrusion rate (Burton *et al.*, 2009). Exceptionally high $SO_2$ emissions have been observed during paroxysmal eruptions, while precursor $SO_2$-flux variations prior to eruptions are more difficult to identify, possibly due to masking by large and persistent $SO_2$ emissions during quiescent phases.

### 6.4.2 $CO_2$

The rates of $CO_2$ release from subaerial volcanism are even less well known than those of $SO_2$. Burton *et al.* (2013) compiled a list of only 33 volcanoes for which $CO_2$ flux information is available. Given the paucity of available data, and the wide range of emission fluxes that have been measured, from $> 20\,000$ t/day for top emitters to $< 50$ t/day (Figure 6.6a), the reported wide variation in the estimates of the global volcanic $CO_2$ output of 65–540 Tg/year is not surprising. Observations have been sporadic and discontinuous at most of the volcanoes listed in Burton *et al.* (2013). The utility of such spot measurements is unclear, especially given the conspicuous $CO_2$-flux variations demonstrated at the few volcanoes where systematic observations have been made (Figure 6.6b).

Hypotheses for the mechanisms underlying large variations in $CO_2$ emissions include changes in magma supply rate from the mantle source (Poland *et al.*, 2012), pulses of $CO_2$-rich magma/gas transfer from deep to shallow magma storage zones (Aiuppa *et al.*, 2007; Patanè *et al.*, 2013; Werner *et al.*, 2012), precursor leakage of a $CO_2$-rich foam from a deeply stored magma (Aiuppa *et al.*, 2010), and assimilation of limestone blocks into shallow magma (Goff *et al.*, 2001). However, the utility of forecasting volcanic activity based on gas emissions is supported by $CO_2$ flux variations reportedly occurring days to several weeks before eruptions of different styles and magnitudes (Aiuppa *et al.*, 2010; Poland *et al.*, 2012; Werner *et al.*, 2012) (Figure 6.5b).

The relatively narrow range of compositions revealed by Multi-GAS observations (Figure 6.2) for the strongest degassing arc volcanoes, all of which have $CO_2/SO_2$ molar ratios in the range of 1–5, might be useful for confining the global $CO_2$ emissions from arc volcanism. Although the strongest $CO_2$ arc emitter (Popocatépetl) is not included in the Multi-GAS data set, its time-average background $CO_2/SO_2$ ratio might be close to the range of 1–5 (Goff *et al.*, 2001). Then, even if the entire 13–18 Tg/year $SO_2$ volcanic emissions derive from arc volcanoes, the $CO_2/SO_2$ ratio range of 1–5 would lead to an arc $CO_2$ flux of 9–61 Tg/year, with a best-guess range of 17.9–24.7 Tg/year based on the average Multi-GAS ratio of $CO_2/SO_2 \sim 2$ in Figure 6.2. This is at the lower end of estimates reported by Burton *et al.* (2013). For comparison, Fischer (2008) calculated a total arc $CO_2$ flux of 85 Tg/year, which partially reflected the use of a higher $SO_2$ arc flux of 20.2 Tg/year. Following the same line of reasoning, the overall $H_2O/CO_2$ ratio of $\sim 35\pm10$ in the Multi-GAS data set would fix the global volcanic arc $H_2O$ flux at 256–354 Tg/year, which is twofold less than that quoted by Fischer (2008).

The large spread of $CO_2/SO_2$ ratios evident in Figure 6.2 means that similar arguments cannot be extended to within-plate/rift volcanism. Burton *et al.* (2013) reported that these volcanoes may have a combined $CO_2$ output of $\sim 30$ Tg/year. This, combined with our above estimate for the arc volcanism, would increase the total $CO_2$ output from subaerial volcanism to 39–91 Tg/year.

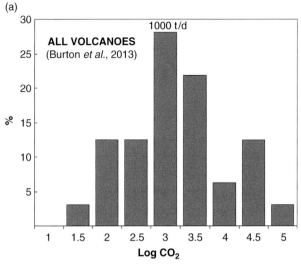

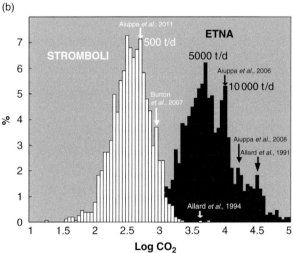

Figure 6.6 (a) Histogram of volcanic $CO_2$ fluxes (log scale) for the 33 volca-noes listed in Burton *et al.* (2013); $CO_2$ fluxes span three orders of magnitude, with a mode at ~ 1000 t/day. (b) Histogram of $CO_2$ fluxes (log scale) for Stromboli (white; data are for 5 years of daily $CO_2$ measurements during 2006–2011; from Aiuppa *et al.*, 2011) and Etna (black; 3 years of daily $CO_2$ measurements during 2009–2011; from Patanè *et al.*, 2013). Previous $CO_2$ flux estimates for both volcanoes are indicated. The large (> two orders of magnitude) temporal variations of $CO_2$ emissions from both volcanoes cast doubts on the validity of campaigns involving spot measurements, and empha-sise the need for continuous long-term observations of $CO_2$ flux using perman-ent networks.

### 6.4.3 Halogens

Halogens have the second highest environmental impact after sulfur (Mather, 2008). Volcanic halogens have long been studied (Symonds *et al.*, 1988), but precise determinations of their emissions are complicated by large heterogeneities of $SO_2/HCl$ and $SO_2/HF$ ratios in volcanic gases. This variability has been attributed to (i) scrubbing of water-soluble Cl and F by hydrothermal/groundwater systems (Symonds *et al.*, 2001) and (ii) degassing-induced fractionations due to the greater affinity of sulfur for volcanic gas (Aiuppa, 2009). Pre-eruptive halogen contents in magmas, which vary with the geological context, are likely additional key factors (Pyle and Mather, 2009). The understanding of Br–I recycling through volcanism is even less complete, but a recent review (Pyle and Mather, 2009) yielded the following best-accepted estimates of halogen degassing fluxes from arc volcanoes: $4.3\pm1.0$ Tg/year HCl, $0.5\pm0.2$ Tg/year HF, 5–15 Gg/year HBr and 0.5–2.0 Gg/year HI.

## 6.5 Conclusions

Volcanic-gas studies have progressed enormously over the last 20 years, and new measurement techniques with higher temporal resolutions combined with greatly improved interpretative degassing models are improving gas-based volcano monitoring. However, large uncertainties remain in the current volcanic-gas global inventories, reinforcing the need for future research into obtaining higher-quality compositional and gas-flux data.

### References

Aiuppa, A. (2009). Degassing of halogens from basaltic volcanism: insights from volcanic gas observations. *Chem. Geol.*, **263**, 99–109.

Aiuppa, A., Federico, C., Giudice, G., Gurrieri, S. (2005). Chemical mapping of a fumarolic field: La Fossa Crater, Vulcano Island (Aeolian Islands, Italy). *Geophys. Res. Lett.*, **32**, L13309.

Aiuppa, A., Federico, C., Giudice, G. *et al.* (2006). Rates of carbon dioxide plume degassing from Mount Etna volcano. *J. Geophys. Res.*, **111**, B09207.

Aiuppa, A., Giudice, G., Gurrieri, S. *et al.* (2008). Total volatile flux from Mount Etna. *Geophys. Res. Lett.*, **35**, L24302.

Aiuppa, A., Moretti, R., Federico, C. *et al.* (2007). Forecasting Etna eruptions by real-time observation of volcanic gas composition. *Geology*, **35**, 1115–1118.

Aiuppa, A., Federico, C., Giudice, G. *et al.* (2009). The 2007 eruption of Stromboli volcano: insights from real-time measurement of the volcanic gas plume $CO_2/SO_2$ ratio. *J. Volcanol. Geotherm. Res.*, **182**, 221–230.

Aiuppa, A., Burton, M., Caltabiano, T. *et al.* (2010). Unusually large magmatic $CO_2$ gas emissions prior to a basaltic paroxysm. *Geophys. Res. Lett.*, **37**, L17303.

Aiuppa, A., Burton, M., Allard, P. *et al.* (2011). First observational evidence for the $CO_2$-driven origin of Stromboli's major explosions. *Solid Earth*, **2**, 135–142

Allard, P., Carbonelle, J., Dajlevic, D. *et al.* (1991). Eruptive and diffuse emissions of $CO_2$ from Mount Etna. *Nature*, **351**, 387–391.

Allard, P., Carbonnelle, J., Métrich, N. *et al.* (1994). Sulphur output and magma degassing budget of Stromboli volcano. *Nature*, **368**, 326–329.

Allard, P., Burton, M.R., Mure, F. (2005). Spectroscopic evidence for a lava fountain driven by previously accumulated magmatic gas. *Nature*, **433**, 407–410.

Allard, P., Burton, M., Sawyer, G. (2012). Remote OP-FTIR sensing of magmatic gases driving Yasur trachyandesitic explosive activity, Vanuatu island arc. *Proc. 2102 EGU Meeting*, GMPV4.3/ NH2.3/AS4.29, EGU2012-14326.

Burton, M., Allard, P., Murè, F., La Spina, A. (2007). Depth of slug-driven strombolian explosive activity. *Science*, **317**, 227–230.

Burton, M.R., Caltabiano, T., Murè, F., Randazzo, D. (2009). $SO_2$ flux from Stromboli during the 2007 eruption: results from the FLAME network and traverse measurements. *J. Volcanol. Geotherm. Res.*, **182**, 214–220.

Burton, M.R., Sawyer, G.M., Granieri, D. (2013). Deep carbon emissions from volcanoes. *Rev. Mineral. Geochem.*, **75**, 323–354.

Caltabiano, T., Romano, R., Budetta, G. (1994). $SO_2$ flux measurements at Mount Etna (Sicily). *J. Geoph. Res.*, **99**, D6, 12.809–12.819.

Carn, S.A., Krueger, A.J., Bluth, G.J.S. *et al.* (2003). Volcanic eruption detection by the Total Ozone Mapping Spectrometer (TOMS) instruments: a 22-year record of sulphur dioxide and ash emissions. *Geol. Soc. Spec. Publ.*, **213**, 177–202.

Chiodini, G. (2009). $CO_2/CH_4$ ratio in fumaroles a powerful tool to detect magma degassing episodes at quiescent volcanoes. *Geophys. Res. Lett.*, **36**, L02302.

Chiodini, G., Marini, L. (1998). Hydrothermal gas equilibria: the $H_2O–H_2–CO_2–CO–CH_4$ system. *Geochim. Cosmochim. Acta*, **62** (15), 2673–2687.

Chiodini, G., Cioni, R., Marini, L. (1993). Reactions governing the chemistry of crater fumaroles from Vulcano Island, Italy, and implications for volcanic surveillance, *Appl. Geochem.*, **8**, 357–371.

Chiodini, G., Caliro, S., De Martino, P. *et al.* (2012). Early signals of new volcanic unrest at Campi Flegrei caldera? Insights from geochemical data and physical simulations, *Geology*, **40**, 943–946.

Christopher, T., Edmonds, M., Humphreys, M.C.S., Herd, R.A. (2010). Volcanic gas emissions from Soufrière Hills Volcano, Montserrat 1995–2009, with implications for mafic magma supply and degassing. *Geophys. Res. Lett.*, **37**, L00E04.

Conde, V., Bredemeyer, S., Duarte, E. *et al.* (2013). $SO_2$ degassing from Turrialba volcano linked to seismic signatures during the period 2008–2012. *Int. J. Earth Sci.*, 1–16, 10.1007/s00531-013-0958-5.

Conde, V., Robidoux, P., Avard, G. *et al.* (in press). Measurements of $SO_2$ and $CO_2$ by combining DOAS, Multi-GAS and FTIR: study cases from Turrialba and Telica volcanoes. *Int. J. Earth Sci.*, in press, DOI: 10.1007/s00531-014-1040-7.

Daag, A.S., Tubianosa, B.S., Newhall, C.G. *et al.* (1996.) Monitoring sulfur dioxide emission at Mount Pinatubo. In *Fire and Mud: Eruptions and Lahars of Mount Pinatubo Philippines*, ed. C.G. Newhall and R.S. Punongbayan, University of Washington Press, Seattle, pp. 409–434.

Delmelle, P. (2003). Environmental impacts of tropospheric volcanic gas plumes. *Geol. Soc. Spec. Publ.*, **213**, 381–399.

Edmonds, M., Gerlach, T.M. (2007). Vapor segregation and loss in basaltic melts. *Geology*, **35**, 751–754.

Edmonds, M., Herd, R.A., Galle, B. *et al.* (2003). Automated, high time-resolution measurements of $SO_2$ flux at Soufrière Hills Volcano, Montserrat. *Bull. Volcanol.*, **65**, 578–586.

Fischer, T.P. (2008). Fluxes of volatiles ($H_2O$, $CO_2$, $N_2$, Cl, F) from arc volcanoes. *Geochem. J.*, **42**, 21–38

Fischer, T.P., Morrissey, M.M., Calvache, V.M.L. *et al.* (1994). Correlations between $SO_2$ flux and long period seismicity at Galeras Volcano. *Nature*, **368**, 135–137.

Francis, P., Burton, M.R., Oppenheimer, C. (1998). Remote measurements of volcanic gas compositions by solar occultation spectroscopy. *Nature*, **396**, 567–570.

Galle, B., Oppenheimer, C., Geyer, A. *et al.* (2003). A miniaturised ultraviolet spectrometer for remote sensing of $SO_2$ fluxes: a new tool for volcano surveillance. *J. Volcanol. Geotherm. Res.*, **119**, 241–254.

Galle, B., Johansson, M., Rivera, C. *et al.* (2010). Network for Observation of Volcanic and Atmospheric Change (NOVAC) – A global network for volcanic gas monitoring: Network layout and instrument description. *J Geophys. Res.*, **115**, D05304.

Gerlach, T.M., Westrich, H.R., Symonds, R.B. (1996). Pre-eruption vapor in magma of 455 the climactic Mount Pinatubo eruption: source of the giant stratospheric sulfur dioxide 456 cloud. In *Fire and Mud: Eruptions and Lahars of Mount Pinatubo, Philippines*, ed. C.G. Newhall and R.S. Punongbayan, University of Washington Press, Seattle, 415–433.

Giggenbach, W.F. (1975). A simple method for the collection and analysis of volcanic gas samples. *Bull. Volcanol.*, **39**, 132–145.

Giggenbach, W.F. (1987). Redox processes governing the chemistry of fumarolic gas discharges from White Island, New Zealand, *Appl. Geochem.*, **2**, 143–161.

Giggenbach, W.F. (1996). Chemical composition of volcanic gases. In *Monitoring and Mitigation of Volcanic Hazards*, ed. M. Scarpa, R.J. Tilling, Springer, Heidelberg, pp. 221–256.

Goff, F., Love, S., Warren, R. *et al.* (2001). Passive infrared remote sensing evidence for large, intermittent $CO_2$ emissions at Popocatepetl volcano, Mexico. *Chem. Geol.*, **177**, 133–156.

Hammouya, G., Allard, P., Jean-Baptiste, P. *et al.* (1998). Pre- and syn-eruptive geochemistry of volcanic gases from Soufrière Hills of Montserrat, West Indies. *Geophys. Res. Lett.*, **25**, 3685–3688.

Hilton, D.R., Fischer, T.P., Marty, B. (2002). Noble gases and volatile recycling at subduction zones, *Rev. Mineral. Geochem.*, **47**.

Holland, P.A.S., Watson, M.I., Phillips, J.C. *et al.* (2011). Degassing processes during lava dome growth: insights from Santiaguito lava dome, Guatemala. *J. Volcanol. Geotherm. Res.*, **202**, 153–166.

La Spina, A., Burton, M.R., Harig, R. *et al.* (2013). New insights into volcanic processes at Stromboli from Cerberus, a remote-controlled open-path FTIR scanner system. *J. Volcanol. Geotherm. Res.*, **249**, 66–76.

Mather, T.A. (2008). Volcanism and the atmosphere: the potential role of the atmosphere in unlocking the reactivity of volcanic emissions. *Phil. Trans. R. Soc. A*, **366**, 4581–4595.

Mori, T., Notsu, K., Tohjima, Y. *et al.* (1993). Remote detection of HCl and $SO_2$ in volcanic gas from Unzen volcano, Japan. *Geophys. Res. Lett.*, **20**, 1355–1358.

Mori, T., and Burton, M. (2006). The $SO_2$ camera: a simple, fast and cheap method for ground-based imaging of $SO_2$ in volcanic plumes. *Geophys. Res. Lett.*, **33**, L24804.

Notsu, K., and Mori, T. (2010). Chemical monitoring of volcanic gas using remote FT-IR spectroscopy at several active volcanoes in Japan, *Appl. Geochem.* **25**, 505–512.

Olmos, R., Barrancos, J., Ivera, C.R. *et al.* (2007). Anomalous emissions of $SO_2$ during the recent eruption of Santa Ana volcano, El Salvador, Central America. *Pure Appl. Geophys.*, **164**, 2489–2506.

Oppenheimer, C., Lomakina, A.S., Kyle, P.R. *et al.* (2009). Pulsatory magma supply to a phonolite lava lake. *Earth Planet. Sci. Lett.*, **284**, 392–398.

Paonita, A., Federico, C., Bonfanti, P. *et al.* (2013). The episodic and abrupt geochemical changes at La Fossa fumaroles (Vulcano Island, Italy) and related constraints on the dynamics, structure, and compositions of the magmatic system. *Geochim. Cosmochim. Acta*, **120**, 158–178.

Patanè, D., Aiuppa, A., Aloisi, M. *et al.* (2013). Insights into magma and fluid transfer at Mount Etna by a multiparametric approach: a model of the events leading to the 2011 eruptive cycle. *J. Geophys. Res. B*, **118**, 3519–3539.

Poland, M.P., Miklius, A., Sutton, J.A. *et al.* (2012). A mantle-driven surge in magma supply to Kīlauea volcano during 2003–2007. *Nature Geosci.*, **5**, 295–300.

Pyle, D.M., Mather, T.A. (2009). Halogens in igneous processes and their fluxes to the atmosphere and oceans from volcanic activity: a review. *Chem. Geol.*, **263**, 110–121.

Shinohara, H. (2005). A new technique to estimate volcanic gas composition: plume measurements with a portable multi-sensor system. *J. Volcanol. Geotherm. Res*, **143**, 319–333.

Shinohara, H. (2008). Excess degassing from volcanoes and its role on eruptive and intrusive activity. *Rev. Geophys.*, **46**, RG4005.

Shinohara, H. (2013). Composition of volcanic gases emitted during repeating Vulcanian eruption stage of Shinmoedake, Kirishima volcano, Japan. *Earth Plan. Space*, **65**, 667–675.

Shinohara, H. and Witter, J.B. (2005). Volcanic gases emitted during mild Strombolian activity of Villarrica volcano, Chile. *Geophys. Res. Lett.*, **32**, L20308.

Stix, J., Zapata, G.J.A., Calvache, V.M. *et al.* (1993). A model of degassing at Galeras Volcano, Colombia, 1988–1993. *Geology*, **21**, 963–967.

Stoiber, R.E., Malinconico, Jr, L.L., Williams, S.N. (1983). Use of the Correlation Spectrometer at volcanoes. In *Forecasting Volcanic Events*, ed. H. Tazieff and J-C. Sabroux, Elsevier, Amsterdam.

Sutton, A.J., Elias, T., Gerlach, T.M. *et al.* (2001). Implications for eruptive processes as indicated by sulfur dioxide emissions from Kilauea volcano, Hawai'i, 1979–1997. *Volcanol. Geotherm. Res.*, **108**, 283–302.

Symonds, R., Rose, W.I., Reed, M.H. (1988). Contribution of Cl- and F-bearing gases to the atmosphere by volcanoes. *Nature*, **334**, 415–418.

Symonds, R.B., Rose, W.I., Reed, M.H. *et al.* (1987). Volatilization, transport and sublimation of metallic and non-metallic elements in high temperature gases at Merapi Volcano, Indonesia. *Geochim. Cosmochim. Acta*, **51** (8), 2083–2101.

Symonds, R.B., Rose, W.I., Bluth, G.J.S., Gerlach, T.M. (1994). Volcanic-gas studies: methods, results and applications, *Rev. Mineral.*, **30**, 1–66.

Symonds, R.B., Gerlach, T.M., Reed, M.H. (2001). Magmatic gas scrubbing: implications for volcano monitoring. *J. Volcanol. Geotherm. Res.*, **108**, 303–341.

Tamburello, G., Aiuppa, A., Kantzas, E.P. *et al.* (2012). Passive vs. active degassing modes at an open-vent volcano (Stromboli, Italy). *Earth Planet. Sci. Lett.*, **359–360**, 106–116.

Tamburello, G., Aiuppa, A., McGonigle, A.J.S. *et al.* (2013). Periodic volcanic degassing behavior: the Mount Etna example. *Geophys. Res. Lett.*, **40**, 4818–4822.

Vaselli, O., Tassi, F., Duarte, E. *et al.* (2010). Evolution of fluid geochemistry at the Turrialba volcano (Costa Rica) from 1998 to 2008. *Bull. Volcanol.*, **72**, 397–410.

Wallace, P.J. (2005). Volatiles in subduction zone magmas: concentrations and fluxes based on melt inclusion and volcanic gas data. *J. Volcanol. Geotherm. Res.*, **140**, 217–240.

Werner, C., Evans, W.C., Kelly, P.J. *et al.* (2012). Deep magmatic degassing versus scrubbing: elevated $CO_2$ emissions and C/S in the lead-up to the 2009 eruption of Redoubt volcano, Alaska. *Geochem. Geophys. Geosyst.*, **13**, Q03015.

# 7

# Remote sensing of volcanic ash and sulfur dioxide

FRED PRATA AND GEMMA PRATA

## 7.1 Introduction

In 1979, the National Oceanic and Atmospheric Administration (NOAA) launched
the TIROS-N satellite carrying an Earth-observing instrument designed principally
for studying the atmosphere and oceans. The Advanced Very High Resolution
Radiometer (AVHRR) was a four-channel visible and infrared (IR) imaging device
that provided global images of the Earth's atmosphere and surface at a spatial scale
of about $1 \times 1$ km$^2$. The data were essentially free: once a simple receiving dish
and some decoding electronics were installed, digital imagery could be received
twice per day (or more at polar latitudes) giving users access to real-time data over
a geographically large region (2,400 km $\times$ 10,000 km). The data were also
radiometrically calibrated and could be used quantitatively to measure sea-surface
temperatures and the health of vegetation, apart from monitoring weather systems,
storms and many other meteorological phenomena. It was immediately apparent
that these data could be used for hazard monitoring and in particular to monitor
volcanoes and volcanic activity. The same satellite also carried an instrument for
measuring solar backscattered ultraviolet (SBUV) radiation in order to measure
ozone. In a landmark paper, Krueger (1983) reported the sighting of SO$_2$ clouds
from El Chichón using TOMS (Total Ozone Mapping Spectrometer) retrievals,
based on the same principles as the SBUV, by noting that bands at 312.5 and 317.5
nm were affected by SO$_2$ absorption. The same year Robock and Matson (1983)
showed the spread of the ash cloud from El Chichón using data from the NOAA
AVHRR instruments. One of the first TOMS measurements of SO$_2$ emissions was
from Cerro Azul volcano on the Galápagos islands, acquired in February, 1979.
When TOMS was launched it was not anticipated that it would be able to monitor
volcanic activity. Exploitation of geostationary meteorological data for monitoring

*Volcanism and Global Environmental Change*, eds. Anja Schmidt, Kirsten E. Fristad and Linda T. Elkins-Tanton.
Published by Cambridge University Press. © Cambridge University Press 2015.

volcanic activity was pioneered in Japan by Y. Sawada (Sawada, 1983) at the Japan Meteorological Agency, and Malingreau and Kaswanda (1986) published a paper showing how weather satellite data (from the AVHRR) could be used to study Indonesian volcanoes.

These early uses of satellite data to monitor mostly volcanic ash and $SO_2$ established the basis for future uses and encouraged researchers to develop innovative methods for exploiting satellite data. At the end of 2013 there is still no satellite instrument in orbit designed specifically to measure volcanic emissions, although as shown in Section 2, many instruments have significant capability to do so.

Studying volcanic emissions is important for many reasons; perhaps the most important being because of their effects on the global atmosphere (see Schmidt and Robock, this volume), on the environment and on health. The eruptions of Mt Pinatubo in June 1991, which placed *c.* 30 Gt of $SO_2$ (Bluth *et al.*, 1992; Guo *et al.*, 2004) and *c.* 100 Gt of fine ash (Guo *et al.*, 2004) into the atmosphere, caused *c.* 0.5 K drop in global surface temperatures in the 3 years following (Robock, 2000). Operational IR and ultraviolet (UV) satellite instruments measured the $SO_2$ from Pinatubo as it spread mostly westwards in the stratospheric winds circling the tropics in about 3 weeks (see Figure 7.1). Satellites are less able to quantify emissions close to the surface, but while developments in technology and algorithmic improvements are addressing this issue, ground-based and airborne instruments are also being developed to monitor $SO_2$ and ash remotely.

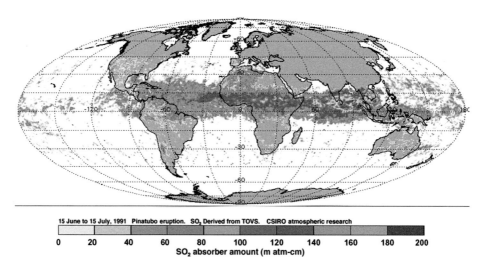

15 June to 15 July, 1991   Pinatubo eruption.   $SO_2$ Derived from TOVS.   CSIRO atmospheric research

0      20      40      60      80      100      120      140      160      180      200
$SO_2$ absorber amount (m atm-cm)

Figure 7.1    The global spread of $SO_2$ from Pinatubo, Philippines observed by the HIRS (High-resolution Infra-Red Sounder), part of the TOVS (TIROS Operational Vertical Sounder) suite of instruments on board the polar orbiting NOAA operational satellites. A black and white version of this figure will appear in some formats. For the colour version, please refer to the plate section.

## 7.2 Remote-sensing instruments and platforms

Remote sensing has the attribute of being able to supply information on a target (a gas or cloud of particles) without touching it. This is especially useful when making measurements at volcanoes where proximity to toxic gases or within range of ejecta is hazardous. It is also useful for monitoring targets that are remote and difficult to get to. Remote-sensing instruments are deployed on satellite platforms, on aircraft, on unmanned aerial vehicles (see also Chapter 9) and, for ground-based activities, they can be mounted on vehicles, on tripods or attached to moveable mounts for scanning purposes. Here we are concerned with remote-sensing measurements of gases, principally $SO_2$ and volcanic ash particles. Table 7.1 provides a summary of the most widely used instruments deployed on satellite platforms and on the ground, together with a few details of the wavelength ranges utilised, the gases measured and the horizontal spatial resolution.

All of the instruments shown in Table 7.1 are passive; the main active instrument useful for measuring volcanic species is the CALIOP (Cloud-Aerosol Lidar with Orthogonal Polarization) lidar on board the Cloud-Aerosol Lidar and Infrared Pathfinder Satellite Observation (CALIPSO) satellite. CALIOP measures backscattered laser light along a narrow strip at nadir as CALIPSO travels along its orbital path. The lidar is insensitive to $SO_2$ gas but can detect sulfate aerosols and ash particles. The recent work by Winker *et al.* (2012) gives a good overview of the capability of CALIOP for measuring volcanic ash. There are some passive microwave instruments (e.g. Special Sensing Microwave Instrument – SSMI – and the Advanced Microwave Sounding Unit – AMSU/B) that might be able to provide information on ash in the early stages of a volcanic eruption when particle sizes are large, radii $> 500$ μm. At the moment the spatial resolution of passive microwave sensors in space is quite large (pixel sizes *c.* 30–100 km) and more research is required before an assessment can be made of the utility of passive microwave data. However, the Microwave Limb Sounder (MLS) has been used to retrieve volcanic $SO_2$ from space (Read *et al.*, 1993; Prata *et al.*, 2007).

## 7.3 Remote sensing of volcanic ash

### 7.3.1 Volcanic ash

Volcanic ash is the term given to anything ejected from a volcano that is less than 2 mm in diameter. Volcanic ash encompasses a wide variety of compositional types, size distributions and shapes. Particle size[1] and the size distribution are

---

[1] For historical reasons, the volcanological and atmospheric science communities differ in the way they typically describe particle size. Volcanologists define particle size in terms of diameter whereas in atmospheric science, particle size is described in terms of radius. The log-based Krumbein $\phi$ scale of Krumbein and Sloss (1963) is also commonly used in the geological sciences; it is defined as $\phi = -log2D$, where $D$ is particle diameter.

Table 7.1 *Some details on the capabilities of various satellite and ground-based instruments for remote-sensing measurements of gases and ash. For ash, Y = yes; AAI = Aerosol Absorbing Index; ? = unproven.*

| Instrument | Bands | Gases | Ash | Spatial resolution (km) |
|---|---|---|---|---|
| **Infrared** | | | | |
| HIRS | 4, 7.3, 9.6, 11, 12, 13.2 μm | $SO_2$, $CO_2$, $O_3$ | Y | *c.* 15–30 |
| VISSR | 7.2, 10.7, 12, 13.4 μm | $SO_2$ | Y | *c.* 5 |
| MODIS | 4.0, 7.3, 8.6, 11, 12, 13.2–14.4 μm | $SO_2$ | Y | *c.* 1 |
| Himawari-8/9 | 4.0, 7.3, 8.6, 11, 12, 13.3 μm | $SO_2$ | Y | *c.* 2 |
| ASTER | 8–12 μm | $SO_2$ | Y | *c.* 0.09 |
| SEVIRI | 4.0, 7.3, 8.6, 11, 12, 13.4 μm | $SO_2$ | Y | *c.* 4 |
| MTSAT-2 | 3.7, 11.0, 12.0 μm | – | Y | *c.* 4 |
| AIRS | 4.0, 7.3, (8.6) μm  Hyperspectral 3–15 μm | $SO_2$, $CO_2$, HCl, $H_2S$, CO, $O_3$ | Y | *c.* 15 |
| TES | 4.0, 7.3, 8.6 μm | $SO_2$, $CO_2$, HCl, $H_2S$ | ? | *c.* 5 |
| IASI | 4.0, 7.3, 8.6 μm  Hyperspectral 3–15 μm | $SO_2$, $CO_2$, HCl $H_2S$, CO, $O_3$ | Y | *c.* 15 |
| VIIRS | 3–15 μm | $SO_2$ | Y | 0.75–1 |
| **Ultraviolet** | | | | |
| TOMS | 300–360 nm | $SO_2$, $O_3$ | AAI | *c.* 50 |
| OMI | 270–500 nm | $SO_2$, BrO, OClO, $NO_2$, $O_3$ | AAI | *c.* 12–48 |
| OMPS | 250–1000 nm | $O_3$, $SO_2$, $NO_2$ | AAI | *c.* 50 |
| GOME | 240–790 nm | $SO_2$, BrO, OClO, $NO_2$, $O_3$ | AAI | *c.* 40–20 |
| GOME-2 | 240–790 nm | $SO_2$, BrO, OClO, $NO_2$, $O_3$ | AAI | 40–80 |
| SCIAMACHY | 0.24–2.4 μm | $SO_2$, BrO, OClO, $NO_2$, $CO_2$ | AAI | *c.* 32–215 |
| **Microwave** | | | | |
| MLS | 0.118–2.5 THz | $SO_2$, BrO, HCl, $NO_2$, $H_2O$, $O_3$, CO, ClO | N | *c.* 2–17 |
| **Ground-based** | | | | |
| COSPEC | 300–315 nm | $SO_2$, $NO_2$ | N | Variable |
| FTIR | 3–15 μm | $CO_2$, CO, HF, $SO_2$, HCl, $H_2SO_4$ | ? | Variable |
| UV spectrometers | 300–315 nm | $SO_2$, HCl, BrO, $NO_2$ | ? | Variable |
| IR cameras | 7–14 μm | $SO_2$, $H_2O$, $H_2SO_4$ | Y | Variable |
| UV cameras | 300–320 nm | $SO_2$ | ? | Variable |

related to the energetics of the eruption (Heiken *et al.*, 1985; Zimanowski *et al.*, 2003). Sub-micron sized particles are present in every eruption and are often found adhering to larger particles. The style of magma fragmentation (e.g. magmatic vs. phreatomagmatic) influences the shape of the resulting ash particles (Sigurdsson *et al.*, 1999). Ash particles are very rarely spherical, and shapes can range from equant blocky fragments to angular cusps formed from broken bubble walls. The composition of volcanic ash can be highly heterogeneous, and typically contains a mixture of glass and crystal fragments, and lithic (rock) components that were entrained in the magma, originating from, for example, the surrounding country rock or walls of the volcanic conduit (De Rosa, 1999). The glass particles may themselves contain phenocrysts (large crystals), microlites (micro-crystals) and bubbles formed by the exsolution of magmatic volatile phases. A backscattered electron (BSE) image of a volcanic ash sample from the May 2010 Eyjafjallajökull eruption (Figure 7.2) shows that the fragments vary in composition, have irregular shapes and occupy a variety of sizes from a few microns up to 0.5 mm.

All of these properties influence how ash is transported and how long it remains in the atmosphere. They also influence how electromagnetic radiation is transmitted, absorbed and scattered when it interacts with a cloud of ash particles. Having an understanding of this relationship allows us to retrieve information on these properties using the remote-sensing techniques previously described.

### 7.3.2 Microphysical properties

The physical attributes relevant to remote sensing of fine ash, defined here as particles with diameters less than *c.* 60 μm, include the size distribution and composition, with less importance on shape. The spectral region between 8 and 13 μm has mainly been exploited to infer fine-ash properties, but limits the range of particle radii that can be retrieved to between *c.* 1 μm to *c.* 16 μm. There are few measurements of the size distribution of ash while resident in the atmosphere and so this often has to be determined from deposits on the ground, which is not a satisfactory situation. Some airborne data exist from the Mt St Helens eruptions of May 1980 (Hobbs *et al.*, 1981) and from the recent eruptions of Eyjafjallajökull (Schumann *et al.*, 2011) and Grímsvötn (Vogel *et al.*, 2012) in Iceland. These data suggest a bimodal size distribution with peaks near to 0.5 μm and 3–6 μm radius and have been modelled mathematically using gamma and log-normal distributions. There is very little experimental evidence of significant amounts of ash with particle radii > 50 μm. There are measurements of the composition of fine ash, e.g. Gislason *et al.* (2011), Lieke *et al.* (2013); for the purpose of remote

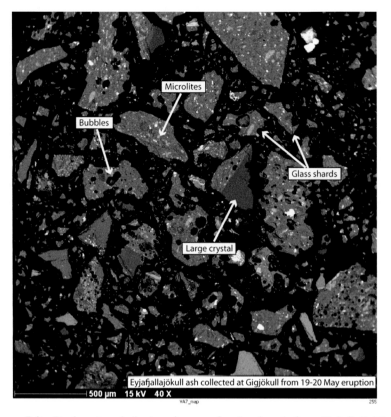

Figure 7.2   Backscattered electron image of volcanic ash from Eyjafjallajökull volcano from May 2010 showing the range of different components ash can be made up of. The greyscale level represents differing densities, with brighter shades indicating higher density (high atomic weight), and darker shades representing lower density (low atomic weight).

sensing in the IR the most salient feature is the per cent composition of $SiO_2$, since this molecule exhibits characteristic absorption features through the range 7–12 μm. Shape does affect the radiative transfer of light interacting with particles. The error caused by assuming spherical particles, rather than more realistic ragged, agglomerates with asperities, is mostly seen in the retrieval of the mass loading.

The optical regime governing how radiation interacts with a particle depends on both the wavelength ($\lambda$) and particle size ($r$). Models of light scattering can be divided into different optical regimes depending on the size of the particle relative to the wavelength of light. To incorporate this, we describe particle size in terms of a dimensionless number known as the size parameter ($X$).

$$X = \frac{2\pi r}{\lambda},$$ (7.1)

where $r$ is the particle radius and $\lambda$ is wavelength. These optical regimes are:

- Rayleigh – the particle is small compared to the wavelength ($X \leq 1$)
- Mie – the particle is of comparable size to the wavelength ($X \approx 1$)
- Geometric – the particle is large compared to the wavelength ($X \geq 1$)

As the size distribution of particles influences the way in which a cloud of particles interacts with radiation, it is necessary to characterise the shape of different size distributions and investigate how this affects optical parameters such as single scattering albedo, mass absorption coefficient and phase function. The size parameter is useful for determining which optical regime describes how electromagnetic radiation will interact with the particles, and guides the choice of wavelength to use for retrievals of ash properties. Distal particle size distributions rarely fit a log-normal distribution and are often polymodal. They typically show a coarser sub-population which shifts to finer sizes and decreases in proportion with increasing distance from the vent as the cloud is transported downwind due to coarser particles settling out. As distance from the volcano increases, settling transitions from inertia-dominated single-particle settling to aggregation-dominated settling of very fine ($< 30$ μm diameter) ash particles (Rose and Durant, 2009). Nevertheless, it is these particles that remain in the atmosphere for longer and present a hazard to air traffic and human health (Horwell and Baxter, 2006). The mass proportion of very fine ash ($< 30$ μm diameter) in explosive eruptions can vary from a few per cent (typical for basaltic eruptions) to greater than 50% (in some silicic eruptions). In distal regions, this finer sub-population typically retains consistent size characteristics but becomes proportionally dominant. The effect of size parameter (and therefore particle size) on various optical parameters can be modelled using Mie theory by varying the mean size parameter and standard deviation from the fitted distribution. This is analogous to shifting the size range of the distribution but maintaining its shape.

### 7.3.3 Optical properties

To retrieve fine-ash properties from remote sensing it is necessary to have knowledge of the spectral refractive index of ash. The amount of absorption of light by the particle is controlled by the imaginary part of the refractive index ($R_i$), which is strongly dependent upon wavelength and composition. The absorption coefficient $k$ (m$^{-1}$) can be expressed as

$$k = \frac{4\pi R_i(\lambda)}{\lambda}. \tag{7.2}$$

The compositional effects are manifest through the spectrally dependent refractive index. Several efforts are underway to improve the measurement database of optical properties of volcanic ash. These data feed directly into Mie scattering models that provide the relevant optical parameters viz. the asymmetry parameter, the single-scattering albedo and extinction efficiencies that are needed as input to the radiative transfer codes.

### 7.3.4 Satellite retrievals

Infrared remote sensing of volcanic ash was first developed by Prata (1989) who recognised that two channels within the wavelength region 8–13 μm could be used to discriminate volcanic ash from meteorological water/ice clouds. Subsequent work by Wen and Rose (1994) showed how to determine mass loadings and effective particle size by using microphysical models and the optical properties of idealised (spherical) ash particles. Further refinements and improvements have been made by Prata and Grant (2001), Pavolonis *et al.* (2006), Pavolonis (2010), Clarisse *et al.* (2010) and Prata and Prata (2012). The determination of mass loading and effective particle radius relies essentially on two pieces of independent information or observations: the brightness temperatures at two wavelengths (most often at 11 μm and 12 μm); and two *a priori* constraints – a microphysical model and a radiative transfer model. Different schemes have been proposed to invert the brightness temperature measurements into the required geophysical parameters and here we give a graphical explanation of the retrieval of ash from IR measurements.

### Methods

The panels of Figure 7.3 show the brightness temperatures[2] at 11 μm ($T_{11}$) and 12 μm ($T_{12}$) and the temperature difference, $\Delta T = T_{11} - T_{12}$, for a MODIS (MODerate resolution Imaging Spectrometer) Terra image acquired on 8 May 2010 during the eruption of Eyjafjallajökull. There is a plume emanating from Eyjafjallajökull dispersing southwards that can be seen in the $T_{11}$ and $T_{12}$ images and more clearly in the difference image (Figure 7.3c). Figure 7.3d shows a retrieval of mass loading based on these data, a microphysical model of the ash and a radiative transfer model. Note that the retrieval is shown for just three levels: 0.2, 2 and 4 g m$^{-2}$.

The mass loading (and concentration for a 1-km-thick cloud) can be determined from the following simplified expression:

$$m_l = \frac{4}{3}\rho\frac{\tau(\lambda)r_e}{Q_{ext}(\lambda)}, \qquad (7.3)$$

---

[2] Brightness temperature is defined as the temperature obtained when using the monochromatic measured radiance in the inverse Planck function.

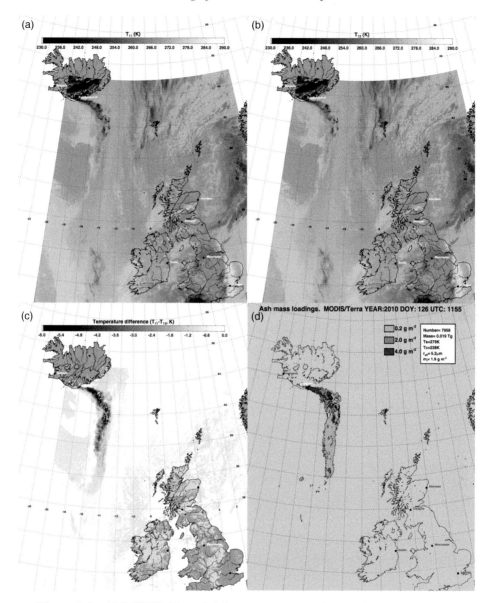

Figure 7.3 (a) MODIS 11 μm brightness temperature image. (b) MODIS 12 μm brightness temperature image. (c) MODIS 11–12 μm brightness temperature difference image. (d) Ash mass-loading retrieval based on the MODIS brightness temperature data. The MODIS/Terra image was acquired on 8 May 2010 at 11:55 UT.

where $\tau$ is the IR optical depth, $r_e$ is the effective particle radius, $\rho$ is the density of the ash, $Q_{ext}$ is the extinction efficiency and $\lambda$ is the wavelength. It is possible to simplify further by assuming that the optical depth is equal at the two IR wavelengths used; then we may write,

$$\tau = -\ln\left[1 - \frac{T_s - T_b}{T_s - T_c}\right],  \qquad (7.4)$$

where $T_c$ is the cloud-top temperature, $T_s$ is the surface temperature and $T_b$ is the brightness temperature in one of the channels. The term $T_s$–$T_c$ is often referred to as the thermal contrast and it can be seen that this should be large, while $T_s$ should be larger than the cloud-top temperature. These conditions may be interpreted as implying that retrieval of $\tau$ is best for semi-transparent clouds where the ash cloud has a temperature different to that of the surface below. For very transparent clouds (i.e. dilute ash clouds), $T_b \rightarrow T_s$ and retrieval becomes problematic. Likewise, as $T_b \rightarrow T_c$ the cloud becomes opaque and $\tau \rightarrow \infty$. An IR optical depth of $\approx$4 corresponds to $T_b$–$T_c \approx 0.5$ K and may be considered a limiting value for most IR satellite sensors. For an optically thick ash cloud with $\rho \approx 2.5 \times 10^6$ g m$^{-3}$, $Q_{ext} \approx 2.5$, $r_e \approx 5$ µm, the mass loading $m_i \approx$20 g m$^{-2}$, while for an optically thin ash cloud, with $T_s$–$T_b \approx 0.5$ K, $\tau \approx 0.01$, and with the same values as before, $m_i \approx 0.15$ g m$^{-2}$. In practice, the lower limit has been found to have the slightly higher value of $m_i \approx 0.2$ g m$^{-2}$ (Prata and Prata, 2012). An important point to note here is that it is the IR opacity that matters; ash clouds with large particles outside the sensitivity range are easily detected.

In summary, these simple calculations show that satellite IR remote sensing of volcanic ash clouds can be used to determine mass loadings within the broad range of 0.2 to 20 g m$^{-2}$ and, for a 1-km-thick ash cloud, this covers the entire range of concentrations of concern to aviation (viz. 0.2, 2 and 4 mg m$^{-3}$ and higher).

## 7.4 Remote sensing of SO$_2$

### 7.4.1 Absorption properties of SO$_2$

The SO$_2$ molecule exhibits a wide range of absorption properties from the UV to the microwave part of the electromagnetic spectrum. In the UV, SO$_2$ exhibits significant absorption features within the region between 240 and 338 nm. There are several satellite-borne sensors that exploit these UV features for SO$_2$ retrieval, including TOMS, GOME (Global Ozone Monitoring Experiment), GOME-2, SCIAMACHY (SCanning Imaging Absorption spectroMeter for Atmospheric CHartographY), OMI (Ozone Monitoring Instrument) and, more recently, OMPS (Ozone Mapping Profiler Suite).

In the IR region, SO$_2$ has three important absorption bands centred near 4, 7.3 and 8.6 µm. The 7.3 µm band (1362 cm$^{-1}$) is the very strong anti-symmetric stretch absorption feature ($v_3$-band) of the SO$_2$ molecule, but the band lies in a region of strong water-vapour absorption, which limits its use in the lower troposphere. This band is used to determine upper troposphere–lower stratosphere (UTLS) SO$_2$ partial column abundance using data from the AIRS (Atmospheric Infrared Sounder) and IASI (Infrared Atmospheric Sounding

Interferometer) polar orbiting sensors and is useful for studies of climate-related effects due to volcanic $SO_2$ (see Schmidt and Robock, this volume) and also for aviation. The weaker bands at 4 and 8.6 µm are also in much more transparent parts of the electromagnetic spectrum and consequently can be used down to the surface. The 4 µm band is generally too weak to be useful for diffuse $SO_2$ emissions but for large emissions, close to the surface, or for systems utilising solar absorption spectroscopy, the band can be used. The 8.6 µm band has been used from space and from the surface to measure $SO_2$ (Realmuto *et al.*, 1994; Prata and Bernardo, 2014).

There are also absorption lines in the microwave region, principally around 204.25, 346.52, 624.34, 624.89, 625.84, 626.17 and 649.24 GHz.

### 7.4.2 Satellite retrievals

#### Methods

Retrieval of $SO_2$ from remote-sensing instruments falls into three main categories: UV, IR and microwave methods. The UV methods essentially adopt the DOAS (differential optical absorption spectroscopy) technique (see Platt and Bobrowski, this volume) to determine the slant-path molecular column density of $SO_2$. Good explanations of the current methods used can be found in Yang *et al.* (2010) and Krotkov *et al.* (2010), including a description of the effects of the $SO_2$ height in the atmosphere and ways to ameliorate these.

#### Examples

To illustrate the capability of satellite instruments to measure $SO_2$ the example of the Nabro eruption in June 2011 is used. This stratovolcano in Eritrea, longitude 41.69° E, latitude 13.36° N, elevation 2219 m (above sea level), erupted on 12 June 2011. Very little was known about the eruptive history of Nabro so the eruption came as a surprise. There are many satellite images of the eruption including high-resolution thermal imagery from the ASTER (Advanced Space-borne Thermal Emission and Reflection Radiometer) instrument, 15-minute geostationary data from the MSG SEVIRI (Meteosat Second Generation Spin-stabilised Enhanced Visible and Infrared Imager) instrument as well as research satellite data from Aura/OMI, MODIS/Terra and Aqua, the AIRS instrument, IASI and GOME-2, among others. The AIRS sensor provided a good assessment of the $SO_2$ in the dispersing emissions from Nabro. Prata and Bernardo (2007) have described the retrieval of $SO_2$ using AIRS and noted that because there is limited sensitivity within the waveband (*c.* 7.3 µm) to $SO_2$ below *c.* 3 km, largely UTLS $SO_2$ is measured. In the cases of large emissions of $SO_2$ into the UTLS,[3] significant

---

[3] For Nabro this has been estimated as *c.*1.5 Tg

chemical conversion of the gas into sulfate aerosols can cause radiative heating of the UTLS and cooling at the surface. A composite of AIRS retrievals on 18 June 2011 for the Nabro eruption, 6 days after the start, shows (Figure 7.4, upper panel) that the cloud had dispersed widely, reaching China. The plot shows 'plumes' or 'streamers' of $SO_2$ winding across the Asian subcontinent, giving a comprehensive overview of the horizontal dispersion. Little height information is available from this retrieval scheme, but coincident CALIOP data can be used to obtain vertical height information. The lower panel of Figure 7.4 shows a backscatter 'curtain' obtained from the CALIOP 532 nm lidar between 18:13 and 18:27 UT on 18 June 2011.

The satellite UV sensors (e.g. TOMS/OMI and GOME/GOME-2) have proven to be extremely valuable for assessing $SO_2$ emissions from both explosive and passively degassing volcanoes. The work by Carn *et al.* (2003, 2008) describe how the TOMS and OMI sensors can be used to study passively degassing volcanoes and more vigorous explosive activity. A useful aspect of the OMI sensor is that it has been operating for many years (9 years by the end of 2013), so that regional composites can be made providing insights into trends in temporal increases or decreases of $SO_2$ emissions. Such information is helping climate scientists to better constrain $SO_2$ effects on the chemistry of the atmosphere and possible radiative impacts.

### *7.4.3 Ground-based retrievals*

#### *Methods*

The COSPEC instrument (Moffat and Millan, 1971) has been widely used to measure and quantify $SO_2$ in volcanic emissions but it has been largely superseded by compact scanning spectrometers utilising the DOAS method (see Platt and Bobrowski, this volume) for retrieving $SO_2$. $SO_2$ cameras (Kern, 2009) are beginning to replace the spectrometers and they are undergoing rapid development. As both spectrometers and cameras seem to be the future technologies for remote sensing of $SO_2$, we describe these in more detail below.

#### *Compact UV spectrometers*

The development of UV spectrometers for $SO_2$ measurement occurred in the late 1990s when it became economically feasible to use small fibre-optic spectrometers with quite high resolution (0.5–1 nm), and good signal-to-noise ratios operating between about 300–360 nm within the UV part of the electromagnetic spectrum. At the heart of these systems is a small, compact spectrometer utilising a fibre-optic probe and focusing optics, and a mirror to allow scanning.

Emission rates can be obtained from these systems by scanning under the plume from either a stationary instrument or by traversing with a mobile unit, and then estimating the wind speed at plume height. The integral of the path concentration

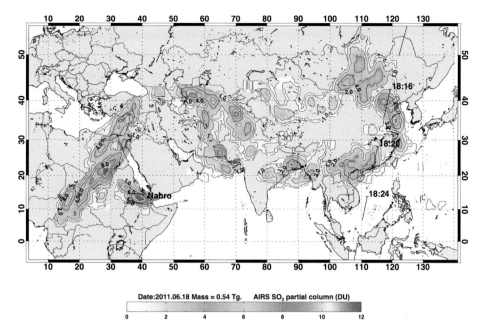

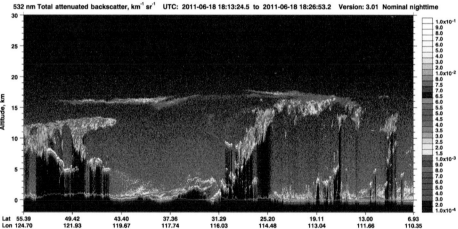

Figure 7.4   *Upper panel*: Upper troposphere $SO_2$ from the eruption of Nabro, Eritrea, determined from the AIRS sensor on board the Aqua platform. AIRS has the capability of distinguishing UTLS $SO_2$ from gas lower down in the troposphere; this feature makes these retrievals useful for investigating the effects of volcanic $SO_2$ emission on climate. One path of the CALIPSO satellite is shown by the black line, with time labels at 4-minute intervals. *Lower panel*: Corresponding CALIOP total attenuated backscatter at 532 nm for the CALIPSO overpass at 18:13–18:27 UT on 18 June 2011. The thin high-backscatter features around 16 km altitude from 112–122 °E longitude, 15–50 °N latitude, coincide well with the $SO_2$ identified in the AIRS data. A black and white version of this figure will appear in some formats. For the colour version, please refer to the plate section.

along the path scanned (or traversed) multiplied by the plume speed (m s$^{-1}$), gives an estimate of the SO$_2$ emission rate in g s$^{-1}$. Wind-speed estimation can often be the limiting factor in the accuracy of emission-rate measurements. However, one desirable aspect of these scanning systems is that multiple time-series of the SO$_2$ path concentration can be assembled as the plume passes overhead. By correlating features between the time-series, the speed of movement of the features can be estimated and hence the scanning mini-DOAS instrument itself can provide all the necessary information to obtain emission rates. A good overview of the developments in UV spectroscopy for applications to measuring SO$_2$ and SO$_2$ emission rates has been provided by McGonigle (2005). The NOVAC (Galle *et al.*, 2010) and FLAME networks are now providing near continuous SO$_2$ measurements at volcanoes around the world based on UV spectroscopy.

### The SO$_2$ camera

The SO$_2$ camera is a natural extension of the compact spectrometer, where high spectral resolution is sacrificed in favour of high spatial sampling. Typically the camera employs filters (or uses two cameras with different filters) to acquire 'on-band' and 'off-band' images at *c.* 308 nm and *c.* 330 nm. The principle of the method assumes only absorption by SO$_2$ and uses the Beer–Bouguer–Lambert law to estimate the path concentration, given knowledge of the band-averaged absorption coefficient. The mass loading, $m_i$ is determined from,

$$m_l = \frac{1}{\overline{k}^a} \ln \left[ \frac{\overline{I}_o(\lambda)}{\overline{I}(\lambda)} \right], \tag{7.5}$$

where $\overline{k}$ is an absorption coefficient averaged over the bandpass of the filter with mean wavelength $\overline{\lambda}$, $\overline{I}(\lambda)$ is the measured plume intensity and $\overline{I}_o(\lambda)$ is the intensity outside the plume. To reduce the effect of dark current,[4] the intensity obtained when viewing a blackened target is usually removed from all measurements. Under good conditions of clear skies, camera–plume distance small ($< 2$ km), and few aerosols, this simple retrieval can work well; however, these conditions are seldom met. Addition of a second off-band filter can help to alleviate some of the problems associated with interference from aerosols. With a second filter, Equation (7.5) can be re-written as:

$$m_l \approx \frac{1}{k^a} \ln \left[ \frac{I_o^a(\lambda)/I^a(\lambda)}{I_o^b(\lambda)/I^b(\lambda)} \right], \tag{7.6}$$

where the superscripts refer to the on-band (*a*) and off-band (*b*) filters and we have omitted the overbar for notational convenience. The retrievals by this method have been explained by Mori and Burton, (2006). Many refinements, both technological

---

[4] The residual current obtained when no light is falling on the detector.

and algorithmic, can be made to the $SO_2$ camera and Kern *et al.* (2010) provide a good overview of these. One quite obvious improvement is to incorporate a compact spectrometer into the camera providing spectral information in one part of the image. The spectrometer can be used as a calibration point for the image data. At a recent workshop (June, 2013) seven different $SO_2$ camera systems were tested and inter-compared, showing a strong convergence towards a system with two cameras (on- and off-band filters), an integrated spectrometer and the use of large (*c.* 50 mm) $SO_2$ cells for calibration.

Measurements from field campaigns and field tests using the $SO_2$ camera are published in the open literature and there is a growing body of data now available to volcanologists. An illustration of the retrieval technique and analysis methodology can be found in the work of Mori and Burton (2006).

Recently, Prata and Bernardo (2014) have developed an $SO_2$ camera operating in the IR region. The system uses filters to isolate radiation in the 8.6 μm band and extra filters for monitoring plume temperateure and to correct for background effects. Measurements from the IR camera and an $SO_2$ compact DOAS instrument have been compared by Lopez *et al.* (2013) at Karymsky volcano.

## 7.5 Future prospects

At the present time (2014) there is a suite of satellite-borne instruments capable of determining important properties of volcanic ash and $SO_2$ gas. These instruments span the UV to the IR and even into microwave regions. None of the instruments was specifically designed to measure volcanic emissions and so it is serendipitous that they can. There are no plans to develop satellite remote-sensing instruments for volcanic applications but we have learned to use existing measurements; since many future instruments will inherit the specifications of their predecessors there is every reason to believe that new instruments will be exploited well for volcanic research. While the situation at the moment is remarkably good, it seems there will be periods in the future where satellite coverage for volcanic applications will be worse than now. There is a need for both geostationary (with very high time resolution) coverage and polar orbiters (two to four would be good) that provide adequate temporal resolution in the polar regions not sampled by the geosensors. Europe's Sentinel-5 geostationary platform carrying a high spectral resolution imager will satisfy many of the needs for the region it images (about a 70° circle centred at 0° latitude and longitude).

In summary, the future looks bright for satellite remote sensing, at least in the passive domain. The great success of the CALIOP lidar on board the polar orbiting CALIPSO platform has shown that active sensing from space in the visible region is of great value to volcanic research. Active microwave instruments (radars) have already shown their worth for deformation studies. A geostationary lidar, preferably with a scanning capability, would provide an excellent complement to the passive suite of sensors, but there are no plans for such a system to be developed.

Ground-based systems are developing rapidly and the trend from *in situ* sensors, to single-pixel remote-sensing systems, scanning systems and on to imaging systems will likely lead to some standardisation and more frequent operational use for gas monitoring. One can envisage ground-based remote-sensing systems consisting of hyper spectral imaging cameras complemented, as in the space-based systems, with active lidars scanning plumes and providing gas emission rates and plume tomography in real-time.

As the technologies converge and become standardised, numerical models will begin to assimilate the measurements with the aim of providing forecasts in times of crises and perhaps more routine information for the public, particularly for forecast air quality. Cities near volcanoes, airports and vulnerable industries will benefit from such forecasts and it is entirely possible to see, in the not too distant future, the daily weather report complemented by a short report on the current air quality due to volcanic emissions. Remote sensing is likely to play an increasingly important role in volcanological research.

## Acknowledgement

The authors are grateful to Sarah Millington for a thoughtful review that helped to improve the manuscript.

## References

Bluth, GJS, Doiron, SD, Schnetzler, CC, Krueger, AJ and Walter, LS. 1992. Global tracking of the $SO_2$ clouds from the June, 1991 Mount Pinatubo eruptions. *Geophysical Research Letters*, **19**(2), 151–154.

Carn, SA, Krueger, AJ, Bluth, GJS *et al.* 2003. Volcanic eruption detection by the Total Ozone Mapping Spectrometer (TOMS) instruments: a 22-year record of sulphur dioxide and ash emissions. *Geological Society, London, Special Publications*, **213**, 177–202.

Carn, SA, Krotkov, NA, Fioletov, V *et al.* 2008. Emission, transport and validation of sulfur dioxide in the 2008 Okmok and Kasatochi eruption clouds. *AGU Fall Meeting Abstracts*, vol. **1**, p.7.

Clarisse, L, Prata, F, Lacour, J-L *et al.* 2010. A correlation method for volcanic ash detection using hyperspectral infrared measurements. *Geophysical Research Letters*, **37**(19).

De Rosa, R. 1999. Compositional modes in the ash fraction of some modern pyroclastic deposits: their determination and significance. *Bulletin of Volcanology*, **61**, 162–173.

Galle, B, Johansson, M, Rivera, C *et al.* 2010. Network for Observation of Volcanic and Atmospheric Change (NOVAC). A global network for volcanic gas monitoring: network layout and instrument description. *Journal of Geophysical Research: Atmospheres*, **115**(D5).

Gislason, SR, Hassenkam, T, Nedel, S *et al.* 2011. Characterization of Eyjafjallajökull volcanic ash particles and a protocol for rapid risk assessment. *Proceedings of the National Academy of Sciences*, **108**, 7307–7312.

Guo, S, Bluth, GJS, Rose, WI, Watson, IM and Prata, AJ. 2004a. Re-evaluation of $SO_2$ release of the 15 June 1991 Pinatubo eruption using ultraviolet and infrared satellite sensors. *Geochemistry, Geophysics, Geosystems*, **5**(4).

Heiken, G and Wohletz, K. 1985. *Volcanic Ash*. University Presses of California, Chicago, Harvard & MIT, London.

Hobbs, PV, Radke, LF, Eltgroth, W and Hegg, DA. 1981. Airborne studies of the emissions from the volcanic eruptions of Mount St. Helens. *Science*, **211**, 816–818.

Horwell, CJ and Baxter, PJ. 2006. The respiratory health hazards of volcanic ash: a review for volcanic risk mitigation. *Bulletin of Volcanology*, **69**, 1–24.

Kern, C. 2009. Spectroscopic measurements of volcanic gas emissions in the ultra-violet wavelength region. *PhD Thesis, University of Heidelberg*.

Kern, C, Deutschmann, T, Vogel, L *et al.* 2010. Radiative transfer corrections for accurate spectroscopic measurements of volcanic gas emissions. *Bulletin of Volcanology*, **72**, 233–247.

Krotkov, NA, Schoeberl, MR, Morris, GA, Carn, S and Yang, K. 2010. Dispersion and lifetime of the $SO_2$ cloud from the August 2008 Kasatochi eruption. *Journal of Geophysical Research: Atmospheres*, **115**(D2).

Krueger, AJ. 1983. Sighting of El Chichon sulfur dioxide clouds with the Nimbus 7 total ozone mapping spectrometer. *Science*, **220**, 1377–1379.

Krumbein, WC and Sloss, L. 1963. *Stratigraphy and Sedimentation*. Freeman, San Francisco, CA.

Lieke, KI, Kristensen, TB, Korsholm, U *et al.* 2013. Characterization of volcanic ash from the 2011 Grímsvötn eruption by means of single-particle analysis. *Atmospheric Environment*, **79**, 411–420.

Lopez, T, Fee, D, Prata, F and Dehn, J. 2013. Characterization and interpretation of volcanic activity at Karymsky volcano, Kamchatka, Russia, using observations of infrasound, volcanic emissions, and thermal imagery. *Geochemistry, Geophysics, Geosystems*, **14**(12), 5106–5127.

Malingreau, JP and Kaswanda, O. 1986. Monitoring volcanic eruptions in Indonesia using weather satellite data: the Colo eruption on July 28, 1983. *Journal of Volcanology and Geothermal Research*, **27**, 179–194.

McGonigle, AJS. 2005. Volcano remote sensing with ground-based spectroscopy. *Philosophical Transactions of the Royal Society A: Mathematical, Physical and Engineering Sciences*, **363**, 2915–2929.

Moffat, AJ and Millan, MM. 1971. The applications of optical correlation techniques to the remote sensing of $SO_2$ plumes using sky light. *Atmospheric Environment (1967)*, **5**, 677–690.

Mori, T and Burton, M. 2006. The $SO_2$ camera: a simple, fast and cheap method for ground-based imaging of $SO_2$ in volcanic plumes. *Geophysical Research Letters*, **33**(24), L24804.

Pavolonis, MJ. 2010. Advances in extracting cloud composition information from space-borne infrared radiances – a robust alternative to brightness temperatures. Part I: theory. *Journal of Applied Meteorology and Climatology*, **4**, 1992–2012.

Pavolonis, MJ, Feltz, WF, Heidinger, AK and Gallina, GM. 2006. A daytime complement to the reverse absorption technique for improved automated detection of volcanic ash. *Journal of Atmospheric and Oceanic Technology*, **23**, 1422–1444.

Prata, AJ. 1989. Infrared radiative transfer calculations for volcanic ash clouds. *Geophysical Research Letters*, **16**(11), 1293–1296.

Prata, AJ and Bernardo, C. 2007. Retrieval of volcanic $SO_2$ column abundance from Atmospheric Infrared Sounder data. *Journal of Geophysical Research: Atmospheres*, **112**(D20).

Prata, AJ and Bernardo, C. 2014. Retrieval of sulphur dioxide from a ground-based thermal infrared imaging camera. *Atmospheric Measurement Techniques Discussions*, **7**, 1153–1211.

Prata, AJ and Grant, IF. 2001. Retrieval of microphysical and morphological properties of volcanic ash plumes from satellite data: application to Mt Ruapehu, New Zealand. *Quarterly Journal of the Royal Meteorological Society*, **127**, 2153–2179.

Prata, AJ and Prata, AT. 2012. Eyjafjallajökull volcanic ash concentrations determined using Spin Enhanced Visible and Infrared Imager measurements. *Journal of Geophysical Research: Atmospheres*, **117**(D20).

Prata, AJ, Carn, SA, Stohl, A and Kerkmann, J. 2007. Long range transport and fate of a stratospheric volcanic cloud from Soufrière Hills volcano, Montserrat. *Atmospheric Chemistry and Physics*, **7**, 5093–5103.

Read, WG, Froidevaux, L and Waters, JW. 1993. Microwave limb sounder measurement of stratospheric $SO_2$ from the Mt. Pinatubo volcano. *Geophysical Research Letters*, **20**(12), 1299–1302.

Realmuto, VJ, Abrams, MJ, Buongiorno, MF, and Pieri, DC. 1994. The use of multi-spectral thermal infrared image data to estimate the sulfur dioxide flux from volcanoes: a case study from Mount Etna, Sicily, July 29, 1986. *Journal of Geophysical Research: Solid Earth*, **99**(B1), 481–488.

Robock, A. 2000. Volcanic eruptions and climate. *Reviews of Geophysics*, **38**, 191–219.

Robock, A and Matson, M. 1983. Circumglobal transport of the El Chichón volcanic dust cloud. *Science*, **221**, 195–197.

Rose, WI and Durant, AJ. 2009. Fine ash content of explosive eruptions. *Journal of Volcanology and Geothermal Research*, **186**, 32–39.

Sawada, Y. 1983. Attempt at surveillance of volcanic activity by eruption cloud image from artificial satellite. *Bulletin of the Volcanological Society Japan II*, **28**, 357–373.

Schumann, U, Weinzierl, B, Reitebuch, O *et al* 2011. Airborne observations of the Eyjafjalla volcano ash cloud over Europe during air space closure in April and May 2010. *Atmospheric Chemistry and Physics*, **11**, 2245–2279.

Sigurdsson, H, Houghton, B, Rymer, H, Stix, J and McNutt, S (eds.). 1999. *Encyclopedia of Volcanoes*. Access online via Elsevier.

Vogel, A, Weber, K, Eliasson, J *et al* 2012. Airborne and ground based measurements of ash particles on Iceland and over Germany during the Grömsvötn eruption May 2011. *EGU General Assembly Conference Abstracts*, vol. **14**, p. 12854.

Wen, S and Rose, WI. 1994. Retrieval of sizes and total masses of particles in volcanic clouds using AVHRR bands 4 and 5. *Journal of Geophysical Research: Atmospheres*, **99**(D3), 5421–5431.

Winker, DM, Liu, Z, Omar, A, Tackett, J and Fairlie, D. 2012. CALIOP observations of the transport of ash from the Eyjafjallajökull volcano in April 2010. *Journal of Geophysical Research: Atmospheres*, **117**(D20).

Yang, K, Liu, X, Bhartia, PK. *et al.* 2010. Direct retrieval of sulfur dioxide amount and altitude from spaceborne hyperspectral UV measurements: theory and application. *Journal of Geophysical Research: Atmospheres*, **115**(D2).

Zimanowski, B, Wohletz, K, Dellino, P and Büttner, R. 2003. The volcanic ash problem. *Journal of Volcanology and Geothermal Research*, **122**, 1–5.

# 8

# Quantification of volcanic reactive halogen emissions

ULRICH PLATT AND NICOLE BOBROWSKI

## 8.1 Introduction

Volcanic gases are composed of many species; in order of abundance these are water vapour (typically 50–90% of the total emissions), $CO_2$ (carbon dioxide, 1–40%), $SO_2$ (sulfur dioxide, 1–25%), $H_2S$ (hydrogen sulfide, 1–10%) and HCl (hydrogen chloride, 1–10 %) (e.g. Textor *et al.*, 2004). Trace species include $CS_2$ (carbon disulfide), COS (carbonyl sulfide), CO (carbon monoxide), HF (hydrogen fluoride), HBr (hydrogen bromide), a number of volatile metal chlorides, mercury compounds (see e.g. Carroll and Holloway, 1994; Symonds *et al.*, 1994) and other heavy metals (e.g. Buat-Menard and Arnold, 1978; Hinkley *et al.*, 1999). Upon mixing with ambient air, water vapour often condenses; the number of droplets is enhanced by the availability of aerosols, both primarily emitted and condensed out of the gas phase (e.g. Varekamp *et al.*, 1986; Mather *et al.*, 2003). Aerosol particles and water droplets provide surfaces for heterogeneous reactions, which are a precondition for reactive halogen chemistry. Generally there are three main motivations to study volcanic plumes and their chemical composition:

(1) It is interesting to learn about the composition of these plumes and to study the (compared to 'usual' atmospheric chemistry) very 'strange' chemical processes during the evolution of such a plume, from the time of emission at the crater until it blends into the atmospheric background.

(2) Volcanic-gas emissions influence the atmosphere in a number of ways and on different timescales (see also Chapters 13 and 14). In particular, on shorter timescales this is true for the budgets of sulfur, ozone and other oxidants in the troposphere and the stratosphere (e.g. Robock, 2000; von Glasow *et al.*, 2009; Kutterolf *et al.*, 2013) as well as the tropospheric background of reactive

*Volcanism and Global Environmental Change*, eds. Anja Schmidt, Kirsten E. Fristad and Linda T. Elkins-Tanton. Published by Cambridge University Press. © Cambridge University Press 2015.

halogens and gaseous mercury. The study of plume chemistry will ultimately allow a precise assessment of these influences.

(3) The composition of volcanic gases gives hints on processes occurring in the Earth's interior, in particular on magma composition and degassing processes. For as long as half a century we have known that halogen/sulfur ratios measured in fumaroles indicate changes in volcanic activity (e.g. Cl/S: Noguchi and Kamiya, 1963).

Today, advances in technology (see Section 8.3), in particular in spectroscopic techniques, allow remote analysis of many species in volcanic plumes (see also Chapters 7 and 9) and even continuous monitoring (Galle *et al.*, 2010). Among other technologies, differential optical absorption spectroscopy (DOAS) and Fourier-transform infrared (FTIR) spectroscopy are popular remote-sensing techniques. The data gained using these techniques provide new possibilities to investigate volcanic volatile compositions with high temporal resolution and also during explosive eruptions. Besides relatively long-lived molecules (such as $SO_2$), transient species, in particular halogen radicals (e.g. bromine monoxide, BrO), can also be routinely detected. Frequently, the relatively stable gases (e.g. $CO_2$, $SO_2$) are simultaneously measured as indicators for dilution of the reactive species by ambient air entrained into the plume.

## 8.2 Composition and halogen chemistry of volcanic plumes

The chemical composition of volcanic plumes is influenced not only by the relative emission source strengths of the individual species, but is also modified by the chemical conversion of volcanic gases in the atmosphere to secondary emission products.

For a long time it was assumed that halogens in volcanic plumes would behave rather passively, being mainly important for the acidity budget of the atmosphere (e.g. acid rain) and, possibly, for stratospheric chemistry. In general, the chemistry in volcanic plumes was believed to be restricted to the oxidation of sulfur. This view changed drastically with the detection of bromine monoxide (BrO) in the plume of Soufrière Hills (Montserrat) by Bobrowski *et al.* (2003) using a passive optical absorption technique (Multi-Axis – DOAS or MAX-DOAS, see Section 8.3). Since then, BrO has been detected at many other volcanoes, including Ambrym, Masaya, Mount Etna, Nyiragongo, Stromboli, Villarica, Popocateptl, Kasatochi, and many more (e.g. Bobrowski and Platt, 2007; Theys *et al.*, 2009; Boichu *et al.*, 2011; Heue *et al.*, 2011; Hörmann *et al.*, 2013; Kelly *et al.*, 2013; see Figure 8.1). Today chlorine radicals (chlorine monoxide, ClO; chlorine dioxide, OClO) are also measured by DOAS and many other halogen species (bromine dioxide, OBrO; iodine monoxide, IO; iodine dioxide, OIO; molecular bromine, $Br_2$; molecular iodine, $I_2$) can in principle be measured by DOAS, which has become a standard measurement technique for volcanic emissions monitoring. This chapter will focus on bromine emissions and chemistry, including some discussion on chlorine and iodine.

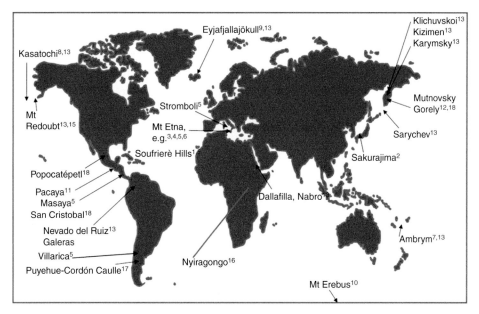

Figure 8.1    World map of sites where spectroscopic BrO measurements were performed. [1]Bobrowski *et al.*, 2003; [2]Lee *et al.*, 2005; [3]Oppenheimer *et al.*, 2006; [4]Bobrowski *et al.*, 2007; [5]Bobrowski and Platt, 2007; [6]Louban *et al.*, 2009; [7]Bani *et al.*, 2009; [8]Theys *et al.*, 2009; [9]Heue *et al.*, 2011; [10]Boichu *et al.*, 2011; [11]Vogel, 2011; [12]Bobrowski *et al.*, 2012; [13]Hörmann *et al.*, 2013; [14]Lübcke *et al.*, 2013; [15]Kelly *et al.*, 2013; [16]Bobrowski *et al.*, in press; [17]Theys *et al.*, 2014; [18]this work.

### 8.2.1 The origin of volcanic halogen species

The source of volcanic halogen emissions can be divided into two categories:

(1) 'deep' volcanic sources, which are associated with melt generation, evolution and exsolution of vapour and/or hydrosaline fluids, and
(2) shallow, more secondary sources, which include, e.g., re-volatilisation of seawater, or other crustal fluids; and thermal decomposition of hydrothermal deposits inside the volcano.

The relative importance of both sources still needs to be investigated. Volcanic halogen fluxes to the atmosphere are most commonly estimated by two approaches:

(1) petrological methods (namely melt inclusion studies) on erupted products, and
(2) measurements of halogen to $SO_2$ ratios combined with $SO_2$-flux measurements at active or quiescent degassing volcanoes. Both methods have advantages and disadvantages.

Global estimates of contemporary volcanic halogen emissions were made by, e.g., Pyle and Mather (2009) who state that arc-related emissions dominate the global halogen degassing budget with 4.3 ($\pm$1) Tg/a for HCl, 0.5 ($\pm$0.2) Tg/a for HF and 5–15 Gg/a for HBr; other authors find somewhat higher values (e.g. Shinohara, 2013). Data on the global volcanic iodine (hydrogen iodide, HI) flux are very uncertain; estimates are in the range of 0.5–2 Gg/a (Pyle and Mather, 2009), 0.2–7.7 Gg/a (Snyder and Fehn, 2002) or 0.11 Gg/a (Aiuppa *et al.*, 2005). In the geological past there were periods for which much higher volcanic halogen emissions are estimated, e.g. during Siberian Traps volcanism (see also Chapter 20).

### 8.2.2 Chemical speciation of halogens in volcanic plumes

Halogens released to the atmosphere by volcanoes are thought to be predominantly degassed as hydrogen halides (see, e.g., Symonds *et al.*, 1994; Schwandner *et al.*, 2004; Carroll and Holloway, 1994), which should be thermodynamically favoured over other halogen species, as suggested by equilibrium model data (e.g. Gerlach, 2004; Martin *et al.*, 2006, 2008). Nevertheless, there may be some conversion to molecular halogens, halogen atoms and even halogen oxides once the plume begins to mix with air and is still at high temperature (in the 'source region'). However, measurements and laboratory kinetic data show that the transformation – at least of HBr to other bromine molecules – can be relatively fast in dilute plumes. Already at plume ages < 2 minutes the contribution of HBr to total bromine might fall below 50% (von Glasow, 2010). Also, recent measurements of ClO and OClO indicate oxidation of chloride in the plume.

   In comparison to bromine and chlorine measurements, volcanic iodine chemistry has been given even less consideration, owing to difficulties in detection. Iodine is probably also released as HI (e.g. Honda, 1970), but no thermodynamic model studies exist. Volcanic iodine could eventually form the reactive halogen species IO that affects plume and atmospheric chemistry. However, to date iodine oxides have not been detected in volcanic plumes.

### 8.2.3 Chemical transformation in plumes

In the following we use the term 'reactive halogen species' (RHS), for X, $X_2$, XO, OXO, HOX, $XNO_x$ where X denotes chlorine, bromine or iodine atoms. Thermodynamic (equilibrium) models of the source region only predict that a very small fraction of Br is initially present as BrO, followed by a period of bromide to BrO conversion. Both predictions are supported by a series of studies: Oppenheimer *et al.* (2006) and Bobrowski *et al.* (2007) found initially low $BrO/SO_2$ ratios in plumes, which generally increase with distance from the emission source and thus atmospheric processing time.

These observations and kinetic considerations, as well as atmospheric chemistry model calculations (e.g. Bobrowski *et al.*, 2007; Roberts *et al.*, 2009; von Glasow, 2010) qualitatively explain BrO formation via self-accelerating, autocatalytic photo-chemical reactions involving multi-phase chemistry. The temporal evolution of BrO in the plume is, however, probably influenced by the bromine speciation at the vent and possibly by the presence of other trace species (e.g. $NO_x$). Interestingly, the currently assumed bromide to BrO conversion mechanism in volcanic plumes was found to be the same as the 'bromine explosion' mechanism (Platt and Lehrer, 1997) occurring in polar regions (see, e.g., von Glasow and Crutzen, 2007). A first step is the uptake of gas phase HOBr and HBr into aerosol particles:

$$HOBr_{(gas)} \rightarrow HOBr_{(aq)} \tag{8.1}$$

$$HBr_{(gas)} \rightarrow HBr_{(aq)} \rightarrow Br^-_{(aq)} + H^+_{(aq)} \tag{8.2}$$

and a pH-dependent reaction in the aqueous phase:

$$HOBr_{(aq)} + Br^-_{(aq)} + H^+_{(aq)} \rightarrow Br_{2\ (aq)} + H_2O \tag{8.3}$$

In a volcanic plume, acidic aerosol particles (e.g. sulfuric acid aerosol) are abundant making reaction (8.3) very efficient. Subsequently $Br_2$ is released in the gas phase, where it is rapidly photolysed during daytime:

$$Br_2 + hv \rightarrow 2Br \tag{8.4}$$

The resulting Br radicals react with $O_3$ producing BrO:

$$Br + O_3 \rightarrow BrO + O_2 \tag{8.5}$$

The BrO resulting from reaction (8.5) has been confirmed by observations of the plume composition. In addition, the presence of BrO in the plume leads to rapid ozone destruction via self reaction:

$$BrO + BrO \rightarrow 2Br + O_2 \tag{8.6a}$$

$$BrO + BrO \rightarrow Br_2 + O_2 \tag{8.6b}$$

The key ozone destruction steps in the reaction scheme above are reactions (8.6a) (and reaction (8.6b) followed by reaction (8.4)) and reaction (8.5). In addition, the reaction of BrO with free radicals ($HO_2$) leads to the formation of HOBr:

$$BrO + HO_2 \rightarrow HOBr + O_2 \tag{8.7}$$

The required atmospheric oxidants $O_3$ and $HO_2$ will be mixed into the plume while it expands into the ambient atmosphere ($HO_2$ may be partly produced in the source region). It is likely that the rate-limiting step of the BrO formation is the mixing of $O_3$ and free radicals into the plume across the 'boundary' of the plume to the ambient 'background' atmosphere. The plume diameter grows according to the square root of the plume age (for constant eddy diffusivity) and therefore the plume surface area grows proportionally to the plume age. Oxidation of halogenides should be more effective at the plume edge, which is indeed seen in the measurements of Bobrowski *et al.* (2007), Louban *et al.* (2009) and General *et al.* (2014).

Chlorine oxides (ClO, OClO) and $Cl_2$ have also been observed in volcanic plumes and fumaroles, though only on few occasions – e.g. Sakurajima, Japan (Lee *et al.*, 2005), Mt Etna, Italy (Bobrowski *et al.*, 2007; General *et al.*, 2014; Gliß *et al.*, 2014), Puyehue–Cordón Caulle, Chile (Theys *et al.*, 2014) and at the Tolbachik volcanic complex, Russia (Zelenski and Taran, 2012). It should be noted that the direct detection of ClO by UV spectroscopy (DOAS, see Section 8.3.2, below) is difficult and the abundance of this species may have been considerably overestimated. In fact, model simulations by Bobrowski *et al.* (2007) and Roberts *et al.* (2009) found ClO levels about a factor of 40 lower than the early ClO measurements. Recent DOAS measurements of OClO, which is in a photostationary state with ClO via:

$$BrO + ClO \quad \rightarrow \quad OClO + Br \tag{8.8}$$

and:

$$OClO + h\nu \quad \rightarrow \quad ClO + O \tag{8.9}$$

do indicate much lower ClO levels, which are more in line with the model calculations mentioned above (e.g. General *et al.*, 2014). The chemistry of chlorine is very different from that of bromine. As (in contrast to Br atoms) Cl atoms react with ubiquitous $CH_4$ (forming HCl) there is no 'chlorine explosion'; thus, most of the emitted HCl stays in the gas phase. This difference in the reactivity of hydrogen halides is probably also the reason why the ratio of $HBr/SO_2$ dramatically decreases downwind of volcanoes, whereas the ratio $HCl/SO_2$ appears to stay fairly constant (e.g. Voigt *et al.*, 2014).

### *8.2.4 Influence of volcanic reactive halogens on the atmosphere*

From the above it becomes clear that volcanic plumes are an interesting environment for homogeneous and heterogeneous chemical processes, which in particular modify the influence of volcanic gases on the atmosphere. For instance, the conversion of (very water-soluble) hydrogen halides to (much less soluble) halogen oxides probably leads to longer atmospheric residence times of RHS and thus wider spatial distribution of these species.

It is well known that stratospheric chemistry can be strongly influenced by halogens released from volcanic eruptions, which reach the stratosphere (see also Chapter 16). Specifically, halogen species can lead to stratospheric ozone destruction for periods of months to years after an eruption (e.g. Coffey, 1996).

However, tropospheric chemistry is also influenced by volcanic emission of RHS in various ways:

(1)  RHS catalyse $O_3$ destruction (see Section 8.2.3);
(2)  RHS inhibit the $O_3$ production in the troposphere (Stutz *et al.*, 1999);
(3)  RHS reactions interfere with the $NO_x$ reaction cycle (Stutz *et al.*, 1999);
(4)  RHS reactions enhance the $OH/HO_2$ ratio;
(5)  Hypohalous acids (HOBr, HOCl) accelerate the formation of sulfate in aqueous particles (e.g. von Glasow and Crutzen, 2007);
(6)  Br can oxidise mercury and therefore reduce its atmospheric lifetime (e.g. von Glasow, 2010);
(7)  Chlorine also plays a key role in trace-metal transport from the magma to the atmosphere (Symonds *et al.*, 1994);
(8)  Chlorine atoms released in volcanic plumes may take part in the degradation of atmospheric methane ($CH_4$). For instance, Platt *et al.* (2004) conclude from $^{13}C/^{12}C$ ratio measurements in atmospheric $CH_4$ that $> 3\%$ of the $CH_4$ in the extra tropical southern hemisphere is removed by reaction with Cl atoms.

Points 1–4 above directly or indirectly influence tropospheric ozone, which is a key atmospheric constituent. Therefore, it is important to find experimental evidence for ozone reduction in tropospheric volcanic plumes by volcanic RHS and to evaluate and quantify related model predictions. Recent field studies on this issue (see, e.g., Vance *et al.*, 2010; Schumann *et al.*, 2011; Surl *et al.*, 2014) still largely suffered from instrumental limitation and cross-sensitivities of $O_3$ measurements with the abundant $SO_2$ in volcanic plumes.

## 8.3 *In situ* measurements and remote sensing of halogens

### 8.3.1 In situ *measurements*

*In situ* measurements have been the basis for monitoring volcanic-gas emissions for many years and they still play an important role. Today, they are complemented by remote-sensing techniques (see Section 8.3.2). Traditionally, several techniques for collecting gas samples have been in use (e.g. Finlayson, 1970; Carroll and Holloway, 1994); evacuated sampling flasks were introduced by Robert W. Bunsen and later refined by Walter Giggenbach (1975, now known as Giggenbach bottles), providing comprehensive information on the sampled gas composition, including halogens.

A clear disadvantage is that *in situ* sampling is dangerous, and sometimes access to the emission source is impossible. A problem when sampling the frequently more easily accessible fumaroles is the possible interaction of their gaseous emissions with ground water and soil on their way to the surface. Therefore, direct plume measurements at open vent volcanoes, if craters are accessible, are of importance. However, in using the currently available *in situ* techniques, a fundamental problem is the high water solubility of some halogen compounds (e.g. HBr) and how to distinguish the contributions of gas and particle phase. The techniques developed to trap and determine volcanic halogen compounds are usually divided into active and passive sampling techniques. The latter usually consist of a vessel containing an alkaline solution or a filter wetted by an alkaline solution that is protected by a ventilated container and exposed near the vent of a volcano. Originally devised by Noguchi and Kamiya (1963), passive alkaline traps are small, cheap and easy to use. Since the sampled gas volume depends on diffusion, measurement times are long and passive alkaline measurements often represent mean values over periods of the order of weeks. Moreover, the success of sampling also depends on the meteorological conditions.

Reduced collection periods and more flexible reaction to local weather (e.g. the wind direction) are enabled by active sampling methods that collect large volumes of data in a short period of time. One possibility is active alkaline traps (filter packs, bubblers or Raschig-ring tubes), which generally consist of a solution exposed to the sample media by pumping air through it (Wittmer *et al.*, 2014). The method can be used to sample highly concentrated gases from fumaroles as well as to probe strongly diluted plumes. Also in use are electro-chemical sensors, which can be deployed in passive as well as active mode, but frequently suffer from relatively high cross-sensitivity between different gaseous species.

### 8.3.2 *Remote-sensing measurements*

Remote-sensing measurements rely on the analysis of radiation emitted by plumes or the determination of the attenuation of external radiation (e.g. from the Sun) to determine remotely the concentration of species of interest inside the plume. In either case the approach is based on Lambert–Beer's law relating the attenuation of an initial radiation intensity $I_0(\lambda)$, ($\lambda$ = wavelength) to the column density $S$ of a gaseous constituent of the plume:

$$I(\lambda) = I_0(\lambda) \cdot \exp\left(-S \cdot \sigma(\lambda)\right) = I_0(\lambda) \cdot \exp\left(-D\right) \tag{8.10}$$

Here $\sigma(\lambda)$ denotes the wavelength-dependent absorption cross section of the constituent of interest. The column density $S$ of a gaseous constituent is related to its concentration $c(x)$ ($x$ denoting the position inside the plume), or mean concentration $\bar{c}$, and the extension of the plume $L$:

$$S = \int_0^L c(x)\,\mathrm{d}x = \bar{c}\cdot L \tag{8.11}$$

More generally, the optical density $D(\lambda)$ is given as:

$$D(\lambda) = \sigma(\lambda)\cdot \int_0^L c(x)\,\mathrm{d}x = \sigma(\lambda)\cdot S \tag{8.12}$$

Thus, in principle, gas column densities $S$ can be determined by solving Equations (8.10) and (8.12) for $S$ (and thus the average gas concentration, if the plume diameter $L$ is known, can be estimated):

$$S = \frac{D(\lambda)}{\sigma(\lambda)} = \frac{1}{\sigma(\lambda)}\cdot \ln\left(\frac{I_0(\lambda)}{I(\lambda)}\right) \tag{8.13}$$

In practice, this simple approach outlined above will not work very well, since processes other than absorption by the gas of interest (such as aerosol extinction and absorption by other gases) will also occur and may cause significant errors. The well-known solution to this problem is to use the 'differential absorption' approach, where differences in optical densities rather than absolute absorptions and optical densities are analysed. In other words, differential optical absorption spectroscopy (DOAS, see Platt and Stutz, 2008) makes use of the fact that $\sigma(\lambda)$ for a given gas (and only gases which fulfil this condition are measurable by DOAS), and thus $D(\lambda)$ (see Equation (8.12)) are frequently a strong function of the wavelength $\lambda$. In particular, $SO_2$, BrO, OClO were thus identified in volcanic plumes (e.g. Bobrowski *et al.*, 2003; Galle *et al.*, 2010; Hörmann *et al.*, 2013). Effectively, the trace-gas abundance is derived from the intensity $I(\lambda)$ of (UV, visible and IR) radiation originating from an external light source (e.g. scattered sunlight) after traversing the plume as shown in Figure 8.2a,b.

An alternative approach is the recording of the thermal emission (in the mid-IR around 8–20 µm) originating from the gas of interest itself, as shown in Figure 8.2c. Neglecting external light sources, the spectral distribution of the intensity radiated by a gas is given by:

$$I(\lambda) = B(\lambda, T)\cdot exp\left(-D(\lambda)\right) \tag{8.14}$$

$$B(\lambda, T) = \frac{2h\frac{c^2}{\lambda^5}}{e^{\frac{hc}{\lambda kT}} - 1} \tag{8.15}$$

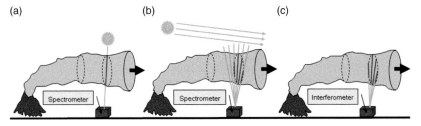

Figure 8.2    Plume scanning schemes: (a) direct light spectroscopy in the UV,
visible or near IR; (b) scattered light spectroscopy in the UV or visible spectral
ranges; and (c) thermal emission is recorded (passive spectroscopy in the thermal
IR). A black and white version of this figure will appear in some formats. For the
colour version, please refer to the plate section.

where $B(\lambda)$ (with $c$ = speed of light, $h$ = Planck constant, $k$ = Boltzmann constant,
$T$ = temperature) denotes the Planck function and $D$ the optical density as given by
Equation (8.12). It is interesting to note that the emission spectrum closely resem-
bles the absorption spectrum but, instead of absorption lines (narrow spectral
regions with reduced intensity), emission lines (spectral regions with enhanced
intensity) are seen. Thus, from measurements of the intensity $I(\lambda)$ in the thermal IR
the optical density of a species and (using Equation (8.13)) its column density can
be determined (see Figure 8.2). Note that in practice $I(\lambda)$ frequently has to be
corrected for absorption between the plume and the detector.

Early measurements of $CO_2$, $H_2O$ and $SO_2$ by recording their absorption of
direct sunlight were reported by Naughton *et al.* (1969), more recently also HCl
(Mori *et al.*, 1993) and HF were observed. Passive, IR-emission measurements of
HCl, HF, $SiF_4$ and $SO_2$ were e.g. reported by Love *et al.* (1998). Recently, two-
dimensional distributions of $SO_2$ and $SiF_4$ were also measured by passive IR
emission spectroscopy (Stremme *et al.*, 2012).

Independent of the particular spectroscopic technique, the plume scanning
approach usually results in a series of gas column densities $S(\alpha)$ as a function of
observation elevation angle $\alpha$ (see Figure 8.3), i.e. integrals along the line of sight
through the plume as given in Equation (8.12).

When the distance $Y$ to the plume is known then the angle can be converted to a
lateral distance $y \approx \alpha \cdot Y$ across the plume and an integral

$$Q = \int_{y_1}^{y_2} S(y)\mathrm{d}y \approx Y \cdot \int_{\alpha_1}^{\alpha_2} \int_0^L c(x)\,\mathrm{d}x\mathrm{d}\alpha \qquad (8.16)$$

can be calculated. Assuming that the plane of the scan is perpendicular to the
direction of plume motion (if it is not, a simple geometric correction has to be
applied) $Q$ denotes the amount of gas in the cross section of the plume. Finally, the
gas flux $J$ can be readily calculated as $J = Q \cdot v$, where $v$ is the plume propagation
speed, i.e. usually the wind speed at the altitude of the plume.

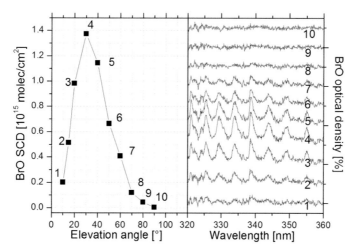

Figure 8.3 Example of DOAS plume scanning (see Figure 8.2b). Left panel: cross section of the BrO column density in the plume of Soufrière Hills volcano on Montserrat, May 2002. Right panel: BrO optical density spectra $(D(\lambda))$ corresponding to the observation; elevation angles are shown on the left (after Bobrowski *et al.*, 2003).

From the above it becomes clear that remote sensing has a number of decisive advantages over *in situ* observations:

(a)  Under most conditions the total amount of gas in the plume can be determined, rather than the gas concentration at the plume edge.
(b)  Measurements can be made from distances of typically a few kilometres; thus, remote sensing is much safer than *in situ* sampling.
(c)  The technology allows for easy automation and thus continuous real-time measurements even during periods of explosive activity.

It should also be kept in mind that spectroscopic techniques are fundamentally different from the analysis schemes applied to *in situ* samples (see Section 8.3.1) in that spectroscopic techniques identify individual molecules and are inherently calibrated (with respect to the column density).

The different spectroscopic technologies (see Figure 8.2) are – in addition to the advantages described above – characterised by the following specific further properties:

Passive IR instruments (Figure 8.2a) have the significant advantage that measurements at night are possible, since they rely on emission from the gas itself. However, at present, instrumentation for recording IR spectra is still rather complicated and comparatively expensive.

In comparison, DOAS instruments have been shown to be rather simple and effective (e.g. 'mini DOAS', Bobrowski *et al.*, 2003) making it, for instance,

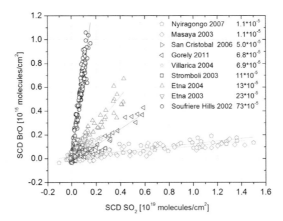

Figure 8.4   BrO column densities as a function of SO$_2$ column densities inside the plume for eight volcanic sites visited between 2002 and 2011; the species are well correlated for each location, but show different slopes for different sites. (Extended version of figure in Bobrowski and Platt, 2007; data were re-evaluated for SO$_2$ to now better account for non-linearity effects at high SO$_2$ concentrations, see also Kern *et al.*, 2010.)

possible to install many instruments within the framework of Network for Detection of Volcanic and Atmospheric Change (NOVAC), presently observing 25 volcanoes with a total of more than 60 individual spectrometers (Galle *et al.*, 2010). Note, however, that this technology only allows measurements during daytime.

### 8.4. Observations at active volcanoes

Recent observations of reactive halogen species (example in Figure 8.3) at several volcanoes worldwide (see Figure 8.1) provided quantitative insight into the variation of emission ratios and evolution of reactive halogens (in particular reactive bromine) in volcanic plumes. These data allow the theoretical concepts presented in Section 8.2.3 to be tested. The relative abundance of halogens in turn gives insight into processes in volcanic plumbing systems (see the review by Aiuppa, 2009). The measurements and observations of RHS made to date at a series of volcanoes can be summarised as follows:

(1) The BrO/SO$_2$ ratios range between $10^{-6}$ to $10^{-3}$ (e.g. Bobrowski *et al.*, 2003; Oppenheimer *et al.*, 2006; Bobrowski and Platt, 2007; Bani *et al.*, 2009; Theys *et al.*, 2012; Hörmann *et al.*, 2013; Bobrowski *et al.*, in press) with most volcanic plumes showing ratios in the range of $10^{-5}$ to $10^{-4}$. Some examples are shown in Figure 8.4.

(2) The monotonous growth of the BrO level as a function of plume age ($t$) or distance from the crater ($L_P$), where $t$ is given by $t = L_P/v$ ($v$ = wind speed at

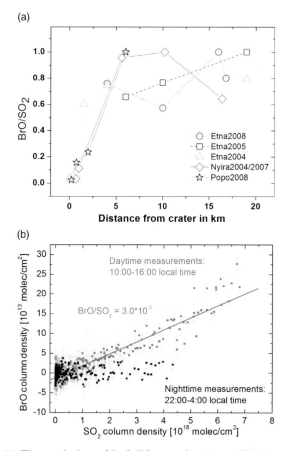

Figure 8.5 (a) The evolution of BrO/SO$_2$ – ratios (normalised) as a function of distance from the crater for three volcanoes (Etna, Nyiragongo, Popocatepetl) visited during 2004 and 2008 (after Bobrowski and Giuffrida, 2012). (b) BrO/SO$_2$ ratios (based on column densities) during daytime and at night at Masaya volcano, Nicaragua. Only during daytime BrO levels significantly exceed zero, thus strongly indicating that BrO-formation is a photochemical process (from Kern *et al.*, 2009). See also colour plates section.

plume level). An example in Figure 8.5a shows data from three volcanoes for distances up to 18 km (*t* up to ≈ 30 minutes). While initially reactive bromine abundances are very low, there appears to be a saturation level in the BrO/SO$_2$ ratio at distances > 5 km (*t* ≈ 10 minutes) (e.g. Vogel, 2011; Bobrowski *et al.*, 2012; Bobrowski *et al.*, in press).

(3) BrO/SO$_2$ ratios – derived from cross sections through the volcanic plume of Mt Etna by passive-imaging DOAS (Louban *et al.*, 2009; General *et al.*, 2014) – are clearly higher at the plume edges than in the centre, again indicating dependence of the bromine activation on species mixed into the plume from the surrounding air.

(4) A series of day- and nighttime observations at Masaya volcano, Nicaragua using active DOAS (Kern *et al.*, 2009) revealed that BrO levels significantly exceed zero only during daytime (see Fig. 8.5b). This observation confirms that BrO formation in volcanic plumes is a photochemical process.

(5) Measurements of the $BrO/SO_2$ ratios at Etna and Nevado del Ruiz volcanoes during eruptive and non-eruptive periods show clearly that during eruptions significantly lower $BrO/SO_2$ ratios prevail than for non-eruptive periods (Bobrowski and Giuffrida, 2012). At Nevado del Ruiz, Lübcke *et al.* (2014) found a drop in the $BrO/SO_2$ ratio from $\approx 4{\cdot}10^{-5}$ to $< 2{\cdot}10^{-5}$ preceding the eruption on 30 June 2012 by about 5 months. After the eruption, the $BrO/SO_2$ ratio reverted to the initial level.

(6) There are indications that bromine might behave differently from chlorine and fluorine. Bobrowski and Giuffrida (2012) proposed, on the basis of a 3-year $BrO/SO_2$ data set and an empirical model, that bromine – in contrast to Cl and F (see, e.g., Aiuppa, 2009) – could be of low effective solubility in the melt and thus could already be emitted by deep degassing magma (similar to $CO_2$). Further indications for this unexpected finding are measurements at the Nyiragongo lava lake, which show simultaneous and correlated variations in the $CO_2/SO_2$ and $BrO/SO_2$ ratios. However, since BrO is largely a secondary species, it has yet to be shown that the $BrO/SO_2$ ratio is directly correlated with the Br/S ratio.

## 8.5 Summary

During the recent decade, modern spectroscopic technology and advanced *in situ* observations have given new insight into volcanic halogen emission and chemical transformation processes in volcanic plumes. In particular, detailed analyses of the temporal and spatial distribution of reactive halogen species in volcanic plumes have revealed that complex multi-phase (photo-) chemistry is taking place there. For instance hydrogen halides (HCl and HBr, probably also HI) can be transformed into reactive halogen species.

Quantitative studies of volcanic emissions and plume chemistry make up a new field, which is still in its infancy. While chemical equilibrium modelling is relatively well developed, there are only a few studies of chemical reaction kinetics in volcanic plumes. Among the many unknowns are the abundance of reactive chlorine and iodine species and their temporal evolution (e.g. of the $XO/SO_2$ ratio) in the plume.

Reactive halogen species can profoundly influence atmospheric chemistry on a regional or even global scale. We are just beginning to understand the influences of volcanic halogens on the budgets of ozone and other oxidants, their contribution to the observed tropospheric background of reactive halogens, and how they might change the atmospheric methane and mercury burden.

The abundance of (reactive) halogen species gives hints on processes occurring in the Earth's interior, in particular on magma composition and on degassing processes. For instance, we have known for half a century that halogen/sulfur ratios measured in fumaroles indicate changes in volcanic activity (e.g. Cl/S: Noguchi and Kamiya, 1963). Recent observations suggest that $BrO/SO_2$ ratios might be an indicator for magma composition changes and/or magma pathway changes inside volcanoes – we have only just begun to decipher these messages.

## References

Aiuppa A., Federico C., Franco A. *et al.* (2005), Emission of bromine and iodine from Mount Etna volcano, *Geochem. Geophys. Geosyst.* **6** (8), doi:10.1029/2005GC000965.

Aiuppa A. (2009), Degassing of halogens from basaltic volcanism: insights from volcanic gas observations, *Chem. Geol.* **263**, 99–109.

Bani P., Oppenheimer C., Tsanev V.I. *et al.* (2009), Surge in sulphur and halogen degassing from Ambrym volcano, Vanuatu, *Bull Volcanol.* **71** (10), 1159–1168, DOI 10.1007/s00445-009-0293-7.

Bobrowski N. and Giuffrida G. (2012), Bromine monoxide/sulphur dioxide ratios in relation to volcanological observations at Mt. Etna 2006–2009, *Solid Earth* **3**, 433–445.

Bobrowski N. and Platt U. (2007), Bromine monoxide studies in volcanic plumes, *J. Volcanol. Geotherm. Res.* **166**, 147–160.

Bobrowski N., Hönninger G., Galle B. and Platt U. (2003), Detection of bromine monoxide in a volcanic plume, *Nature* **423**, 273–276.

Bobrowski N., Vogel L. Platt U. *et al.* (2012), Bromine monoxide evolution in early plumes of Mutnovsky and Gorely (Kamchatka, Russia), EGU General Assembly Meeting, Vienna.

Bobrowski N., von Glasow R., Aiuppa A. *et al.* (2007), Reactive halogen chemistry in volcanic plumes, *J. Geophys. Res.* **112**, D06311, doi:10.1029/2006JD007206.

Bobrowski N., von Glasow R., Giuffrida G.B. *et al.* (in press), Gas emission strength and evolution of the molar ratio of $BrO/SO_2$ in the plume of Nyiragongo in comparison to Etna, *J. Geophys. Res.*

Boichu M., Oppenheimer C., Roberts T.J., Tsanev V. and Kyle P.R. (2011), On bromine, nitrogen oxides and ozone depletion in the tropospheric plume of Erebus volcano (Antarctica), *Atmospheric Environment* **45**, 3856–3866.

Buat-Menard P. and Arnold M. (1978), The heavy metal chemistry of atmospheric particulate matter emitted by Mount Etna Volcano, *Geophys. Res. Lett.* **5**, 245–248.

Carroll M.R. and Holloway J.R (1994), *Volatiles in Magmas*. Washington: Mineralogical Society of America.

Coffey M.T. (1996), Observation of the impact of volcanic activity on stratospheric chemistry, *J. Geophys. Res.* **101**, 6767–6780.

Finlayson J. (1970), The collection and analysis of volcanic and hydrothermal gases. *Geothermics* **2**, 1344–1354.

Galle B., Johansson M., Rivera C. *et al.* (2010), Network for Observation of Volcanic and Atmospheric Change (NOVAC) – a global network for volcanic gas monitoring: network layout and instrument description, *J. Geophys. Res.* **115**, D05304, doi:10.1029/2009JD011823.

General S., Bobrowski N., Pöhler D. et al. (2014), Airborne I-DOAS measurements at Mt. Etna BrO and OClO evolution in the plume, *J. Volcanol. Geotherm. Res.*, doi: 10.1016/j.jvolgeores.2014.05.012.

Gerlach T.M. (2004), Volcanic sources of tropospheric ozone-depleting trace gases, *Geochem. Geophys. Geosyst.* **5**, Q09007.

Gliß J., Bobrowski N., Vogal L., and Platt U. (2014), OClO and BrO observations of the volcanic plume of Mt. Etna – implications in the chemistry of chlorine and bromine species in volcanic plumes. *Atmos. Chem. Phys. Discuss.* **14**, 25213–25280.

Giggenbach W.F. (1975), A simple method for the collection and analysis of volcanic gas samples, *Bull. Volcanol.*, **39**, S132–145.

Heue K.-P., Brenninkmeijer C.A.M., Baker A.K *et al.* (2011), $SO_2$ and BrO observation in the plume of the Eyjafjallajökull volcano 2010: CARIBIC and GOME-2 retrievals, *Atmos. Chem. Phys.* **11**, 2973–2989.

Hörmann C., Sihler H., Bobrowski N. *et al.* (2013), Systematic investigation of bromine monoxide in volcanic plumes from space by using the GOME-2 instrument, *Atmos. Chem. Phys.* **13**, 4749–4781.

Hinkley T.K., Wilson S.A., Finnegan D.L. and Gerlach T.M. (1999), Metal emissions from Kilauea, and a suggested revision of the estimated worldwide metal output by quiescent degassing of volcanoes, *Earth Planet. Sci. Lett.* **170**, 315–325.

Honda F. (1970), Geochemical study of iodine in volcanic gases. II. Behaviour of iodine in volcanic gases, *Geochem. J.* **3**, S201–211.

Kelly P.J., Kern C., Roberts T. *et al.* (2013), Rapid chemical evolution of tropospheric volcanic emissions from Redoubt volcano, Alaska, based on observations of ozone and halogen-containing gases, *J. Volcanol. Geothermal Res.* **259**, 317–333.

Kern C., Sihler H., Vogel L. *et al.* (2009), Halogen oxide measurements at Masaya volcano, Nicaragua using active long path differential optical absorption spectroscopy. *Bull. Volcanol.* **71**, 659–670.

Kern C., Deutschmann T., Vogel L. *et al.* (2010), Radiative transfer corrections for accurate spectroscopic measurements of volcanic gas emissions, *Bull. Volcanol.* **72**, 233–247.

Kutterolf S., Hansteen T.H., Appel K. *et al.* (2013), Combined bromine and chlorine release from large explosive volcanic eruptions: a threat to stratospheric ozone?, *Geology* **41**, 707–710.

Lee C., Tanimoto H., Bobrowski N. *et al.* (2005), Detection of halogen oxides in a volcanic plume and observation of surface ozone depletion, *Geophys. Res. Lett.* **32**, L21809, doi: 10.1029/2005GL023785.

Love S.P., Goff F., Counce D., Siebe C. and Degado H. (1998), Delgado passive infrared spectroscopy of the eruption plume at Popocatépetl volcano, Mexico, *Nature*, **396**, 563–567.

Louban I., Bobrowski N., Rouwet D., Inguaggiato S. and Platt U. (2009), Imaging DOAS for volcanological applications, *Bull. Volcanol.* **71**, 753–765.

Lübcke P., Bobrowski N., Arellano S. *et al.* (2014), $BrO/SO_2$ molar ratios from scanning DOAS measurements in the NOVAC network, *Solid Earth*, **5**, 409–424.

Martin R.S., Mather T.A. and Pyle D.M., (2006), High-temperature mixtures of magmatic and atmospheric gases, *Geochem. Geophys. Geosyst.* **7**, doi: 10.1029/2005GC001186.

Martin R.S., Mather T.A., Pyle D.M. *et al.* (2008), Composition-resolved size distributions of volcanic aerosols in the Mt. Etna plumes. *J. Geophys. Res.* **113**(D17).

Mather T.A., Pyle D.M. and Oppenheimer C. (2003), Tropospheric volcanic aerosol. *Geophys. Monogr. Ser.*, **139**, 189–212.

Mori T., Notsu K., Tohjima Y. and Wakita H. (1993), Remote detection of HCl and $SO_2$ in volcanic gas from Unzen volcano, Japan, *Geophys. Res. Lett.* **20**, 1355–1358.

Naughton J.J., Derby J.V. and Glover R.B. (1969), Infrared measurements on volcanic gas and fume: Kilauea eruption, 1968, *J. Geophys. Res.* **74**, 3273–3277.

Noguchi K. and Kamiya H. (1963), Prediction of volcanic eruption by measuring the chemical composition and amounts of gases, *Bull. Volcanol.* **26**, 367–378.

Oppenheimer C., Tsanev V.I., Braban C.F. *et al.* (2006), BrO formation in volcanic plumes, *Geochim. Cosmochim. Acta* **70**, 2935–2941.

Platt U. and Lehrer E. (1997), Arctic tropospheric ozone chemistry, ARCTOC, Final Report of the EU-Project No. EV5V-CT93-0318, Heidelberg.

Platt U., Allan W. and Lowe D. (2004), Hemispheric average Cl atom concentration from $^{13}C/^{12}C$ ratios in atmospheric methane, *Atmos. Chem. Phys.* **4**, 2393–2399.

Platt U. and Stutz J. (2008), *Differential Optical Absorption Spectroscopy, Principles and Applications, XV*, Heidelberg: Springer.

Pyle D. and Mather T. (2009), Halogens in igneous processes and their fluxes to the atmosphere and oceans from volcanic activity: a review, *Chem. Geol.* **263**, 110–121.

Roberts T.J., Braban, C.F., Martin R.S. *et al.* (2009), Modelling reactive halogen formation and ozone depletion in volcanic plumes, *Chem. Geol.* **263**, 131–142.

Robock A. (2000), Volcanic eruptions and climate, *Rev. Geophys.* **38**, 191–219.

Schumann U., Weinzierl B., Reitebuch O. *et al.* (2011), Airborne observations of the Eyjafjalla volcano ash cloud over Europe during air space closure in April and May 2010, *Atmos. Chem. Phys.* **11**, 2245–2279.

Schwandner F.M., Seward T.M., Gize A.P., Hall P.A. and Dietrich V.J. (2004), Diffuse emission of organic trace gases from the flank and crater of a quiescent active volcano (Vulcano, Aeolian Islands, Italy), *J. Geophys. Res.* **109**, D04301.

Shinohara H. (2013), Volatile flux from subduction zone volcanoes: insights from a detailed evaluation of the fluxes from volcanoes in Japan, *J. Volcanol. Geotherm. Res.* **268**, 46–63.

Snyder G.T. and Fehn U. (2002), Origin of iodine in volcanic fluids: $^{129}I$ results from the Central American Volcanic Arc, *Geochim. Cosmochim. Acta*, **66**, S3827–S3838.

Stremme W., Krueger A., Harig R. and Grutter M. (2012), Volcanic $SO_2$ and $SiF_4$ visualization using 2-D thermal emission spectroscopy – Part 1: slant-columns and their ratios, *Atmos. Meas. Tech.* **5**, 275–288.

Stutz J., Hebestreit K., Alicke B. and Platt U. (1999), Chemistry of halogen oxides in the troposphere: comparison of model calculations with recent field data, *J. Atm. Chem.* **34**, 65–85.

Surl L., Donohoue D., Aiuppa A., Bobrowski N. and von Glasow R. (2014), Quantification of the depletion of ozone in the plume of mount Etna. *Atmos. Chem. Phys. Discuss.* **14**, 23639–23680.

Symonds R.B., Rose W.I., Bluth G.J.S. and Gerlach T.M. (1994), Volcanic-gas studies; methods, results, and applications, *Rev. Mineral Geochem.* **30**, 1–66.

Textor C., Graf H.-F., Timmreck C. and Robock A. (2004), Emissions from volcanoes. In Granier, C., Artaxo, P. and Reeves, C. (eds.), *Emissions of Atmospheric Trace Compounds.* Dordrecht: Kluwer, pp. 269–303.

Theys N., Van Roozendael M., Dils B. *et al.* (2009), First satellite detection of volcanic bromine monoxide emission after the Kasatochi eruption, *Geophys. Res. Lett.* **36**, L03809, doi:10.1029/2008GL036552.

Theys N., De Smedt I., Van Roozendael M. *et al.* (2014), First satellite detection of volcanic OClO after the eruption of Puyehue-Cordón Caulle, *Geophys. Res. Lett.*, **41**, doi:10.1002/ 2013GL058416.

Varekamp J.C., Thomas E., Germani M. and Buseck P.R. (1986), Particle geochemistry of volcanic plumes of Etna and Mount St. Helens. *J. Geophys. Res.* **91**, B12, 12233–12248.

Vogel L. (2011), Volcanic plumes: evaluation of spectroscopic measurements, early detection and bromine chemistry, Doctoral Thesis, Combined Faculties for the Natural Sciences and for Mathematics, Ruperto Carola University of Heidelberg, Germany; available at: http://www.ub.uni-heidelberg.de/archiv/13219.

Voigt C., Jessberger P., Jurkat T. *et al.* (2014), Evolution of $CO_2$, $SO_2$, HCl and $HNO_3$ in the volcanic plumes from Etna, *Geophys. Res. Lett.* **41**, 2196–2203.

Vance A., McGonigle A.J.S., Aiuppa A. *et al.* (2010), Ozone depletion in tropospheric volcanic plumes, *Geophys. Res. Lett.* **37**, L22802, doi:10.1029/2010GL044997.

von Glasow R., Bobrowski N. and Kern C. (2009), The effects of volcanic eruptions on atmospheric chemistry, *Chem. Geol.* **263**, 131–142.

von Glasow R. (2010), Atmospheric chemistry in volcanic plumes, *Proc. Natl. Acad. Sci. USA* **107**, 6594–6599.

von Glasow R. and Crutzen P.J. (2007), Tropospheric halogen chemistry. In Holland H.D. and Turekian K.K. (eds.), *Treatise on Geochemistry*, Oxford: Elsevier-Pergamon, pp. 1–67.

Wittmer J., Bobrowski N., Liotta M. *et al.* (2014), Active alkaline traps to determine acid–gas ratios in volcanic plumes: sampling technique and analytical methods, *J. Geochem., Geophys., Geosyst.* **15**, 2979–2820.

Zelenski M. and Taran Y. (2012), Volcanic emissions of molecular chlorine, *Geochim. Cosmochim. Acta* **87**, 210–226.

# 9

# Satellite and aircraft-based techniques to measure volcanic emissions and hazards

DAVID PIERI

## 9.1 Introduction

The outstanding utility of volcanological remote sensing techniques is their ability to provide synoptic data at a variety of spatial scales, often coupled, as in the case of orbital data, with relatively easy repeat access to address dynamic processes, mitigated only by atmospheric obscuration. Volcanic phenomena on the Earth span a range of spatial scales, characteristic activity frequencies, compositions, mass eruption rates, and emitted energies, a catalog well known to volcanologists: explosive and effusive eruptions, passive emissions, and submarine eruptions (often at enormous hydrostatic pressures), and outliers like natrocarbonatite eruptions, mud volcanoes, and hydro-geothermal field emissions. This expansive phenomenological range of volcanism, across a range of temperatures (e.g. ~ 1450K (lavas) to ~ 200K (lofted stratospheric ash)), and its relentless dynamism, poses substantial challenges for the application of remote sensing technology.

Spatial scales and recurrence frequencies of contemporary volcanic environmental effects (see also Chapters 13 and 14) are demanding in terms of remote sensing techniques. For example, within the planetary boundary layer (1–3 km above sea level (ASL) generally), eruption columns and plumes can inject ash and other aerosols (e.g. $H_2SO_{4(liquid)}$, $CaSO_{4(solid)}$) on timescales from minutes to hours, and can persist for days (continuously), or months (intermittently). Above the tropopause, post-explosive-eruption stratospheric (~ 10,000 m above sea level (ASL)) ash and gas clouds (especially sulfate nano-aerosols) can drift thousands of kilometers, occasionally circumnavigating the globe, playing havoc with aviation, and altering the Earth's surface insolation while heating the stratosphere, sometimes for months, or even years. Depletion of the planetary ozone shield by almost 10% occurred after the eruption of Mt. Pinatubo (1991). Repose periods between such large events can

*Volcanism and Global Environmental Change*, eds. Anja Schmidt, Kirsten E. Fristad and Linda T. Elkins-Tanton. Published by Cambridge University Press. © Cambridge University Press 2015.

typically span hundreds of years; however, some volcanoes can erupt, albeit at somewhat less intensity, into the troposphere (e.g. Mt. Etna) or even directly reach the tropopause (e.g. Kliuchevskoi Volcano) with almost annual regularity. Surface effects (e.g. lava flows, ash falls, pyroclastic flows, lahars, debris flows) can occur at relatively small characteristic spatial scales (e.g. of order 1–100 m for dynamic cores; of order 1–10 km in areal extent), and can occur on timescales of just a few minutes (especially pyroclastic flows); or they can can persist for months (e.g. Mauna Loa 1984 eruption) or for years (e.g. Kilauea Pu'u O'o, since 1983 and ongoing). Ancient eruptions (e.g. Yellowstone Lava Creek eruption, 640 Ka BP; Toba eruption, ~73 Ka BP; Santorini eruption, 3.6 Ka BP; Deccan Traps eruptions, 60–68 Ma BP; Siberian Traps eruptions, ~250–251 Ma BP; Altiplano–Puna ignimbrites, 10–11 Ma BP), and even more recent large eruptions (e.g. Krakatoa, 1883; Tambora, 1812–1814; Ksudach, 1907; Katmai, 1912), dwarf our contemporary experience with respect to the application of remote sensing techniques.

Fortunately, remote sensing has undergone astonishing growth in platforms (now including unmanned aerial vehicles (UAVs)) and spectral range. In addition, new large (of order $10^2$–$10^3$ Tbytes) online digital archives of multi-to-hyper spectral satellite images exist, spanning the 40 years since the first Landsat images were acquired, including digital elevation models for much of that time (e.g. http:// ava.jpl.nasa.gov). Thus, there are now detailed systematic synoptic data documenting volcano emissions and deposits over that period.

In this chapter, I provide a catalog of remote sensing strategies and techniques as related to the characteristics of contemporary volcanic activity. Hopefully, examination of current approaches will suggest areas of improvement, as well as strategies to observe and study the largest geologically known eruptions, which modern humans have yet to directly experience, but which lie in our future, and which will have profound implications for weather and climate.

## 9.2 Worldwide capability to detect and monitor volcanic emissions from satellites and aircraft

### *9.2.1 Orbital observations*

Modern orbital satellite remote sensing techniques have been used to collect image data over volcanoes for many years (e.g. Table 9.1). Orbital multispectral missions typically fall into two classes: "mapping" missions and "sampling" missions, with multispectral observations often being the most useful for volcanological applications. Most of these missions have involved national or multi-national civilian assets and data access is relatively economical for researchers. Commercial broadband or panchromatic data can provide very high spatial (1–5 m/pixel) and

Table 9.1 *Representative summary of historically most used orbital remote sensing instruments for volcanological observations.*

| Instrument/platform | Spectral coverage | Spatial resolution/pixel | Nadir repeat interval | Primary volcanological utilization | Pointing |
|---|---|---|---|---|---|
| ASTER VNIR/Terra | 0.5–0.9 µm; 3 bands plus stereo band | 15 m (daylight) | 16 days | Land surface processes; land use | Up to 24° off-nadir |
| ASTER SWIR/Terra (operational from January 2000 to April 2008) | 1–2.5 µm (6 bands) | 30 m (day/night) | 16 days | Mineral mapping (electronic transition bands); high temperature targets (lava flows; summit domes); cloud detection | Up to 8.5° off-nadir |
| ASTER TIR/Terra | 8–12 µm (5 bands) | 90 m (day/night) | 16 days | Emissivity mapping (resstrahlen); temperature and $SO_2$ detection | Up to 8.5° off-nadir |
| MODIS /Terra, Aura, Aqua | 0.4–14.5 µm (36 bands) | 250 m (2 bands) 500 m (5 bands) 1 km (29 bands) | 1 day | Eruption/hotspot detection; ash-plume tracking | N/A |
| OMI/Aura | 350–500 nm 270–314 nm 306–380 nm | 13 × 24 km footprint (day) | 1 day | $SO_2$ tracking | N/A |
| MISR/Terra | Visible red, green, blue, near-IR | 250–275 m | 2–9 days | Aerosol measurements (ash clouds) | 9 view angles (nadir and 26.1°, 45.6°, 60°, 70.5° off-nadir) |
| AVHRR/POES/ NPOESS | 0.5–12.5 µm; 4, 5, 6 bands depending on launch year | 1.1 km | Multiple times per day | Aerosol tracking (ash clouds) | N/A |
| Geostationary imagers GOES (METEOSAT, MTSAT are similar) | 0.5–14 µm | 1 km (visible), 4 km (SWIR-TIR), 8 km (moisture band) | Quasi-continuous | Plume tracking | N/A |

Table 9.1 (*cont.*)

| Instrument/platform | Spectral coverage | Spatial resolution/pixel | Nadir repeat interval | Primary volcanological utilization | Pointing |
|---|---|---|---|---|---|
| Landsat series (1–8) | Visible, near-IR; TIR | 40–80 m multispectral; 15 m pan; 60–120 m TIR | 16 days | Geological mapping; hotspot detection | N/A |
| EO-1 Advanced Land Imager (ALI) | 0.4–2.35 μm (7 bands) 10 m pan | 30 m; 37 × 42 km FOV 10 m | 16 days | Feature mapping; hotspot detection | 6°–17° East–west |
| EO-1 Hyperion | 0.4–2.5 μm (220 bands) | 7.7 × 185 km FOV 30 m | 16 days | Hotspot detection | 6°–17° East–west |

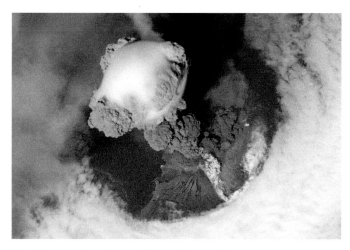

Figure 9.1 Spectacular view of Sarychev Volcano (Matua Island, Kuriles) June 12, 2009 (Astronaut photograph ISS020-E-9048, Nikon D2XS digital camera, 400 mm lens). Photo by NASA Earth Observatory, with permission (http://earthobservatory.nasa.gov/NaturalHazards/view.php?id=38985). A black and white version of this figure will appear in some formats. For the colour version, please refer to the plate section.

temporal (daily) resolution data, and can be useful for observing episodes of heightened activity at specific volcanoes. A unique archive of hand-held film and digital images of eruptions, taken from the International Space Station (ISS, 52° orbital inclination), with resolution of a few meters, is an under-exploited, if qualitative, volcanological asset. Typically, ISS data opportunities cluster in episodes of 3–5-day intervals for any given target volcano, depending on latitude (Figure 9.1).

### 9.2.2 Mapping missions

In orbital mapping missions, earth-observing instruments typically operate at a nearly 100% duty cycle. That is, they are nearly always on, and nadir-looking, thus foregoing the need to aim observing instruments at particular targets, and greatly simplifying operations. Synoptic mapping missions in polar orbit with low (1 km/pixel) to moderate (250 m/pixel) spatial resolutions permit relatively frequent repeat observations, at least one image per day, and often many more, depending on geocentric latitude. Synoptic mapping missions in geostationary (GEO), circular high Earth orbit (HEO), or high elliptical orbits (Molniya), may return data on a minute-by-minute basis across a hemisphere; however, data are returned at relatively low (e.g. 1–4 km/pixel) spatial resolutions for unclassified civil applications.

For example, the National Oceanic and Atmospheric Administration (NOAA) Advanced Very High Resolution Radiometer (AVHHR) has provided daily synoptic (1 km/pixel) coverage of the entire globe and has been effective in detecting and tracking volcanic clouds since the early 1980s. The NOAA Geostationary Operational Environmental Satellite (GOES) series has provided reliable weather forecasting data for North America, and has been useful for tracking volcanic plumes and drifting ash clouds. In Europe, the METEOSAT geostationary platform has served in an effectively identical capacity for EUMETSAT, as has the MTSAT series over Eastern Asia for the Japan Meteorological Agency. Work on airborne volcanic ash (e.g. Prata, 1989a; Schneider *et al.*, 1995; Dean *et al.*, 2004; Pavolonis *et al.*, 2006) and sulfur dioxide (Prata and Grant, 2001; Watson *et al.*, 2004; Krotkov *et al.*, 2010; Carn *et al.*, 2011; Clarisse *et al.*, 2012) has exploited one or more of these classic instruments, and newer ones, for studying volcanic plumes and clouds at regional scale (see also Chapter 7).

### 9.2.3 Sampling missions

Multispectral orbital sampling missions, typically in Sun-synchronous polar orbit and operating at substantially higher spatial resolutions (e.g. 15–120 m/pixel), have revisit intervals of days to weeks. Their relatively narrow swath widths (60–180 km) require them to point off-nadir at target volcanoes, requiring constant comprehensive planning and scheduling of competing observations. This adds enormous mission complexity and, combined with other operational constraints, results in a duty cycle typically an order of magnitude smaller than for "always-on" mapping missions. Early work with Landsat data at relatively high spatial resolution (40 m/pixel visible, 120 m/pixel thermal infrared) focused on detecting thermal anomalies (Rothery *et al.*, 1988; Oppenheimer *et al.*, 1993; Denniss *et al.*, 1998). Later, Landsat-based analyses and modeling focused on assessing the thermal properties of lava flows (e.g. Pieri *et al.*, 1990) and other deposits.

More recently, the Advanced Spaceborne Thermal Emission and Reflection radiometer (ASTER), along with Landsat, has compiled a comprehensive land surface data set (now over 3 million images or "granules") since the year 2000, including a systematic acquisition of ASTER data over the world's volcanoes that have been active since the Holocene (Siebert and Simkin, 2002). ASTER currently acquires 15 m/pixel data in three visible–near-infrared bands (0.5–0.9 um; VNIR). In addition, between 2000 and 2008 it acquired data at 30 m/pixel in six short-wave-infrared bands (1–2.4 μm; SWIR), and, uniquely among current civilian orbital remote sensing instruments, acquires data in five thermal infrared bands (8–12 μm; 90 m/pixel; TIR). ASTER has an additional backward-viewing VNIR band that allows the creation of digital elevation models (DEMs) for each granule.

Over its mission lifetime, ASTER has released several Global Digital Elevation Models (GDEMs; http://asterweb.jpl.nasa.gov/gdem.asp) covering the Earth's land surface at 8–12-m vertical resolution at 30-m postings.

ASTER data has been used for a wide variety of volcanological investigations (e.g. Pieri and Abrams, 2005). A few examples of these include the seminal work on the detection and measurement of passively emitting tropospheric $SO_2$ plumes (e.g. Realmuto *et al.*, 1994), the systematic time-series monitoring of summit crater thermal emissions (e.g. Buongiorno *et al.*, 2013), the monitoring of dynamic summit crater domes (e.g., Ramsey *et al.*, 2012), and the detection of low-temperature thermal eruption precursors (Pieri and Abrams, 2005).

### *9.2.4 Ash- and gas-cloud detection*

No other volcanic hazard has the regional (and even global) impact of drifting volcanic clouds. Their impact on aviation, especially within Europe after the 2010 Icelandic eruptions, was profound, inflicting an estimated $5 billion economic damage globally (Oxford Economics, 2012). Thus there has been a recent enhanced effort within Europe and the United States to improve our understanding of the properties of such clouds, and to enlarge the collection of now scant *in situ* data on volcanogenic gas and aerosols (especially ash), as well as their effect on aircraft.

Data from synoptic weather satellites, the NASA Terra and Aqua orbital platforms (e.g. the Moderate Resolution Imaging Spectroradiometer (MODIS)) and high spatial resolution data (e.g. ASTER) in the TIR have been utilized for investigation of ash and gas clouds at a variety of local-to-regional spatial scales. Typical approaches exploit the reverse absorption effect (Prata, 1989a; 1989b) of dry silicate ash within TIR bands. A negative radiance differential between 10.6 and 12 μm indicates fine ash (1–12 μm effective radius; Rose *et al.*, 2000). When ash clouds diffuse, their radiant emission drops beneath instrument sensitivity thresholds in remote sensing data (e.g. Schneider *et al.*, 1995); however, these limits have only tentatively been correlated between remote sensing and very rare validating *in situ* data (Pieri *et al.*, 2002; Schumann *et al.*, 2010; e.g., particle size-frequency, cloud height-thickness-extent, concentration variations with time; see also Chapter 7).

Likewise, orbital image data from ASTER, MODIS, AIRS (Atmospheric Infrared Sounder; all in the infrared – observations in the troposphere and stratosphere, day and night) and OMI (Ozone Montoring Instrument; ultraviolet – observations primarily in the stratosphere, daytime only) are employed in $SO_2$ detection and tracking. $SO_2$ and ash typically track together – the 2008 Kasatochi eruption is a good recent example (Prata *et al.*, 2010; Krotkov *et al.*, 2010), although there are exceptions (e.g. El Chichon 1982 eruption; Schneider *et al.*, 1999). Still, $SO_2$ has a low (4–5 parts per billion by volume (ppbv) global ambient background, and thus is often viewed as a proxy for volcanic ash (e.g. Kreuger *et al.*, 2009). Better knowledge of how solid aerosols (e.g. ash) and $SO_2$ are related and the conditions under which they are, or are not, spatially correlated, is central to understanding volcanogenic cloud chemistry (e.g. Rose *et al.*, 2000). UAV-based *in situ* observations for calibration and validation of orbital remote sensing data hold out important promise for improving our specific knowledge of such parameters, and for illuminating these problems generally.

Typically, hazard responders rely on volcanic ash transport and dispersion (VATD) models (Stunder *et al.*, 2007), used for research and past event analyses (D'Amours *et al.*, 2010; Webley *et al.*, 2010). Two current Lagrangian trajectory models, one for ash trajectory tracking developed at the University of Alaska, Fairbanks (Searcy *et al.*, 1998), called PUFF, and the other, a NOAA-sponsored Hybrid Single Particle Lagrangian Integrated Trajectory (HYSPLIT) model (Stunder *et al.*, 2007) are in common use. HYSPLIT is a successor to the well-known Volcanic Ash Forecast Transport and Dispersion (VAFTAD) model, pioneered by NOAA (Hefter and Stunder, 1993). Initial boundary conditions at the source vent, e.g. total erupted mass and eruption rate, solid aerosol size-frequency distributions, plume-top altitudes, $SO_2$ flux, amount of ambient air ingested, and the vertical distribution of ash within an eruption column, exert crucial influence on model outcomes.

## 9.3  Airborne observations

### 9.3.1  Manned aircraft

Manned airborne observations employ sophisticated state-of-the-art optical, micro-wave, and electro-magnetic instrumentation (e.g. Table 9.2), particularly effective when correlated with seismic, ground-based geodetic and surface radiometric observations. They have been useful for mapping volcanic deposits and features, capturing dynamic inflation and deflation events, and for elucidating faults and fractures in three dimensions. Capturing eruption activity, including *in situ* sampling of aerosol and gas emissions, particularly for large explosive events, is somewhat more problematic for manned aircraft. While comprehensive obser-vations of volcanic plumes have been made by manned aircraft, these have mainly been from above (where possible) or from stand-off positions (e.g. recent observations of an artificial ash plume with the Airborne Volcanic Object Imaging Detector (AVOID); see also Chapter 7). Potential airborne ash ingestion into aircraft engines, while deemed a marginally acceptable risk by European authorities,

Table 9.2 *NASA airborne remote sensing facility instruments with which the author is familiar or has used (after http://airbornescience.nasa.gov/instrument/facility).*

| Title | Acronym | Aircraft | Type |
|---|---|---|---|
| Airborne Visible/Infrared Imaging Spectrometer | AVIRIS | ER-2, Proteus, Twin Otter, WB-57 | Passive, spectrometer |
| Next-Generation Airborne Visible/Infrared Imaging Spectrometer | AVIRIS-ng | ER-2, Proteus, Twin Otter, WB-57 | Passive, spectrometer |
| Digital Camera System | DCS | B-200, DC-8, ER-2, Twin Otter, WB-57 | Camera, passive |
| Digital Mapping System | DMS | DC-8, P-3 Orion | Camera, passive |
| MODIS Airborne Simulator | MAS | ER-2 | Passive, spectrometer |
| MODIS/ASTER Airborne Simulator | MASTER | B-200, Caravan, Cessna, DC-8, ER-2, J-31, WB-57 | Passive, spectrometer |
| UAV/Gulfstream Synthetic Aperture Radar | UAVSAR | G-III | Active, radar |

disqualifies flight into ash-contaminated airspace under International Civil Aviation Organization (ICAO) operating guidelines. Thus, near-source sampling of volcanic ash plumes where ash concentrations are highest are almost always precluded for manned aircraft, though widely viewed as critical (e.g. Guffanti, 2012).

### 9.3.2 Unmanned aircraft and aerostats

UAV technology and related instrumentation (e.g. Table 9.3; Figure 9.2; Pieri *et al.*, 2013) is undergoing rapid maturation and provides promising new avenues for volcano observations. Such robotic aircraft can be dispatched under hazardous conditions, in bad weather, in close proximity to hazardous terrain, and within volcanic clouds – limited only by their operational envelopes, payload risk–cost–benefit considerations, and regulations. Overall, UAVs are of high utility for *in situ* time-series concentration measurements, and for sampling of gases (e.g. $SO_2$, $CO_2$, $CH_4$, $H_2S$, OCS, CO, $H_2O$) and volcanogenic aerosols (e.g. ash, sulfuric acid), while recording altitudinal profile data of temperature, pressure, humidity, and wind velocity (Pieri *et al.*, 2013).

Systematic monthly *in situ* aerostat (meteorological balloon) $SO_2$ sampling, supporting ASTER, is underway at Turrialba Volcano in Costa Rica (Pieri *et al.*, 2013). Such observations complement those from free-flying small UAVs, providing at-a-station observations at altitudes up to 13,000 ft ASL.

Table 9.3 *Examples of instrument payloads for UAVs and aerostats used by the author (after Pieri et al., 2013).*

| UAV platforms | Payload mass (kg) | Instruments | Operating altitudes (ft) |
|---|---|---|---|
| Tethered balloon | < 1 | SO$_2$, temperature, pressure and humidity detectors; GPS, telemetry | 0–5,000 ft AGL |
| Kite | <1 | SO$_2$, temperature, pressure and humidity detectors; GPS, pan-camera | 0–5,000 ft AGL |
| Micro-UAV, Dragon Eye (by Aerovironment Inc.) | <0.5 | SO$_2$, temperature, pressure and humidity detectors; GPS, optical particle counters (OPCs), nano-PC, color VIS camera, Lo Light VIS, thermal IR, evacuated sampling bottle | > 12,000 ASL |
| Micro-UAV, Vector Wing 100 (by Maryland Aerospace, Inc.) | < 1 | SO$_2$, temperature, pressure and humidity detectors; GPS, color VIS camera | 15,000 ASL |
| Medium-UAV, SIERRA, SIERRA II (operated by NASA Ames Research Center) | < 45 | ULISSES mass spectrometer; SO$_2$, temperature, pressure and humidity detectors; GPS, multispectral imaging, bolometer, optical particle counters, liquid aerosol probe, mini-nephelometer, aerosol drum impactors | 15,000 ASL |

Figure 9.2   Picture of the SIERRA on display at Palo Alto Airport, California, USA. Note the substantial payload capability (up to 100 lb, 45 kg) in the interchangeable nose. This increases its flexibility for multiple instrument missions. Photo by Dan Dawson, with permission.

Effective UAVs represent a compromise: capable, but not too expensive to risk. Sizes of current UAVs have a strong direct correlation with payload capability and platform performance. Small UAVs (e.g. < 50 kg) may be less capable, but are more expendable than medium to large UAVs (e.g. > 50 kg), which are more capable, but economically less expendable. State-of-the-art downsizing is rapidly increasing micro-UAV capability range. Thus, optical particle counters, aerosol impactors, and aerosol optical absorption analyzers are possible autonomous payloads. Likewise, UV–visible–infrared multispectral and hyperspectral imagers, radiometers, lidars, and radars could be flown autonomously in situations where flights over imminently erupting volcanoes are too dangerous, especially where very high spatial resolution (< 1m/pixel) data are required. Within the next few years, small to medium UAVs (~ 100 kg, e.g. the NASA SWIFT), may be able to carry payloads to altitudes above 10 km ASL, which are now solely the realm of very large (and usually prohibitively expensive) UAVs, such as the NASA Global Hawk.

## 9.4 Data archives and global volcanology

Over the last decade or two there has been a rise in accessible global online data bases that are now providing the first global volcano remote sensing catalogs. Some data sets were primarily meteorological in original intent (e.g. AVHRR, GOES, METEOSAT, MTSAT, OMI, AIRS; see also Chapter 7), at relatively low spatial resolution, but with high temporal sampling frequency. Some contain data sets acquired for geological mapping (e.g. Landsat, ASTER) or as an engineering test (EO-1). All are useful to volcanology, which has both meteorological and geological data requirements. We address the latter US archives here.

The Land Processes Distributed Active Archive Center (LP DAAC) is a component of NASA's Earth Observing System Data and Information System. To extend the utility of the LP DAAC specifically for volcanological remote sensing data analyses, the ASTER Volcano Archive (AVA: http://ava.jpl.nasa.gov) was created by the author. A discipline-specific specialty archive, AVA is the world's largest full resolution, fully searchable, volcano image archive (14 ASTER spectral bands – visible to thermal infrared, acquired from 2000 to the present; ~160,000 images ("granules") covering 1546 volcanoes, searchable by spatial, temporal, and spectral parameters). It provides all bands of ASTER data at full native spatial resolution in .jpgw (jpeg geo-located worldfile) or .geoTIFF format. All image products are .kml and .kmz enabled for display and projection within Google Earth™ and Google Maps™. In addition, AVA houses 1,216 digital elevation maps, 4,715 thermal anomaly detections, as well as a variety of other associated data products. Within the next year, AVA will undergo expansion to include comparable Landsat data. AVA's overall goal is to improve our ability to forecast the timing, effects, and extent of volcanic activity, by providing ASTER and other data in a readily accessible format.

## 9.5 Summary

Over the past half century, the increased availability of global multispectral remote sensing data, recording the activity of the world's volcanoes, has contributed to the globalization of volcanology. The global data base enables comparisons of volcanic activity styles, intensities, timing, eruption precursors, as well as mapping surveys of the distributions of erupted products, alteration zones, and thermal anomalies, to name just a few aspects. The collation of relevant meteorological data, as they apply to studies of volcanic activity, provides a way to study the composition and movement of volcanic airborne emissions across the globe. New *in situ* airborne sampling techniques, especially as related to unmanned airborne vehicles, provide an unprecedented, less risky opportunity for sampling key volcanogenic emissions in airspace that would be hazardous to manned aircraft. Comprehensive discipline-specific specialty data archives, such as the ASTER Volcano Archive, can provide real analyses payoffs. As tools for mining massive data sets are applied to ever accumulating remote sensing data, comparisons and contrasts between volcanic systems over time should prove fruitful for science and hazard mitigation. While such data inform contemporary process and transport models, they also enable estimations of boundary conditions and environmental responses for massively larger eruptions in our past, and for the similar-scale eruptions that lie ahead.

## Acknowledgments

Work represented here was carried out under contract to NASA at the Jet Propulsion Laboratory of the California Institute of Technology in Pasadena, California, USA. The author thanks Fred Prata for a very helpful review.

## References

Buongiorno, M. F., Pieri, D. C., Silvestri, M., 2013. Thermal analysis of volcanoes based on 10 years of ASTER data on Mt. Etna. *Remote Sensing and Digital Image Processing*, **17**, 409–428.

Carn, S. A., Froyd, K. D., Anderson, B. E. *et al.*, 2011. *In situ* measurements of tropospheric volcanic plumes in Ecuador and Colombia during TC4. *Journal of Geophysical Research*, **116**, D00J24.

Clarisse, L., Hurtmans, D., Clerbaux, C. *et al.*, 2012. Retrieval of sulphur dioxide from the infrared atmospheric sounding interferometer (IASI). *Atmospheric Measurement Techniques*, **5**, 581–594.

D'Amours, R., Malo, A., Servranckx, R. *et al.*, 2010. Application of the atmospheric Lagrangian particle dispersion model MLDP0 to the 2008 eruptions of Okmok and Kasatochi volcanoes. *Journal of Geophysical Research*, **115**, D00L11.

Dean, K. G., Dehn, J., Papp, K. R. *et al.*, 2004. Integrated satellite observations of the 2001 eruption of Mt. Cleveland, Alaska. *Journal of Volcanology and Geothermal Research*, **135**, 51–73.

Denniss, A. M., Harris, A. J. L., Rothery, D. A., Francis, P. W., Carlton, R. W., 1998. Satellite observations of the April 1993 eruption of Lascar volcano. *International Journal of Remote Sensing*, **19**, 801–821.

Guffanti, M., 2012. Volcanic-ash hazards to aviation in the post-world: a status report. Presented at TETS 2012. Proceedings of the Turbine Engine Technology Symposium 2012, Dayton, OH, 10–13 September.

Hefter, J. L., Stunder, B. J. B., 1993. Volcanic Ash Forecast Transport and Dispersion (VAFTAD) model. *Weather and Forecasting*, **8**, 533–541.

Kreuger, A., Yang, K., Krotkov, N., 2009. Enhanced monitoring of sulfur dioxide sources with hyperspectral UV sensors. *Proceedings of SPIE*, **7475**. See http://dx.doi.org/10.1117/12.830142.

Krotkov, N. A., Schoeberl, M. R., Morris, G. A., Carn, S., Yang, K., 2010. Dispersion and lifetime of the $SO_2$ cloud from the August 2008 Kasatochi eruption. *Journal of Geophysical Research*, **115**, D00L20.

Oppenheimer, C., Francis, P. W., Rothery, D. A., Carlton, R. W. T., Glaze, L. S., 1993. Infrared Image-Analysis of Volcanic Thermal Features – Lascar Volcano, Chile, 1984–1992, *Journal of Geophysical Research*, **98** (B3), 4269–4286.

Oxford Economics, 2012. *Volcanic Ash Impact on Air Travel*, report for Airbus Industries. Oxford Economics, Oxford. See http://www.airbus.com/company/environment/documentation/?docID=10262&eID=dam_frontend_push.

Pavolonis, M. J., Feltz, W. F., Heidinger, A. K. and Gallina, G. M., 2006. A daytime complement to the reverse absorption technique for improved automated detection of volcanic ash. *Journal of Atmospheric and Oceanic Technology*, **23**, 1422–1444.

Pieri, D., Abrams, M., 2005. ASTER observations of thermal anomalies preceding the April 2003 eruption of Chikurachki volcano, Kurile Islands, Russia. *Remote Sensing of Environment*, **99**, 84–94.

Pieri, D., Diaz, J. A., Bland, G. *et al.*, 2013. In situ observations and sampling of volcanic emissions with NASA and UCR unmanned aircraft, including a case study at Turrialba Volcano, Costa Rica. *Geological Society, London, Special Publications*, **380**, 321–352.

Pieri, D. C., Glaze, L. S., Abrams, M. J., 1990. Thermal radiance observations of an active lava flow during the June 1984 eruption of Mount Etna. *Geology*, **18**, 1018–1022.

Pieri, D. C., Ma, C., Simpson, J. J. *et al.*, 2002. Analyses of *in situ* airborne volcanic ash from the Feb 2000 eruption of Hekla. *Geophysical Research Letters*, **29**, 19-1–19-4.

Prata, A. J., 1989a. Infrared radiative transfer calculations for volcanic ash clouds. *Geophysical Research Letters*, **16**, 1293–1296.

Prata, A. J., 1989b. Observations of volcanic ash clouds in the 10–12 μm window using AVHRR/2 data. *International Journal of Remote Sensing*, **10**, 751–761.

Prata, A. J., Grant, I. F., 2001. Retrieval of microphysical and morphological properties of volcanic ash plumes from satellite data: application to Mt Ruapehu, New Zealand. *Quarterly Journal of the Royal Meteorological Society*, **127**, 2153–2179.

Prata, A. J., Gangale, G., Clarisse, L., Karagulian, F., 2010. Ash and sulfur dioxide in the 2008 eruptions of Okmok and Kasatochi: insights from high spectral resolution satellite measurements. *Journal of Geophysical Research*, **115**, D00L18.

Ramsey, M. S., Wessels, R. L., Anderson, S. W., 2012. Surface textures and dynamics of the 2005 lava dome at Shiveluch volcano, Kamchatka. *Geological Society of America Bulletin*, **124**, 5–6, 678–689.

Realmuto, V. J., Abrams, M. J., Boungiorno, M. F., Pieri, D. C., 1994. The use of multispectral thermal infrared image data to estimate the sulfur-dioxide flux from volcanoes – a case study from Mount Etna, Sicily, July 29, 1986. *Journal of Geophysical Research*, **99**, 481–488.

Rose, W. I., Bluth, G. J. S., Ernst, G. G. J., 2000. Integrating retrievals of volcanic cloud characteristics from satellite remote sensors: a summary. *Philosophical Transactions of the Royal Society of London*, **A358**, 1585–1606.

Rothery, D. A., Francis, P. W., Wood, C. A., 1988. Volcano monitoring using short wavelength infrared data from satellites. *Journal of Geophysical Research*, **93** (B7), 7993–7999.

Schneider, D. J., Rose, W. I., Kelley, L., 1995. Tracking of 1992 eruption clouds from Crater Peak vent of Mount Spurr Volcano, Alaska, using AVHRR. *United States Geological Survey Bulletin*, **2139**, 27–36.

Schneider, D. J., Rose, W. I., Coke, L. R. *et al.*, 1999. Early evolution of a stratospheric volcanic eruption cloud as observed with TOMS and AVHRR. *Journal of Geophysical Research*, **104**, 4037–4050.

Schumann, U., Weinzierl, B., Reitebuch, O. *et al.*, 2010. Airborne observations of the Eyjafjallajökull volcano ash cloud over Europe during air space closure in April and May 2010. *Atmospheric Chemistry and Physics Discussions*, **11**, 2245–2279.

Searcy, C., Dean, K. G., Stringer, W., 1998. PUFF: a volcanic ash tracking and prediction model. *Journal of Volcanology and Geothermal Research*, **80**, 1–16.

Siebert, L., Simkin, T., 2002. Volcanoes of the world: an illustrated catalog of Holocene volcanoes and their eruptions. *Smithsonian Institution, Global Volcanism Program Digital Information Series*, GVP-3. See http://www.volcano.si.edu.

Stunder, B. J. B., Heffter, J. L., Draxler, R. R., 2007. Airborne volcanic ash forecast area reliability, weather and forecasting. *Weather and Forecasting*, **22**, 1132–1139.

Watson, I. M., Realmuto, V. J., Rose, W. I. *et al.*, 2004. Thermal infrared remote sensing of volcanic emissions using the Moderate Resolution Imaging Spectroradiometer (MODIS). *Journal of Volcanology and Geothermal Research*, **135**, 75–89.

Webley, P. W., Dehn, J., Lovick, J. *et al.*, 2009. Near real time volcanic ash cloud detection: experiences from the Alaska volcano observatory. *Journal of Volcanology and Geothermal Research*, **186**, 79–90.

Webley, P. W., Dean, K. G., Dehn, J., Bailey, J. E., Peterson, R., 2010. Volcanic-ash dispersion modeling of the 2006 eruption of Augustine Volcano using the Puff model. *United States Geological Survey, Professional Papers*, **1769**, 507–526.

# 10

## The origin of gases that caused the
## Permian–Triassic extinction

ALEXANDER V. SOBOLEV, NICK T. ARNDT, NADEZHDA A. KRIVOLUTSKAYA,
DIMITRY V. KUZMIN AND STEPHAN V. SOBOLEV

### 10.1 Introduction

Even though a causal relationship between the emplacement of the Siberian large igneous province (LIP) and the end-Permian mass extinction remains unproven, the synchronicity of the two events, each the largest of its kind, makes it extremely likely that the magmatism was linked directly to the biological crisis. Recent dating has placed the eruption of the flood basalts in the interval 252.4 to 251.2 Ma (Kamo *et al.*, 2003; White and Saunders, 2005; Svensen *et al.*, 2009; and other chapters in this volume), which almost entirely overlaps that of the mass extinction. Song *et al.* (2012) identified two distinct extinction events: one at the end-Permian and the other at the early Triassic, each of which contributed to the mass extinction, and both within the time frame of the eruptions.

The cause of the mass extinctions was probably the release of gases that dramatically increased global temperatures, upset the chemical balance within the oceans, destroyed the ozone layer or were sufficiently toxic to destroy life both on land and in the oceans (e.g. Wignall *et al.*, 2009). There are two probable sources of these gases, both related with the Siberian LIP. First, the mantle source – the Siberian plume could potentially release large amounts of $CO_2$ and Cl, if it contains altered recycled oceanic crust (Sobolev *et al.*, 2011). These gases are released either by explosive eruptions that typically occurred early in the magmatic episode, or by degassing of erupted or intruded magmas during the major magmatic event.

The second possible source of gases lies within the sedimentary pile beneath the volcanic plateau. These sediments consist, in large part, of carbonates and evaporites, which are overlain by terrestrial sediments containing coal measures. When heated at the margins of the intrusions that fed the flood volcanism, these sediments released large amounts of $CO_2$, $SO_2$ and halocarbons; when organic matter or coal

*Volcanism and Global Environmental Change*, eds. Anja Schmidt, Kirsten E. Fristad and Linda T. Elkins-Tanton.
Published by Cambridge University Press. © Cambridge University Press 2015.

was present, toxic reduced gases such as CO and CH$_4$ were also released (Ganino and Arndt, 2009; Svensen *et al.*, 2009). In addition, contamination of magmas by sediments could generate significant fluxes of Cl, S and F when the lavas degassed upon eruption (Sobolev *et al.*, 2009a; Black *et al.*, 2012).

   In this paper we reappraise the quantities and types of gas released from the two sources during the emplacement of the Siberian LIP and re-evaluate their potential contribution to the end-Permian mass extinction.

## 10.2  Magmatic volatiles – data from melt inclusions in phenocrysts

The only reliable sources of information on the magmatic volatiles in ancient volcanic or intrusive rocks are volcanic glasses and melt or fluid inclusions in phenocrysts. Fresh glasses of basaltic composition are very rare, however, in continental flood volcanic rocks. The compositions of glasses and inclusions, including volatiles, can be measured using modern *in situ* techniques such as electron microprobe, ion microprobe and laser-source inductively coupled plasma mass spectrometry (LA-ICP-MS).

   Various processes can change the concentrations of volatiles in both glasses and inclusions. The most important is degassing during magma ascent. This process mostly affects CO$_2$, which is much less soluble in the melt at crustal pressures, but it also may affect H$_2$O if degassing takes place at shallow depths. The most resistant to degassing are Cl, F and S, which remain relatively soluble in basaltic melts at low pressures close to the surface. Because inclusions in phenocrysts normally are trapped and partially isolated in relatively deep magma chambers well before final eruption, melt inclusions are usually less degassed than glasses and better retain the original volatile contents (Sobolev, 1996). However, certain elements may exchange by diffusion between the included melt and enclosing crystal and magma. Portnyagin *et al.* (2008) showed that the H$_2$O contents of melt inclusions could be significantly modified by diffusion of hydrogen either into or out of the inclusion. The timescale of this process varies from hours to minutes. As temperature decreases in slowly cooled olivine cumulates, Fe–Mg exchange between included melt and host olivine results in Fe loss from the melt (Sobolev and Danyushevsky, 1994). This, in turn, decreases the sulfur solubility and leads to precipitation of a sulfide melt, thus decreasing the sulfur content of the included melt (Danyushevsky *et al.*, 2002). In addition, CO$_2$ can be stored in the gas bubbles in inclusions, thus compromising data on the C concentration in the glass (e.g. Bucholz *et al.*, 2013).

### *10.2.1  Samples and methods*

As shown in Figure 10.1, our samples cover most of the stratigraphic sequence and almost the entire period of eruption of lavas of the Siberian LIP. Here we report

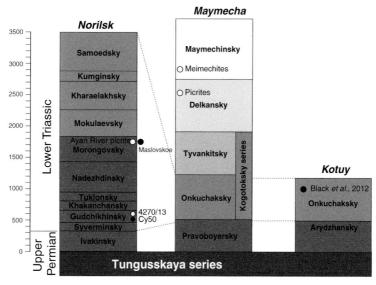

Figure 10.1   Stratigraphy of volcanic sections of Norilsk and Maymecha River modified after Kamo *et al.* (2003). Symbols indicate stratigraphic position of samples: open symbols, uncontaminated by crust; filled symbols, contaminated by crust.

new data for melt inclusions in olivine from alkaline picrites from two localities: Ayan River and the Delkansky suite of the Maymecha–Kotuy province. The Ayan picrites are exposed in a single 100-m-thick flow in the Putorana plateau at a level that stratigraphically corresponds to the upper part of the Morongovsky suite in the middle of the main, tholeiitic, flood-basalt sequence (Vasiliev, 1988; Ryabchikov *et al.*, 2001a, b). The Maymecha–Kotuy region is dominated by mafic–ultramafic alkaline lavas whose composition ranges from trachyte to meimechite. They are probably younger than the main flood-basalt sequence (Fedorenko *et al.*, 1996).

We also report data for melt inclusions in clinopyroxene phenocrysts from the border of the Southern Maslovskoe intrusion, a tholeiitic mafic–ultramafic sill in the Norilsk region (Krivolutskaya *et al.*, 2012). This sample provides volatile contents (Cl, F, B, S, $H_2O$) of tholeiitic basalts of the main volcanic sequence.

Inclusion-bearing minerals were separated and heated for 20 minutes at 1250 °C (olivine) and 1200 °C (clinopyroxene) then quenched to produce glassy inclusions. The samples were prepared in a vertical quenching furnace at controlled oxygen fugacity corresponding to the quartz–fayalite–magnetite (QFM) buffer at the Vernadsky Institute of Geochemistry, Moscow. Quenched grains were mounted in epoxy and polished to expose glass inclusions on the surface.

Electron probe microanalyses were conducted on the Jeol Jxa 8200 Superprobe at the Max-Planck Institute for Chemistry, Mainz, Germany (Sobolev *et al.*, 2009a,b).

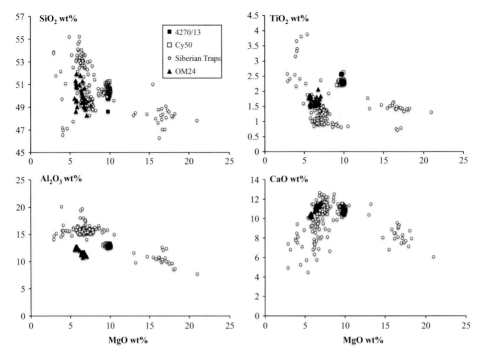

Figure 10.2   Compositions of melt inclusions and Siberian Traps. 4270/13 and Cy50 from Gudchikhinsky picrites after Sobolev *et al.* (2009a); Maslovskoe melt inclusions in clinopyroxene from the Maslovskoe intrusion (OM24). Siberian Traps after Fedorenko *et al.* (1996).

Ion microprobe analyses were conducted on the IMS-4f ion probe at the Institute of Microelectronics, Yaroslavl, Russia (Sobolev, 1996).

We compared our new data with published analyses of melt inclusions in olivine from two samples of Gudchikhinsky picrite (Cy50 and 4270/13) which lie near the base of the stratigraphic sequence (Sobolev *et al.*, 2009a), the Maymechinsky suite of the Maymecha–Kotuy province (Black *et al.*, 2012; Sobolev *et al.*, 2009b) and inclusions in plagioclase from the Onkuchaksky suite of the Maymecha–Kotuy province (Black *et al.*, 2012).

### 10.2.2  Major and trace elements of melt inclusions

The compositions of melt inclusions in clinopyroxene of the Maslovskoe intrusion match well the compositions of typical low-Ti tholeiites for most major and trace elements (Figures 10.2, 10.3). Like most Siberian flood basalts, they possess a clear trace-element signature of contamination by continental crust (positive Pb, U and Rb anomalies and negative Nb anomalies) and no evidence of garnet in the source (flat heavy rare earth elements (HREE)). In contrast, melt inclusions from

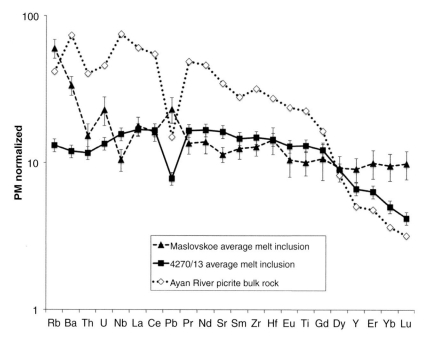

Figure 10.3   Trace-element patterns of studied melt inclusions and rocks, normalized to primitive mantle (PM) after Hofmann (1988).

the Gudchikhinsky (Sobolev *et al.*, 2009a) and Ayan picrites lack the contamination signature and have sloping HREEs, which signals derivation from a deep, garnet-bearing source (Figure 10.3). The alkaline mafic–ultramafic lavas of the Maymecha–Kotuy province also possess strong garnet signatures and no compositional indication of crustal contamination (Carlson *et al.*, 2006; Sobolev *et al.*, 2009b).

### 10.2.3  H₂O contents

Data on $H_2O$ contents were previously reported for melt inclusions in olivine from Gudchikhinsky picrites and from meimechites (Sobolev *et al.*, 2009a,b). Here we report $H_2O$ concentrations in melt inclusions in clinopyroxene of the Maslovskoe intrusion. Water contents are low and variable (0.08–0.45 wt%) for tholeiites and higher (0.7–1.6 wt%) for meimechites. Low concentrations suggest low-pressure degassing and/or H exchange between included and external melt. The highest concentrations from both tholeiites and meimechites yield $H_2O/Ce$ =100–160, close to the typical range of ocean-island basalt (OIB) and mid-ocean-ridge basalt (MORB; Koleszar *et al.*, 2009), suggesting that the water contents are primary and are not influenced by degassing or diffusion exchange.

### *10.2.4 Cl contents*

Concentrations of Cl in melt inclusions are minimally affected by degassing and post-trapping processes and thus should reliably represent magmatic contents. Chlorine is a highly incompatible element with bulk partitioning between crystal-line phases and melt similar to potassium (Michael & Schilling, 1989). This means that the Cl/K ratio in the melt should not change during melting or crystal fractionation and should thus represent the ratio of the mantle source.

In Gudchikhinsky and Maslovskoe melt inclusions, chlorine displays significant excesses (Figure 10.4) yielding $Cl/K_2O = 0.1–0.40$, which is several times higher than for typical MORB and OIB (Koleszar *et al.*, 2009). Extreme Cl excess ($Cl/K_2O$ up to 1.6) has been reported in the highly evolved melt inclusions in minerals from Siberian sills (Black *et al.*, 2012). In sample Cy50 from the Gudchikhinsky suite (Figure 10.5), $Cl/K_2O$ correlates with Nb/U, a proxy of crustal contamination (Hofmann, 2002) indicating assimilation of crustal Cl, presumably from evaporites (Sobolev *et al.*, 2009a). Likewise, inclusions in clinopyroxene from the tholeiitic Maslovskoe intrusion are strongly contaminated by continental crust, as indicated by their low Nb/U. The $Cl/K_2O$ of these samples is moderate to high and the source of the Cl is probably also crustal. In contrast, inclusions in sample 4270/13 from the Gudchikhinsky suite are strongly enriched in Cl but possess no indication of crustal contamination – their trace-element ratios are in the mantle range (Figures 10.3 and 10.5). This feature was interpreted to indicate a Cl-rich source in the mantle plume that yielded the Siberian flood volcanics (Sobolev *et al.*, 2009a).

Chlorine–potassium ratios of alkaline lavas from Ayan River and the Maymecha–Kotuy province are similar and relatively low ($Cl/K_2O = 0.03$, Figure 10.4). They do not show a significant excess over typical mantle values (Koleszar *et al.*, 2009).

### *10.2.5 Sulfur contents*

The tholeiitic melts are undersaturated in sulfide, which yields a positive correl-ation between sulfur and K (Figure 10.4) and $S/Dy < 200$, which is markedly lower than for MORB and OIB (250–300, Koleszar *et al.*, 2009). The sulfur content increases with the concentration of K and other strongly incompatible elements, approaching high contents in strongly alkaline lavas of the Maymecha–Kotuy region (Black *et al.*, 2012). These lavas do not show a correlation between K and S and are probably sulfide-saturated (Figure 10.4).

### *10.2.6 CO₂ contents*

The concentration of carbon in melt inclusions rarely represents original $CO_2$ contents due to the degassing of magmas and gas-bubble segregation within melt

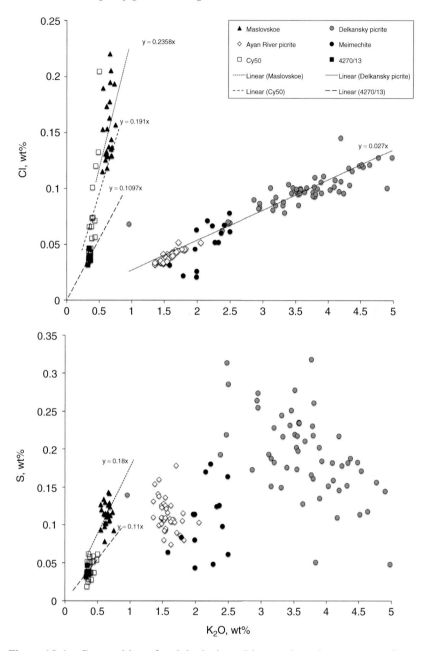

Figure 10.4   Composition of melt inclusions. Lines and numbers represent linear correlations and slopes with zero intercept for particular samples. For sample positions in stratigraphy of the Siberian LIP see Figure 10.1.

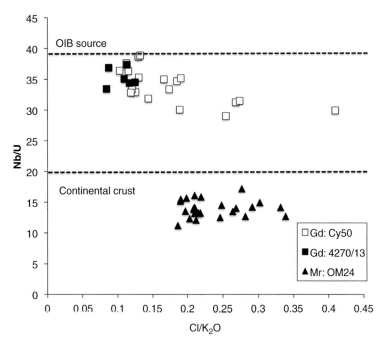

Figure 10.5   Composition of melt inclusions in olivine from the Gudchikhinsky suite (Cy50 and 4270/13) and clinopyroxene from the Maslovskoe intrusion (OM24).

inclusions. Nevertheless, the minimum $CO_2$ contents in parental melt can be estimated if inclusions of dense $CO_2$ fluid are associated with the inclusions. Such estimates for alkaline lavas from Ayan River yield 1.5–2.0 wt% $CO_2$ (Ryabchikov *et al.*, 2001b), which corresponds to 2.0–2.5 wt% $CO_2$ predicted from the Nd contents of these melts (Ryabchikov *et al.*, 2001a; Cartigny *et al.*, 2008).

### 10.2.7  Sources of magmatic volatiles

Presented data and published results suggest a dramatic change in the Cl/K ratio of the mantle source of magmas of the Norilsk section. Considering only crustal uncontaminated magmas, the highest Cl/K ratio is found in the Gudchikhinsky melt inclusions from the lowest part of the Norilsk section. The younger Ayan River picrites, corresponding to the middle part of the Norilsk section and latest in the sequence lavas from the Maymecha–Kotuy province (Delkansky picrites and meimechites), display three times lower Cl/K ratios.

Based on Ni access over Mg, and Mn depletion over Fe of olivine, Sobolev *et al.* (2009a) suggested that the source of Gudchikhinsky picrites is almost pure olivine-free pyroxenite produced by reaction of melted recycled oceanic crust and mantle peridotite. Using the same criteria Sobolev *et al.* (2009b, 2011) suggested that the Ayan River picrites were generated from a pyroxenite-dominated source (80% pyroxenite) while the meimechites were generated from an almost pure peridotite. It was further proposed that altered recycled crust in the Siberian plume was a major source of volatiles, and that release of these volatiles caused both the end-Permian mass extinction and the carbon isotope excursion (Sobolev *et al.*, 2011). If this hypothesis is true, the new data suggest that the pyroxenite source had been exhausted in Cl and thus most probably in other volatiles at the early stage of the Siberian Traps eruption.

Crustal contaminated lavas possess variable Cl/K ratios from very low values, as for the basalts of the Onkuchaksky suite of the Maymacha–Kotuy province (Black *et al.*, 2012) to very high or even extreme as for the Maslovskoe intrusion or sills (Black *et al.*, 2012). Where available, data show strong correlation of Cl excess and crustal contamination. Like Sobolev *et al.* (2009a) and Black *et al.* (2012), we propose a crustal source for the Cl excess in these melts.

### 10.2.8  Calculation of total magmatic $CO_2$, $HCl$ and $SO_2$ release

Following Sobolev *et al.* (2011), we assume that the mantle source for the Siberian flood basalts consisted of two components: (1) pyroxenite derived from melting of the recycled oceanic crust that constituted 15 wt% of the thermochemical mantle plume; and (2) peridotite that made up the remainder of the mantle plume. We obtain concentrations of Cl, S and $H_2O$ of the pyroxenitic source from melt inclusions in olivine from uncontaminated Gudchikhinsky picrites (sample 4270/13). Inclusions in olivine from these magmas have been shown to represent primary melts and thus their Cl, S and $H_2O$ concentrations can be used to estimate the contents of these volatiles in the mantle source. This yields the following values for the composition of recycled oceanic crust: $Cl = 137$ ppm, $S = 135$ ppm after normalization of the values to K. In the deep mantle, these amounts of Cl and S could reside in chloride and sulfide. In contrast with Cl, S and $H_2O$, the amount of $CO_2$ in relatively shallow melts does not represent the primary concentration, because of almost complete degassing at high pressures. Thus, for an assessment of $CO_2$ in the recycled oceanic crust we use global estimations of 3000 ppm $CO_2$ for the bulk 7-km-thick oceanic crust and its maximum outgassing rate through arc volcanism of 70% (Dasgupta and Hirschmann, 2010). This gives a conservative minimum estimate of 900 ppm $CO_2$ in the deeply recycled oceanic crust. The maximum estimate would be about 1800 ppm using the same initial bulk concentrations of $CO_2$ and minimal outgassing of 40% during subduction (Dasgupta and Hirschmann, 2010). In the deep mantle this amount of $CO_2$ could reside in carbonates or diamond.

We consider two end-member models for the extraction of volatiles. In the first model we assume that $CO_2$ and HCl are fully extracted from the plume if the temperature approaches that of the carbonatite solidus (Dasgupta and Hirschmann, 2007). This model gives an upper bound for melt mobility, assuming that melts produced by an infinitely low degree of melting can move out of the plume. In the second model we assume that $CO_2$ and HCl are fully extracted from both peridotitic and pyroxenitic components only if 1% melting is achieved. This model gives a lower boundary for melt mobility, assuming that only 1% carbonate–silicate melts can move out of the plume.

According to the thermomechanical model (Sobolev *et al.*, 2011) the $CO_2$, Cl and S in the plume are fully extracted during its interaction with the lithosphere, and a major part is extracted before the main phase of magmatism. The total amount of released volatiles is proportional to the volume of the plume, which is poorly constrained. Nevertheless, using a 2.5% scaling between the amount of magma intruded into the crust and plume volume (Sobolev *et al.*, 2011), and a total volume of Siberian magmas in the crust (including deep portions of the crust) of 4 to 8 million $km^3$, we estimate the volume of the Siberian plume to be 160 to 320 million $km^3$. This volume is equivalent to a sphere with a radius of 340–420 km. The mass of extracted $CO_2$ for such a plume, which comes mostly from the recycled component of the plume, is more than 100 000–200 000 Gt. We also estimate that such a plume generates 10 000–20 000 Gt of Cl and about the same amount of S.

The ratio of Cl/K in uncontaminated melts could be used as a proxy of mantle source degassing. Reported here are data for Ayan River picrites (middle of the Norilsk section) and alkaline picrites and meimechites of the Maymecha–Kotuy province (latest stage of Siberian LIP), which indicate almost 70% degassing of the mantle source since the beginning of the Norilsk section (the Gudchikhinsky suite). This is consistent with the major conclusion of the model explained above – that the majority of volatile components from the plume should have been extracted before the main magmatic stage. This also allows us to estimate chlorine emission from the lavas (excluding crustal volatiles) as a function of their potassium content. For an average $K_2O$ between 0.5 and 1.0 wt% (Fedorenko *et al.*, 1996) and $Cl/K_2O = 0.03$ (Figure 10.4) a $4 \times 10^6$ $km^3$ of Siberian flood volcanics contains 1500–3000 Gt of Cl. This estimate does not include Cl from crustal sources and thus could be considered as a minimum amount.

## 10.3 Thermogenic gases

Svensen *et al.* (2004) advanced the hypothesis that heating of sedimentary rocks adjacent to intrusions in LIPs released the gases that changed global climate and contributed to mass extinctions. The first paper (Svensen *et al.*, 2004) focused on the relationship between the North Atlantic Magmatic Province and Eocene global warming, but subsequent papers dealt with other LIPs, including the Siberian Traps (Svensen *et al.*, 2007, 2009). This work, summarized in Chapter 12, deals mainly with processes that take place at relatively low temperatures, mainly by the destabilization of hydrocarbons. Iacono-Marziano *et al.* (2012a) subsequently showed that high-temperature interaction between magma and organic matter (hydrocarbons or coal) can lead to a dramatic reduction of the magma redox state, and significant release of reduced CO-dominated gas mixtures. The impact on the environment of these toxic gases was explored in climate models developed by Iacono-Marziano *et al.* (2012b).

Ganino *et al.* (2008) described how the amount of gas released by high-temperature contact metamorphism could be quantified, using the aureole surrounding the Panzhihua intrusion in China as an example. This intrusion is part of the Emeishan LIP and its emplacement coincides with the end-Guadalupian mass extinction, which occurred at 261 Ma, only about 10 Myr before the Permian–Triassic extinction. Ganino *et al.* (2008) used two approaches to quantify gas release: (1) the stoichiometry of metamorphic reactions and (2) differences in the measured volatile contents of the metamorphosed rocks and their protoliths.

A large proportion of the Panzhihua aureole consists of brucite marble, produced by metamorphism of Proterozoic dolostones. Although Ganino *et al.* (2008) assumed that the brucite formed through retrograde hydration of periclase, the petrographic and geochemical observations summarized by Ganino *et al.* (2013) demonstrate that it resulted from a prograde alteration of dolomite to brucite:

$$MgCa(CO_3)_2 \text{ (dolomite)} + H_2O \rightarrow Mg(OH)_2 \text{ (brucite)} \\ + CaCO_3 \text{ (calcite)} + CO_2.$$

Assuming an excess of water, which is justified by the presence of abundant amphibole and biotite in metamorphosed mafic dykes intruding the aureole (Ganino *et al.*, 2013), the proportion of $CO_2$ released by this reaction is 23.8% of the mass of the original dolostone. The measured loss on ignition measured in brucite marbles is about 38% (Ganino *et al.*, 2008), which, when the 12% water in brucite is subtracted, gives 25% $CO_2$, a value very similar to that calculated from the stoichiometry.

Pang *et al.* (2012) applied the same approach to impure limestones, pelites and evaporites in the aureoles surrounding intrusions of the Siberian LIP. The calculated proportion of released $CO_2$ varied from negligible amounts in the pelitic rocks to values like those of Ganino *et al.* (2008) in the carbonates. Pang *et al.* (2012) also evaluated the amount of $SO_2$ that could potentially be released from anhydrite in the evaporites. They noted that this mineral melts only at temperatures well above those monitored from metamorphic assemblages in the aureoles and proposed, following Li *et al.* (2009) and Ripley *et al.* (2003, 2009), that S was liberated during interaction with aqueous fluid that circulated through the aureole. Firm evidence that sulfur from evaporites is transferred from wall rock to magma is found in the heavy S isotopic compositions of the sulfide ores of the Norilsk–Talnakh deposits (Grinenko, 1985; Ripley *et al.*, 2003, 2010). Sulfur dissolved in the magma would be degassed on eruption and additional sulfur would be released during transfer of fluids to the surface, in part through the vents described by Svensen *et al.* (2009). As discussed by Pang *et al.* (2012) it is difficult to quantify the amount of $SO_2$ released by the fluid transfer mechanism and transferred to the surface.

### 10.3.1 Calculation of total thermogenic $CO_2$ and $SO_2$ release

Is it possible to use these estimates of the amount of $CO_2$ and $SO_2$ released by the mass of sedimentary rock to estimate the total amount of gas that escapes to the atmosphere during the course of an eruption? To do this requires some knowledge of the mass of rock that was heated to temperatures above the breakdown temperatures for the minerals involved. Svensen *et al.* (2008) estimated the total volume of aureoles surrounding intrusions in the sedimentary basin beneath the Siberian Traps as 800 000 $km^3$ and proposed that a fixed proportion (0.5 to 1.5%) of the organic matter in the original rocks reacted with the gas. On this basis they calculated that between 9000 and 28 000 Gt of C would be released. Applying the same approach to the gas released by high-temperature reactions, and supposing (1) that the rocks in the aureoles comprise 25% carbonate of which 20% breaks down and (2) that the amount of $CO_2$ released is 25% of the mass of carbonate, we calculate that the aureoles yielded an additional 1% of their mass as $CO_2$, a value similar to that from the organic matter.

The proportion of $SO_2$ is even more difficult to quantify, but if we suppose that evaporite constitutes 10% of the rocks in the aureoles and that these rocks lose 5% of their sulfur to circulating fluids and then to the atmosphere, then the mass of $SO_2$ released is about one-tenth of that of $CO_2$, or about 1000 Gt in round numbers.

To conclude, the quantities of thermogenic gases may be comparable to magmatic outputs. Reduced gases resulting from interaction with coal or organic matter are more toxic than the other species and are likely to have had a significant impact on the biosphere (Iacono-Marziano *et al.*, 2012).

## 10.4 Discussion

Sobolev *et al.*'s (2011) thermomechanical model of interaction of thermochemical mantle plume with lithosphere, which is also adopted in this paper, predicts that most of the $CO_2$ and HCl in the recycled-crust component of the plume is extracted during its interaction with the lithosphere before the main phase of magmatism. The new data on the melt inclusions in olivine from Ayan River picrites fully agree with this prediction. Compared to inclusions in olivine from earlier uncontaminated Gudchikhinsky picrites the Cl/K ratio of melt of Ayan River picrites falls by 70%. The same low Cl/K ratio is recorded in later picritic magmas from the Maymecha–Kotuy province.

From the modeling and melt-inclusion data, we infer that the early magmas were rich in Cl and would have liberated this phase at an early stage of the emplacement of the Siberian province. The amounts of $CO_2$ and Cl emitted at this early stage may have been 100 000–200 000 Gt and 10 000–20 000 Gt respectively. These huge amounts of isotopically light carbon from recycled altered oceanic crust are quite enough to explain the carbon isotope excursion and significant depletion of the ozone layer (Sobolev *et al.*, 2011). We thus associate the first extinction pulse identified by Song *et al.* (2012) with the degassing of these magmatic gases. These magmas contained low $SO_2$ contents and this species did not contribute much to the early extinction event (Figure 10.6).

The eruption of early magmas including Gudchikhinsky picrites, and a major degassing event, was followed by eruption of the much more voluminous main series of magmas. However, the estimate of mantle-sourced Cl released by the main series of magmas is only around 2000 Gt. This number is three to four times less than the estimate of Black *et al.* (2012). The likely reason for this difference is that our estimate does not include Cl assimilated in the crust, while Black *et al.* (2012) use the average concentrations of contaminated and uncontaminated melts. In any case, the estimated Cl emissions are at least two to three times lower than those predicted for the degassing before the main magmatic events. On the other hand, the main series eruptions coincided with the emplacement of the intrusive complex beneath the flood basalts and therefore with the release of thermogenic gases from the metamorphic aureoles surrounding the intrusions. As described by Svensen *et al.* (2009), Pang *et al.* (2012) and Iacono- Marziano *et al.* (2012), contact metamorphism of the sedimentary rocks released large quantities of a toxic cocktail of gases; not only $CO_2$ and halocarbons from the carbonates, but also $SO_2$ (when evaporites were assimilated), and reduced species such as $CH_4$ and CO (when hydrocarbons or coal were assimilated). We thus associate Song *et al.*'s (2012) second pulse of extinction with the release of these gases (Figure 10.6).

## 10.5 Conclusions

1. The mantle source of magmas during the initial stages of the Siberian LIP was dominated by pyroxenite formed from altered recycled oceanic crust with an unusually high Cl content, a relatively low content of S and elevated

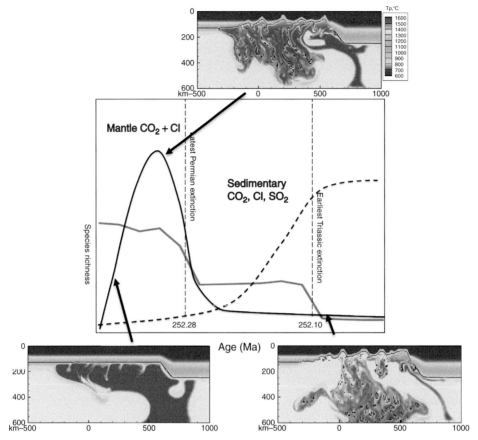

Figure 10.6   Model cartoon. The blue solid curve shows temporal evolution of species richness near Permian–Triassic boundary with two pronounced extinction events (Song *et al.*, 2012). The black curves schematically show suggested temporal evolution of magmatic (solid curve) and thermogenic (dashed curve) volatile fluxes. Colored cross-sections show snapshots of temperature distribution in the Siberian mantle in the model by Sobolev *et al.* (2011) corresponding to different stages of degassing. The most extensive degassing from the plume occurs during the destruction of the mantle lithosphere by the plume, before the main magmatic phase. A black and white version of this figure will appear in some formats. For the colour version, please refer to the plate section.

concentrations of $CO_2$. At the middle and late stages the mantle source evolved towards an increasing fraction of peridotite, and the pyroxenite component became depleted in Cl and $CO_2$. The initial pulse in the end-Permian mass extinction can be linked to the release of magmatic $CO_2$, Cl and other gases from this source.

2. Typical tholeiitic magmas of the Siberian LIP indicate a low amount of Cl of magmatic origin; the Cl that some magmas contained was probably related to the assimilation of sedimentary evaporites.

3. The second extinction pulse can be related to the release of thermogenic gases from contact aureoles surrounding intrusions comagmatic with the main sequence of tholeiitic basalts.

## Acknowledgements

This study was partially supported by Agence Nationale de la Recherche, France, Chair of Excellence grant (ANR-09-CEXC-003–01), Russian Foundation of Basic Research grant (12-05-01154-a) and Earth Sciences Department of Russian Academy grants to A.V.S and the U.S. National Science Foundation Continental Dynamics Program grant 807585 to NTA. We are grateful to Y. R. Vasilev for providing samples of Ayan River and Delkan section picrites and Stephen Self and Kirsten Fristad for constructive reviews.

## References

Black, B. A., Elkins-Tanton, L. T., Rowe, M. C. and Peate, I. U. (2012). Magnitude and consequences of volatile release from the Siberian Traps. *Earth and Planetary Science Letters*, **317**, 363–373.

Bucholz, C. E., Gaetani, G. A., Behn, M. D. and Shimizu, N. (2013). Post-entrapment modification of volatiles and oxygen fugacity in olivine-hosted melt inclusions. *Earth Planet. Sci. Lett.*, **374**, 145–155.

Carlson, R. W., Czamanske, G., Fedorenko, V. and Ilupin, I. (2006). A comparison of Siberian meimechites and kimberlites: implications for the source of high-Mg alkalic magmas and flood basalts. *Geochem. Geophys. Geosys.* **7**, Q11014.

Cartigny, P., Pineau, F., Aubaud, C. and Javoy, M. (2008). Towards a consistent mantle carbon flux estimate: insights from volatile systematics ($H_2O/Ce$, $\delta D$, $CO_2/Nb$) in the North Atlantic mantle (14° N and 34° N). *Earth Planet. Sci. Lett.*, **265**, 672–685.

Danyushevsky, L. V., McNeil, A.W. and Sobolev, A.V. (2002). Experimental and petrological studies of melt inclusions in phenocrysts from mantle-derived magmas: an overview of techniques, advantages and complications. *Chem. Geol.*, **183**, 5–24.

Dasgupta, R. and Hirschmann, M. M. (2007). Effect of variable carbonate concentration on the solidus of mantle peridotite. *Am. Mineral.*, **92**, 370–379.

Dasgupta, R. and Hirschmann, M. M. (2010). The deep carbon cycle and melting in Earth's interior. *Earth Planet. Sci. Lett.*, **298**, 1–13.

Fedorenko, V. A., Lightfoot, P. C., Naldrett, A. J. *et al.* (1996). Petrogenesis of the flood-basalt sequence at Noril'sk, north central Siberia. *Int. Geol. Rev.*, **38**, 99–135.

Ganino, C. and Arndt, N. T. (2009). Climate changes caused by degassing of sediments during the emplacement of large igneous provinces. *Geology*, **37**, 323–326.

Ganino, C., Arndt, N. T., Chauvel, C., Jean, A. and Athurion, C. (2013). Melting of carbonate wall rocks at the margins of mafic intrusions at Panzhihua, China. *Geosci. Front.*, **4**, 535–546.

Ganino, C., Arndt, N. T., Zhou, M. F., Gaillard, F. and Chauvel, C. (2008) Interaction of the magma with the sedimentary wall rock and magnetite ore genesis in the Panzhihua mafic layered intrusion, SW China. *Mineral. Deposita*, **43**, 677–694.

Grinenko, L. N. (1985). Sources of sulfur of the nickeliferous and barren gabbro–dolerite intrusions of the northwest Siberian platform. *Int. Geol. Rev.*, **28**, 695–708.

Hofmann, A. W. (1988). Chemical differentiation of the Earth: the relationship between mantle, continental crust, and oceanic crust. *Earth Planet. Sci. Lett.*, **90**, 297–314.

Hofmann, A. W. (2002). Sampling mantle heterogeneity through oceanic basalts: isotopes and trace elements. *Treat. Geochem.*, **2**, 61–101.

Kamo, S. L., Czamanske, G. K., Amelin, Yu. *et al.* (2003). Rapid eruption of Siberian flood-volcanic rocks and evidence for coincidence with the Permian-Triassic boundary and mass extinction at 251 Ma. *Earth Planet. Sci. Lett.*, **214**, 75–91.

Koleszar, A. M., Saal, A. E., Hauri, E. H. *et al.* (2009). The volatile contents of the Galapagos plume; evidence for $H_2O$ and F open system behavior in melt inclusions. *Earth Planet. Sci. Lett.*, **287**, 442–452.

Krivolutskaya, N. A., Sobolev, A. V., Snisar, S. G. *et al.* (2012). Structure and mineralogical–geochemical characteristics of the Maslovskoe Pt–Cu–Ni deposit, Norilsk area. *Mineral. Deposita*, **47**, 69–88.

Iacono-Marziano, G., Gaillard, F., Scaillet, S. *et al.* (2012a) Extremely reducing conditions reached during basaltic intrusion in organic matter-bearing sediments. *Earth Planet Sci Lett.*, **357**, 319–326.

Iacono-Marziano, G., Gaillard, F., Scaillet, S. *et al.* (2012b) Gas emissions due to magma-sediment interactions during flood magmatism at the Siberian Traps: gas dispersion and environmental consequences. *Earth Planet Sci Lett.*, **357**, 308–318.

Li, C., Ripley, E. M., Naldrett, A. J., Schmitt, A. K. and Moore, C. H. (2009). Magmatic anhydrite–sulfide assemblages in the plumbing system of the Siberian Traps. *Geology* **37**, 259–262.

Michael, P. J. and Schilling, J. G. (1989). Chlorine in mid-ocean ridge magmas – evidence for assimilation of seawater-influenced components. *Geochimica et Cosmochimica Acta*, **53** (12), 3131–3143.

Pang, K. W., Arndt, N. T., Svensen, H. *et al.* (2012). Contact metamorphism and degassing of Siberian evaporites: a mineralogical, whole-rock geochemical and Sr–Nd isotopic study. *Geochim. Cosmochim. Acta*, doi 10.1007/s00410-012-0830-9.

Portnyagin, M., Almeev, R., Matveev, S. and Holtz, F. (2008). Experimental evidence for rapid water exchange between melt inclusions in olivine and host magma. *Earth Planet. Sci. Lett.*, **272** (3–4), 541–552.

Ryabchikov, I. D., Ntaflos, T., Buchl, A. and Solovova, I. P. (2001a). Subalkaline picrobasalts and plateau basalts from the Putorana plateau (Siberian continental flood basalt province). I. Mineral compositions and geochemistry of major and trace elements. *Geochem. Int.*, **39**, 415–431.

Ryabchikov, I. D., Solovova, I. P., Ntaflos, T., Buchl, A. and Tikhonenkov, P. I. (2001b). Subalkaline picrobasalts and plateau basalts from the Putorana plateau (Siberian continental flood basalt province). II. Melt inclusion chemistry, composition of 'primary' magmas and P–T regime at the base of the superplume. *Geochem. Int.*, **39**, 432–446.

Ripley, E. M., Lightfoot, P. C., Li, C., Elswick, E. R. (2003). Sulfur isotopic studies of continental flood basalts in the Norilsk region: implications for the association between lavas and ore-bearing intrusions. *Geochim. et Cosmochim. Acta*, **67**, 2805–2817.

Ripley, E. M., Li, C., Moore, C. H., Schmidt, A. K. (2010). Micro-scale S isotope studies of the Kharaelakh intrusion, Norilsk region, Siberia: constraints on the genesis of coexisting anhydrite and sulfide minerals. *Geochim. et Cosmochim. Acta*, **74**, 634–644.

Sobolev, A. V. (1996). Melt inclusions in minerals as a source of principal petrologic information. *Petrology*, **4**, 209–220.

Sobolev, A. V. and Danyushevsky, L. V. (1994). Petrology and geochemistry of boninites from the north termination of the Tonga Trench: constraints on the generation conditions of primary high-Ca boninite magmas. *J. Petrol.*, **35**, 1183–1213.

Sobolev, A. V., Krivolutskaya, N. A. and Kuzmin, D. V. (2009a). Petrology of the parental melts and mantle sources of Siberian Trap magmatism. *Petrology*, **17**, 253–286.

Sobolev, A. V., Sobolev, S. V., Kuzmin, D. V., Malitch, K. N. and Petrunin, A. G. (2009b). Siberian meimechites: origin and relation to flood basalts and kimberlites. *Russ. Geol. Geophys.*, **50**, 999–1033.

Sobolev, S. V., Sobolev, A. V., Kuzmin, D. V. *et al.* (2011). Linking mantle plumes, large igneous provinces and environmental catastrophes. *Nature*, **477**, 312–316.

Song, H., Wignall, P. B., Tong, J. and Yin, H. (2012). Two pulses of extinction during the Permian–Triassic crisis. *Nature Geosci.*, **6**, 52–56.

Svensen, H., Planke, S., Malthe-Sørenssen, A. *et al.* (2004). Release of methane from a volcanic basin as a mechanism for initial Eocene global warming. *Nature*, **429**, 542–545.

Svensen, H., Planke, S., Chevallier, L. *et al.* (2007) Hydrothermal venting of greenhouse gases triggering Early Jurassic global warming. *Earth Planet. Sci. Lett.*, **256**, 554–566.

Svensen, H., Planke, S. and Polozov, A. G. *et al.* (2009). Siberian gas venting and the end-Permian environmental crisis. *Earth Planet. Sci. Lett.*, **277**, 490–500.

Wignall, P. B., Sun, Y., Bond, D. P. G. *et al.* (2009).Volcanism, mass extinction and carbon isotope fluctuations in the Middle Permian of China. *Science*, **324**, 1179–1182.

White, R. V. and Saunders, A. D. (2005). Volcanism, impact and mass extinctions: incredible or credible coincidences? *Lithos*, **79**, 299–316.

Vasilev, Y. R. (1988). Plagioclase bearing picrites of Ayan River. *Russ. Geol. Geophys.*, **29**, 68–75.

# 11

## Volatile release from flood basalt eruptions: understanding the potential environmental effects

STEPHEN SELF, LORI S. GLAZE, ANJA SCHMIDT AND TAMSIN A. MATHER

### 11.1 Introduction

Continental flood basalt (CFB) eruptions were huge lava-producing events that dwarf the rates of basaltic volcanism today. Fire-fountaining explosive activity accompanied these eruptions (Brown *et al.*, 2014; see also Chapter 1), and a series of such eruptions leads to the formation of a CFB province. Research summarized by Courtillot and Renne (2003) suggests that during the formation of some CFB provinces (e.g. the Columbia River Basalt Group (CRBG), Figure 11.1a) the climax of volcanism, in terms of erupted lava volume, can be brief, often much less than 1 million years in duration.

There is an intriguing age correlation between at least five CFB provinces emplaced in the past 300 Myr with mass extinction events and other environmental changes (Wignall, 2001; Courtillot and Renne, 2003; Kelley, 2007). The link between CFB volcanism, extinctions and environmental change is thought to be due to eruptive gas release from the magmas leading, amongst other processes, to atmospheric aerosol formation and its radiative and depositional effects (see also Chapters 13, 14 and 20). Perhaps most striking is the age correlation of the emplacement of the Siberian Traps, the largest-volume CFB province, with the greatest loss of floral and faunal diversity in Earth's history (the end-Permian mass extinction, ~ 252 Ma), with CFB volcanism proposed as a trigger for this event (e.g. Wignall, 2001; Reichow *et al.*, 2009). However, the mechanisms leading to the association between CFB volcanism and extinctions have yet to be robustly and qualitatively demonstrated.

In this chapter we discuss two aspects of the relationship between CFB volcanism and the potential, detrimental environmental effects it causes. These are (1) the heights of volcanic plumes from CFB episodes, and (2) the emplacement timescales and volatile release rates of CFB eruptions (we focus on volatiles released directly from erupted magma). These are important source parameters for atmospheric

*Volcanism and Global Environmental Change*, eds. Anja Schmidt, Kirsten E. Fristad and Linda T. Elkins-Tanton. Published by Cambridge University Press. © Cambridge University Press 2015.

(a)

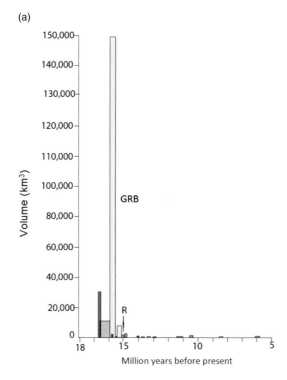

(b)

Figure 11.1 (a) Volume of lava erupted plotted against time for formations of the Columbia River Basalt Group (CRBG), showing a large 'spike' of main pulse volcanism, GRB (Grande Ronde Basalt) Formation. Other formations of CRBG lava are identified in Barry *et al.* (2013) and are not mentioned in this chapter, except for the Roza eruption (R; 1,300 km$^3$). The width of bars represents the age range for each formation (based on latest $^{40}$Ar/$^{39}$Ar age dates). (b) Stack of lava flows, typical of the 'trap' or stepped topography characteristic of flood basalt lavas, exposed along Snake River in Washington State (USA). The ~ 300-m-thick pile of lavas seen here is part of the GRB Formation of the CRBG.

modelling efforts to test whether, and to what degree, CFB eruptions caused or contributed to environmental changes in the geological past.

## 11.2  Volcanic plume rise during CFB volcanism

Studies of modern eruptions, such as Pinatubo in 1991, suggest that the short-term climatic effect of an eruption is greatest if gases such as sulfur dioxide ($SO_2$) reach the stratosphere (see also Chapters 10, 12 and 14 for discussion of other gases released). In general, basaltic lava compositions that characterize CFB volcanism erupt less explosively than their more silicic counterparts, resulting in lower eruption columns. It is therefore important to understand whether, and under what circumstances, basaltic eruption columns from both modern-day and CFB eruptions can reach the stratosphere. Early attempts were made by Stothers *et al.* (1986) and Woods (1993), suggesting that near-stratospheric heights could be attained by high-intensity basaltic eruptions. More recent studies in this area have built upon and improved the formulation of these earlier models, refining plume-height estimates for linear vent sources (Glaze *et al.*, 2011).

   In addition to $SO_2$ and other volcanogenic volatiles, volcanic eruptions are capable of redistributing large volumes of water from the lower atmosphere into the stratosphere (Glaze *et al.*, 1997). A 25-km-high eruption column rising through a wet, tropical atmosphere can transport up to 4 Mt of $H_2O$ per hour up through the plume (we use the term 'plume' to describe the vertical eruption column and downwind ash/gas cloud). This water, if injected at the top of the plume at stratospheric heights, may be important in generating stratospheric aerosols (Textor *et al.*, 2003). Carbon dioxide ($CO_2$) is also released during basaltic volcanism but, due to its longer atmospheric lifetime, the height to which it is injected is not critical for its atmospheric residence time and environmental impacts.

### *11.2.1  Modern observations of buoyant plumes from basaltic volcanism*

Observations in the literature have occasionally noted sustained buoyant plumes associated with historic basaltic fire-fountain events (Figure 11.2). While the overall scale is much smaller than CFB volcanism, basalt magma should behave similarly in terms of the individual eruptions. An important and well-documented example is the Pu`u `Ō`ō eruption, Kīlauea, Hawai`i. Early episodes in 1983 and 1984 generated multiple fire-fountains, with typical heights of 100–200 m, occasionally up to ~ 400 m, and sustained plumes of 5–7-km height above sea level (ASL). The 1984 eruption of Mauna Loa, Hawai`i, produced somewhat larger fire-fountains (up to 500 m high) along a 2-km-long active fissure that generated

Figure 11.2    Etna, Sicily, on 10–11 November 2013, during which the volcano produced new basaltic scoria cones, intense lava fountains, plumes and lava flows. (Photo: Tom Pfeiffer, VolcanoDiscovery).

a buoyant plume estimated to rise to 11 km ASL (7.5 km above the vent) (Smithsonian Institution, 1984).

Much larger fire-fountains were documented during the 1986 basaltic fissure eruption of Izu-Oshima volcano, Japan, and, indirectly, the 1783–1784 eruption of Laki, Iceland (the closest historic analogue to CFB volcanism). At Izu-Oshima, lava fountains 1.6 km high were observed to feed an ashy sub-Plinian plume that reached 16 km altitude (Endo *et al.*, 1988; Mannen and Ito, 2007). Thordarson and Self (1993) estimated that Laki's fire-fountains reached 0.8–1.4 km in height, sustaining eruption columns of up to 15-km altitude, based on historical reports from contemporary observers. Glaze *et al.* (in press) have modelled the observed maximum plume heights (12–16 km ASL) for the 1986 Izu-Oshima eruption (Figure 11.3). The model predicts maximum plume heights of 13.1–17.4 km for source vent widths of between 4 and 16 m (consistent with expected fire-fountain widths) when 32% (by mass) of the erupted magma is fragmented and involved in the buoyant plume (effective volatile content of 6 wt%). The amount of ash in the plume is important, in part because solid particles store heat more effectively than gas.

### 11.2.2 *Flood basalt eruption plumes*

Field observations suggest that the Miocene-age Roza eruption (14.7 Ma, part of the CRBG) could have had sustained fire-fountains of similar height to Izu-Oshima (~ 1.6 km above the vent; Thordarson and Self, 1996; 1998). Assuming 5-km-long

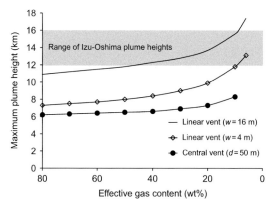

Figure 11.3   Maximum predicted plume heights as a function of effective gas content in buoyant ash and gas plume above a fire-fountain. Grey shading indicates the range of observed buoyant ash plume heights (12–16 km ASL) during the explosive phase of the 1986 Izu-Oshima basaltic fissure eruption. All cases assume eruption temperature of 1350 K and initial velocity of 100 m/s at the buoyant plume source 1000 m above the fissure vent (point of gas separation from fire-fountain, approximately 2/3 of maximum fire-fountain height). Curves indicate buoyant plumes from a circular source with a diameter $d = 50$ m (solid circles), a linear source with width $w = 4$ m (open diamonds), and a linear source with $w = 16$ m (solid line). Figure modified from Glaze *et al.* (in press).

active fissure segments, this suggests that the ~ 180 km of known Roza fissure length supported ~ 36 explosive events/phases (Brown *et al.*, 2014). Fissure segments could have had one to several vents, which later produced lava without significant fountaining for a longer period within a decades-long eruption (Brown *et al.*, 2014).

To estimate the likely plume rise height for the Roza eruptions, Glaze *et al.* (in press) assume parameters analogous to the Izu-Oshima eruption. Based on glass inclusion data reported by Thordarson and Self (1996), a bulk Roza volatile content of 2 wt% is assumed. The range of estimated maximum plume heights for the Roza eruption is 13.1–17.4 km ASL for fissure widths of 4–16 m (Glaze *et al.*, in press). At Roza's ~ 45° N palaeo-latitude, the tropopause is between 10 and 13 km ASL, depending on season. Thus, a plume from a 16-m-wide linear source, analogous to Izu-Oshima, can easily drive a plume into the stratosphere (Figure 11.3), even for ash-poor cases, and this indicates that flood basalt eruptions, such as Roza, were capable of repeatedly injecting large amounts of $SO_2$ into the stratosphere over many years to tens of years. Based on the ~ 9000 Mt of $SO_2$ estimated by Thordarson and Self (1996) to be released mainly with the explosive phases, Glaze *et al.* (in press) estimate that each fissure segment might have injected 62 Mt/day of $SO_2$ into the stratosphere while actively fountaining, each with a duration of 3–4 days. Assuming the CRBG is typical of CFB emplacement characteristics, it thus seems likely that individual CFB eruptions could have influenced climate on timescales of decades to centuries if the eruptions were sustained.

## 11.3 Emplacement characteristics and volatile release rates of flood basalt eruptions

The most abundant volatile species released during a basaltic eruption is $H_2O$ (50% to 90% by volume of the gas phase) followed by $CO_2$ and $SO_2$, either of which may contribute up to 40% by volume. Halogens are released in minor quantities mainly in the form of HCl, HF, and with even lower abundance of HBr and HI. When attempting to constrain volatile fluxes from CFB provinces, eruption characteristics of individual CFB eruptions are important considerations. During the emplacement of CFB provinces, hundreds to possibly thousands of eruptions produce immense lava flow-fields with volumes on the order of 1,000–5,000 km$^3$. All flood basalt lava flow-fields are dominated by basaltic lava flows (see Figure 11.1b), and when examined in detail all CFB provinces are dominated by compound pāhoehoe lava flow-fields (Bryan *et al.*, 2010). In terms of eruption style, these lava-forming eruptions produce fire-fountains at the vents, just like recent Hawai`ian or Etnean eruptions as discussed above.

Constraining the timing and length of hiatus periods within a province-forming sequence of eruptions is as important as knowing the durations of eruptions in order to assess the environmental effects of CFB volcanism (Self *et al.*, 2014). For the Grande Ronde Formation of the CRBG (Figure 11.1a), there were ~ 100 separate eruptions within a maximum of ~ 400,000 years (Reidel and Tolan, 2013). Assuming that each eruption lasted 100 years, then the average hiatus would have been ~ 4,000 years. In reality, both the number of eruptions and the total duration of their emplacement are highly uncertain and in any case this average hiatus may not be typical of the durations between any two individual eruptions.

Basaltic magmas (including those forming flood basalts) are usually rich in dissolved sulfur (commonly with sulfur concentrations of $\geq$ 1,500 ppm, Wallace, 2005; Black *et al.*, 2012). Therefore, the release of sulfur-rich gases from a large basaltic eruption can be much greater than that from an explosive silicic eruption of equal size. It is also known that the erupted basaltic magma is not the only source of volatiles during the emplacement of CFB provinces. In particular, for the Siberian Traps, several authors note the potentially important role of gases released via interactions between the magma and country rock (e.g. Ganino and Arndt, 2009; Svensen *et al.*, 2009, Chapters 10 and 12).

Thordarson and Self (1996) estimated that the Roza flow released about 1,200 Mt of $SO_2$ per year for a decade or longer. The Grande Ronde Basalts (CRBG) appear to have released $\sim 10^6$ Mt of $SO_2$ in intermittent bursts of $< 1 \times 10^3$ to $30 \times 10^3$ Mt, separated by long-lasting non-eruptive intervals represented by thick soil horizons in the lava sequence (Blake *et al.*, 2010). For the Siberian Traps, estimates of the total $SO_2$ emissions are in the range $10–20 \times 10^6$ Mt of $SO_2$ (Black *et al.*, 2012) over the duration of emplacement of the whole province. Basically, every 1 km$^3$ of basaltic magma emplaced during a CFB eruption releases about 3.5–6.5 Mt of $SO_2$ (Self *et al.*, 2014). For context, studies of the Laki lava flows and ash indicate that the magma originally contained about 1,700 ppm of sulfur, and that the eruption could have released $\sim$ 120 Mt of $SO_2$ over 8 months (with two-thirds released during the first 2 months of intense activity; Thordarson and Self, 1993).

There are fewer estimates of the halogen content of CFB magmas than for sulfur. However, studies on the Deccan Traps and the Siberian Traps suggest that emissions of chlorine and fluorine might be comparable to those of sulfur (Black *et al.*, 2012), or possibly higher (Sobolev *et al.*, 2011; see also Chapter 10). Lower halogen emissions are suggested from the Paraná–Etendeka CFB province (Marks *et al.*, 2014).

Estimates of $CO_2$ released from a CFB province are challenging because direct determinations of $CO_2$ degassing from flood basalt magmas upon eruption are not available. Self *et al.* (2006) used 0.5 wt% as a high, but reasonable, value for pre-eruptive $CO_2$ concentration in flood basalt magmas to estimate that approximately 14 Mt of $CO_2$ could be released for every 1 km$^3$ of CFB lava erupted (assuming a density of 2,750 kg/m$^3$). Therefore, the total release from an erupted lava volume of $10^3$ km$^3$ (the approximate volume of one Deccan eruption) would be about $14 \times 10^3$ Mt of $CO_2$, or $1.4 \times 10^7$ Mt $CO_2$ from the emplacement of the entire Deccan Traps assuming a total lava volume of $10^6$ km$^3$. Although this estimate suggests that a single eruption can release a very large mass of $CO_2$, this mass equates to only 40% of the current anthropogenic $CO_2$ flux ($\sim 35 \times 10^3$ Mt/yr; Le Quéré *et al.*, 2013). If we assume a realistic release timescale of 10 years, then the volcanic $CO_2$ flux from a single eruption equates to only $\sim$ 4% of the current anthropogenic flux.

## 11.4 Towards quantifying the environmental effects of CFB volcanism

The proxy record clearly suggests significant environmental changes that are temporally coincident (within the uncertainty range of radiometric age dating) with some periods of CFB volcanism. Figure 11.4 shows proposed cause-and-effect mechanisms attributed to CFB volcanism including, amongst others,

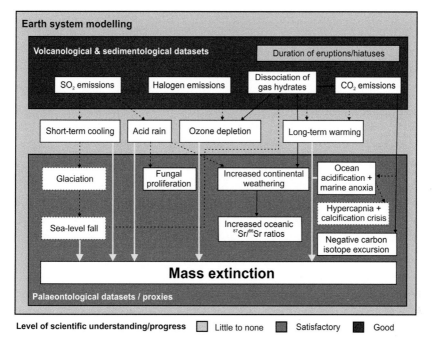

Figure 11.4    Schematic illustrating some of proposed environmental effects and Earth system feedbacks resulting from CFB volcanism (adapted from Wignall, 2001). Not all volatile emissions and therefore feedbacks shown here are applicable to all CFB provinces. Dashed lines and dashed box outlines indicate effects and feedbacks that are possible in theory but have not yet been fully quantified or are still debated in the authors' opinion. Bold light-grey arrows indicate main mechanisms thought to cause a mass extinction (either terrestrial or marine or both). Grey shadings indicate level of scientific understanding, and progress of each discipline, illustrating that whilst quantification of volatile fluxes from CFB volcanism and palaeontological proxies are available (dark grey and grey), some key mechanistic understanding that links the two is missing (light grey). Figure modified from Self *et al.* (2014).

a short-term cooling effect from sulfuric acid aerosol particles lasting years to decades, and acid deposition leading to damage of vegetation and ecosystems (Black *et al.*, 2014). Proposed long-term effects include global warming from $CO_2$ released by the eruptions lasting tens to thousands of years (Sobolev *et al.*, 2011). The dashed lines in Figure 11.4 illustrate that there is substantial uncertainty concerning the exact mechanisms, extent and Earth system feedbacks associated with many of the proposed effects. For example, there is debate about the link between Deccan volcanism $CO_2$ emissions and the effect on Late-Cretaceous temperatures (e.g. Caldeira and Rampino, 1990; Chenet *et al.*, 2008). Volcanologists have made considerable progress constraining volatile fluxes during individual CFB eruptions (see also Chapter 20) and, similarly, palaeontologists have gathered detailed proxy records of the environmental changes. However, extensive further work is required to develop a mechanistic understanding of the links between CFB volcanism and environmental changes. Earth system modelling likely has an important role to play to quantify these linkages.

As a minimum, information on the volcanic volatile flux, the altitude of emissions, and the eruption duration is needed as inputs for atmospheric modelling studies (see Self *et al.*, 2014, for further discussion). To date, the Roza lava flow-field is the only individual CFB eruption where it has been possible to constrain eruptive vent conditions based on the geological record (Brown *et al.*, 2014). Uncertainties in such parameters for the majority of CFB provinces hamper a full assessment of the length and severity of the potential environmental effects and Earth system feedbacks. One such key uncertainty is hiatus length: if these periods of quiet between eruptions outlasted the duration of the volcanic forcing of climate, then the Earth system would have had time to recover after an eruptive phase, weakening the feedback chain through to mass extinctions in Figure 11.4.

For the assessment of the environmental effects of CFB volcanism, the Laki eruption is seen as the best historical analogue, but this was much shorter-lived (8 months) compared with the possible length of a CFB eruption (a decade or more based on the Roza case). Modelling studies suggest that Laki's gas and aerosol clouds dispersed widely across the northern hemisphere (e.g. Chenet *et al.*, 2005; Stevenson *et al.*, 2003; Oman *et al.*, 2006a; Schmidt *et al.*, 2010). The winter of 1783–1784 was up to –0.5°C colder than average in central Europe (e.g. Thordarson and Self, 2003; Oman *et al.*, 2006b; Schmidt *et al.*, 2012). However, the differences in eruptive volume, duration and eruption style mean that scaling the climatic effects of historical and present-day volcanic activity to CFB eruption scale may be flawed. Evidence from both observations and modelling of short-lived explosive eruptions has also suggested that a simple scaling between eruption magnitude and climatic impact has limited validity (Rampino and Self, 1982; see also Chapter 13).

To date there have only been a few modelling studies directly tailored to CFB volcanism. Black *et al.* (2014) used a three-dimensional chemistry–climate model to suggest that pulsed episodes of Siberian Traps volcanism resulted in global-scale ozone depletion and intense episodes of acid rain (Black *et al.*, 2014), which could have contributed to the end-Permian mass extinction. Schmidt *et al.* (unpublished) used a global aerosol–climate model to quantify the climatic and environmental effects of sulfur released during a decade-long Roza flow eruption (emitting 1,200 Mt of $SO_2$ per year into 9–13 km altitude) and a Deccan-scale eruption (emitting 2,400 Mt of $SO_2$ per year into 9–13 km altitude). They concluded that when the buffering capacities of soils is taken into account, a decade-long episode of acid rain from sulfur alone is too short-lived to significantly acidify soils or to damage vegetation on a global scale. In their model, global surface temperatures change between –3 °C and –4.5 °C when averaged over a decade of eruptive activity and return to background values within less than 50 years. These cooling estimates are at the lower end of previous estimates but nonetheless represent a substantial climate perturbation (see also Chapter 13; Thordarson and Self, 1996). Schmidt *et al.* (unpublished) conclude that better constraints on the frequency and duration of individual eruptions as well as hiatus periods are needed to fully quantify the magnitude and duration of the climatic and environmental effects.

## 11.5 Summary

CFB provinces appear to form within 1 to 3 million years, with many showing evidence for one or more pulses of increased magma production and eruption lasting from 1 million to as little as a few 100,000 years. Each of these pulses features tens to hundreds of individual eruptions, each potentially producing up to 10,000 km$^3$ (more usually 1,000–5,000 km$^3$) of predominantly pāhoehoe lava. The eruption of the Roza flow-field produced a 1,300 km$^3$ lava flow-field over a period of at least 10–15 years (Thordarson and Self, 1998) and more recent work on other CRBG flow-fields generally agree with this duration (Vye-Brown *et al.*, 2013). For other CFB provinces, we can only infer eruption durations of decades to centuries by analogy, backed up by the dominance of basaltic–andesitic pāhoehoe lava flow-fields.

CFB eruptions could have injected gases into the stratosphere, if not semi-continuously, then intermittently during decade-long eruptions. Measurements suggest that the Roza eruption released about 12,000 Mt of $SO_2$ over its course (Thordarson and Self, 1996) and, in general, it appears that every 1 km$^3$ of magma emplaced during a CFB eruption releases 6 ± 1.7 Mt of $SO_2$. The majority of magmatic gas species released during CFB volcanism have potential environmental impacts. However, the fate of these volatiles and therefore the magnitude of their environmental effects is poorly constrained at present. For example, further work is needed to understand the atmospheric burdens, atmospheric lifetimes and climatic impact of the magmatic sulfur- and halogen-bearing species released during a typical CFB eruption. Recent progress has included the use of global climate and aerosol models (Black *et al.*, 2014; Schmidt *et al.*, unpublished). Better CFB magmatic volatile release scenarios are needed to facilitate further such modelling efforts (Self *et al.*, 2014) but it must also be remembered that interactions of magma with sedimentary bedrock during the emplacement of some CFB provinces may be an important additional flux of volatile species into the atmosphere (see also Chapters 10 and 20). These interactions are specific to each different province and further work is needed to constrain their magnitude on a case-by-case basis.

The coincidence, to within the limits of radiometric age determinations, of several CFB provinces with times of major biotic change has been well documented; however, the mechanisms by which these episodes might perturb the environment to this extent are not well understood. Figure 11.4 summarizes proposed causal mechanisms between CFB volcanism and environmental change and gives our assessment of the level of current scientific understanding in each area shown. It is very likely that the Earth system responded differently to volcanic perturbations during different periods of geological history; therefore, we suggest that future research efforts should be tailored to each CFB province on a case-by-case basis. In particular, we suggest that knowledge of the number and length of hiatuses during each individual case of CFB volcanism is as important as knowledge of the eruptive phases, when attempting to quantify the severity and duration of any potential environmental effects it caused.

## Acknowledgements

SS was funded by UK-NERC grants NER/B/S/2003/00246 and GR3/11474, and The Open University Research Development Fund. LSG is funded by the NASA Planetary Geology and Geophysics Program (WBS 811073.02.01.05.80). AS is funded by an Academic Research Fellowship from the School of Earth and Environment, University of Leeds. TAM acknowledges the Leverhulme Trust and NERC (NE/G01700X/1) for financial support. Thanks to an anonymous reviewer for their helpful comments.

## References

Barry, T.L., Kelley, S.P., Camp, V. *et al.* (2013). A review of radiometric age constraints for the stratigraphy and eruptions of the Columbia River Basalt Group lavas. *Geological Society of America Special Paper*, **497**, 45–66.

Black, B.A., Elkins-Tanton, L.T., Rowe, M.C. and Peate, I.U. (2012). Magnitude and consequences of volatile release from the Siberian Traps. *Earth and Planetary Science Letters*, **317–318**, 363–73.

Black, B.A., Lamarque, J-F., Shields, C., Elkins-Tanton, L.T. and Kiehl, J. (2014). Acid rain and ozone depletion from pulsed Siberian Traps magmatism. *Geology*, **42**, 67–70.

Blake, S., Self, S., Sharma, K. and Sephton, S. (2010). Sulfur release from the Columbia River Basalts and other flood lava eruptions constrained by a model of sulfide saturation. *Earth and Planetary Science Letters*, **299**, 328–38.

Brown, R.J., Blake, S., Thordarson, T. and Self, S. (2014). Pyroclastic edifices record vigorous lava fountains during emplacement of a flood basalt flow field, Roza Member, Columbia River Basalt Province, USA. *Geological Society of America Bulletin* (in press).

Bryan, S.E., Ukstins Peate, I.A., Self, S. *et al.* (2010). The largest volcanic eruptions on Earth. *Earth-Science Reviews*, **102**, 207–29.

Caldeira, K. and Rampino, M.R. (1990). Carbon dioxide emissions from Deccan volcanism and a K/T boundary greenhouse effect. *Geophysical Research Letters*, **17**, 1299–302.

Chenet, A.L., Fluteau, F. and Courtillot, V. (2005). Modelling massive sulphate aerosol pollution, following the large 1783 Laki basaltic eruption. *Earth and Planetary Science Letters*, **236**, 721–31.

Chenet, A.L., Fluteau, F., Courtillot, V., Gerard, M. and Subbarao, K.V. (2008). Determination of rapid Deccan eruptions across the Cretaceous–Tertiary boundary using paleomagnetic secular variation: results from a 1200-m-thick section in the Mahabaleshwar escarpment. *Journal of Geophysical Research*, **113**, B04101.

Courtillot, V.E. and Renne, P.R. (2003). On the ages of flood basalt events. *Comptes Rendus Geoscience*, **335**, 113–40.

Endo, K., Chiba, T., Taniguchi, H. *et al.* (1988). Tephrochronological study on the 1986–1987 eruptions of Izu-Oshima volcano. *Journal of Volcanological Society Japan*, **2**, 32–51.

Ganino, C. and Arndt, N.T. (2009). Climate changes caused by degassing of sediments during the emplacement of large igneous provinces. *Geology*, **37**, 323–26

Glaze, L.S., Baloga, S.M. and Wilson, L. (1997). Transport of atmospheric water vapor by volcanic eruption columns. *Journal of Geophysical Research*, **102**(D5), 6099–108.

Glaze, L.S., Baloga, S.M. and Wimert, J. (2011). Explosive volcanic eruptions from linear vents on Earth, Venus, and Mars: comparisons with circular vent eruptions. *Journal of Geophysical Research*, **116**(1), E01011.

Glaze, L.S., Self, S., Schmidt, A. and Hunter, S.J. (2014). Assessing eruption column height in ancient flood basalt eruptions. *Earth and Planetary Science Letters*.

Kelley, S.P. (2007). The geochronology of large igneous provinces, terrestrial impact craters, and their relationship to mass extinctions on Earth. *Journal of the Geological Society*, **164**, 923–36.

Le Quéré, C., Andres, R.J., Boden, T. *et al.* (2013). The global carbon budget 1959–2011. *Earth System Science Data*, **5**, 165–85.

Mannen, K. and Ito, T. (2007). Formation of scoria cone during explosive eruption at Izu-Oshima volcano, Japan. *Geophysical Research Letters*, **34**, L18302.

Marks, L., Keiding, J., Wenzel, T. *et al.* (2014). F, Cl, and S concentrations in olivine-hosted melt inclusions from mafic dikes in NW Namibia and implications for the environmental impact of the Paraná–Etendeka Large Igneous Province. *Earth and Planetary Science Letters*, **392**, 39–49.

Oman, L., Robock, A., Stenchikov, G.L. *et al.* (2006a). Modeling the distribution of the volcanic aerosol cloud from the 1783–1784 Laki eruption. *Journal of Geophysical Research*, **111**, D12209.

Oman, L., Robock, A., Stenchikov, G.L. and Thordarson, T. (2006b). High-latitude eruptions cast shadow over the African monsoon and the flow of the Nile. *Geophysical Research Letters*, **33**, L18711.

Rampino, M.R. and Self, S. (1982). Historic eruptions of Tambora (1815), Krakatau (1883), and Agung (1963), their stratospheric aerosols, and climatic impact. *Quaternary Research*, **18**, 127–43.

Reichow, M.K., Pringle, M.S., Al'Mukhamedov, A.I. *et al.* (2009). The timing and extent of the eruption of the Siberian Traps Large Igneous Province: implications for the end-Permian environmental crisis. *Earth and Planetary Science Letters*, **277**, 9–20.

Reidel, S.P. and Tolan, T.L. (2013). The Grande Ronde Basalt, Columbia River Basalt Group. *Geological Society of America Special Paper* **497**, 117–54.

Schmidt, A., Carslaw, K.S., Mann, G.W. *et al.* (2010). The impact of the 1783–1784 AD Laki eruption on global aerosol formation processes and cloud condensation nuclei. *Atmospheric Chemistry and Physics*, **10**, 6025–41.

Schmidt, A., Thordarson, T., Oman, L.D., Robock, A. and Self, S. (2012). Climatic impact of the long-lasting 1783 Laki eruption: inapplicability of mass-independent sulfur isotopic composition measurements. *Journal of Geophysical Research: Atmospheres*, **117**, D23116.

Self, S., Widdowson, M., Thordarson, T. and Jay, A.E. (2006). Volatile fluxes during flood basalt eruptions and potential effects on the global environment: a Deccan perspective. *Earth and Planetary Science Letters*, **248**, 518–32.

Self, S., Schmidt, A. and Mather, T.A. (2014). Emplacement characteristics, timescales, and volatile release rates of continental flood basalt eruptions on Earth. *Geological Society of America Special Paper*, **505**, doi:10.1130/2014.2505(16).

Smithsonian Institution (1984). Mauna Loa. *Scientific Event Alert Network (SEAN) Bulletin*, **9**(3), 2–9.

Sobolev, S.V., Sobolev, A.V., Kuzmin, D.V. *et al.* (2011). Linking mantle plumes, large igneous provinces and environmental catastrophes. *Nature*, **477**, 312–16.

Stevenson, D.S., Johnson, C.E., Highwood, E.J. *et al.* (2003). Atmospheric impact of the 1783–1784 Laki eruption: part I – chemistry modeling. *Atmospheric Chemistry and Physics*, **3**, 487–507.

Stothers, R.B., Wolff, J.A., Self, S. and Rampino, M.R. (1986). Basaltic fissure eruptions, plume heights, and atmospheric aerosols. *Geophysical Research Letters*, **13**(8), 725–28.

Svensen, H., Planke, S., Polozov, A.G. *et al.* (2009). Siberian gas venting and the end-Permian environmental crisis. *Earth and Planetary Science Letters*, **277**, 490–500.

Textor, C., Graf, H.-F., Herzog, M. and Oberhuber, J.M. (2003). Injection of gases into the stratosphere by explosive volcanic eruptions. *Journal of Geophysical Research*, **108**(D19), 4606.

Thordarson, T. and Self, S. (1993). The Laki (Skaftar Fires) and Grimsvötn eruptions in 1783–1785. *Bulletin of Volcanology*, **55**, 233–63.

Thordarson, T. and Self, S. (1996). Sulfur, chlorine and fluorine degassing and atmospheric loading by the Roza eruption, Columbia River Basalt Group, Washington, USA. *Journal of Volcanology and Geothermal Research*, **74**, 49–73.

Thordarson, T. and Self, S. (1998). The Roza Member, Columbia River Basalt Group: a gigantic pahoehoe lava flow field formed by endogenous processes? *Journal of Geophysical Research*, **103**(B11), 27411–45.

Thordarson, T. and Self, S. (2003). Atmospheric and environmental effects of the 1783–1784 Laki eruption: a review and reassessment. *Journal of Geophysical Research*, **108**(D1), 4011.

Vye-Brown, C.L., Self, S. and Barry, T.L. (2013). Physical volcanology and emplacement of flood basalt flow fields: case studies from the Columbia River flood basalts, USA. *Bulletin of Volcanology*, **75**, 697.

Wallace, P.J. (2005). Volatiles in subduction zone magmas: concentrations and fluxes based on melt inclusion and volcanic gas data. *Journal of Volcanology and Geothermal Research*, **140**(1–3), 217–40.

Wignall, P.B. (2001). Large igneous provinces and mass extinctions. *Earth-Science Reviews*, **53**, 1–33.

Woods, A.W. (1993). A model of the plumes above basaltic fissure eruptions. *Geophysical Research Letters*, **20**(12), 1115–18.

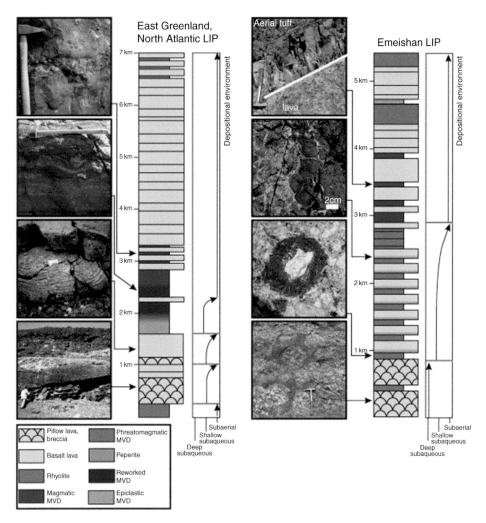

Figure 1.1    For figure caption, see text, p.7.

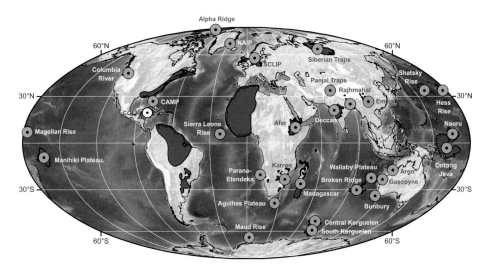

Figure 3.1    For figure caption, see text, p.31.

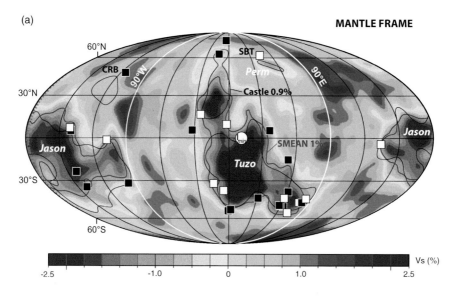

60°N

SBT

CRB

*Perm*

Castle 0.9%

30°N

*Jason*

SMEAN 1%

*Jason*

Tuzo

30°S

60°S

Vs (%)

-2.5    -1.0    0    1.0    2.5

60°N

SBT

CRB

Perm

30°N

Jason

Jason

Tuzo

30°S

60°S

5  3  1  0  Cluster analysis (1000-2800 km)    1% slow contour SMEAN (2800 km)

Figure 3.2    For figure caption, see text, p.37.

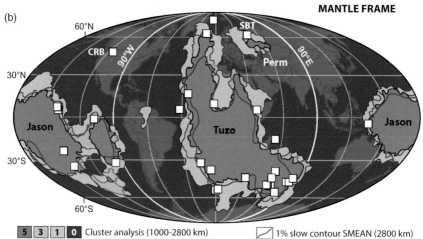

(a)    (b)

(c)

Figure 4.1    For figure caption, see text, p.50.

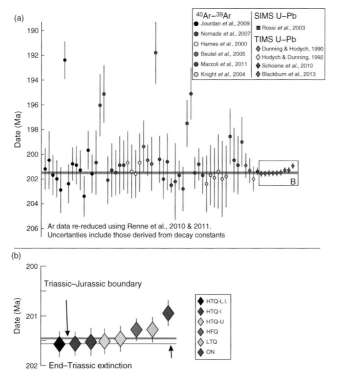

Figure 4.6    For figure caption, see text, p.57.

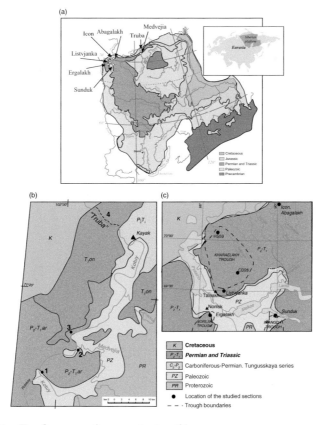

Figure 5.1    For figure caption, see text, p.64.

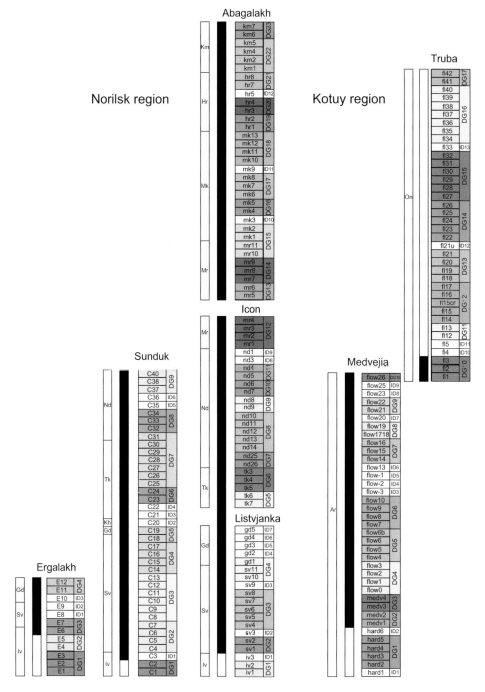

Figure 5.4    For figure caption, see text, p.70.

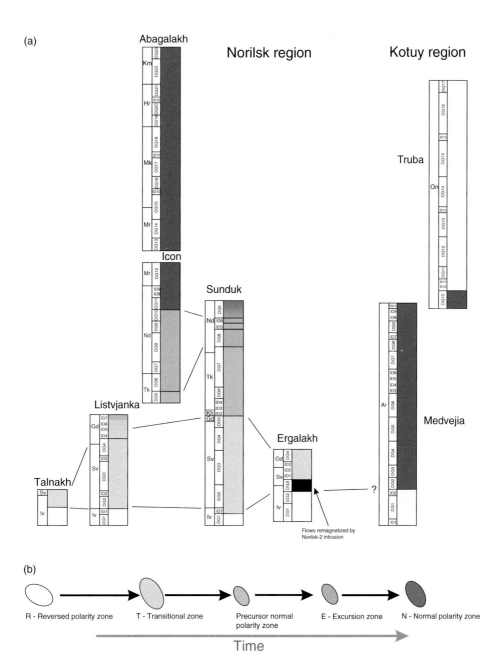

Figure 5.5   For figure caption, see text, p.74.

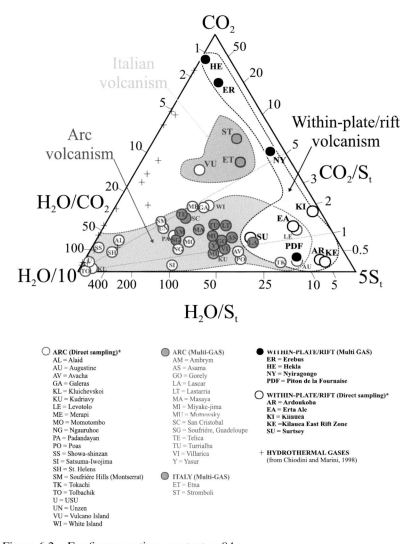

Figure 6.2    For figure caption, see text, p.84.

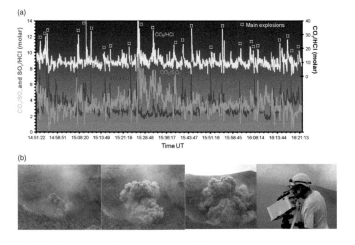

Figure 6.3    For figure caption, see text, p.86.

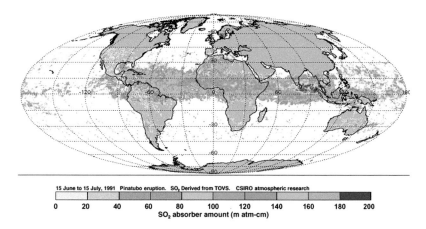

15 June to 15 July, 1991   Pinatubo eruption.   SO₂ Derived from TOVS.   CSIRO atmospheric research

0        20        40        60        80        100        120        140        160        180        200
SO₂ absorber amount (m atm-cm)

Figure 7.1    For figure caption, see text, p.98.

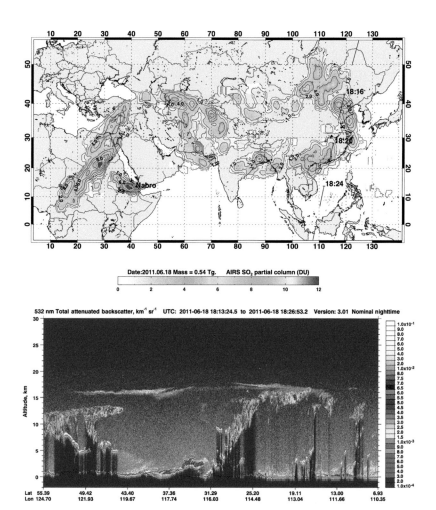

Date:2011.06.18 Mass = 0.54 Tg.    AIRS SO₂ partial column (DU)

0          2          4          6          8          10          12

532 nm Total attenuated backscatter, km⁻¹ sr⁻¹   UTC: 2011-06-18 18:13:24.5 to 2011-06-18 18:26:53.2   Version: 3.01 Nominal nighttime

Figure 7.4    For figure caption, see text, p.109.

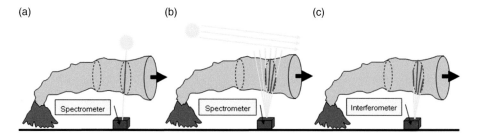

Figure 8.2    For figure caption, see text, p.124.

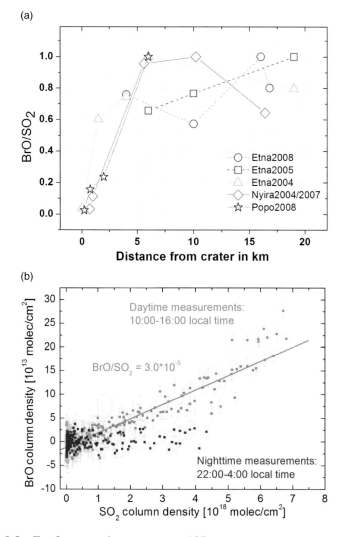

Figure 8.5    For figure caption, see text, p.127.

Figure 9.1   For figure caption, see text, p.137.

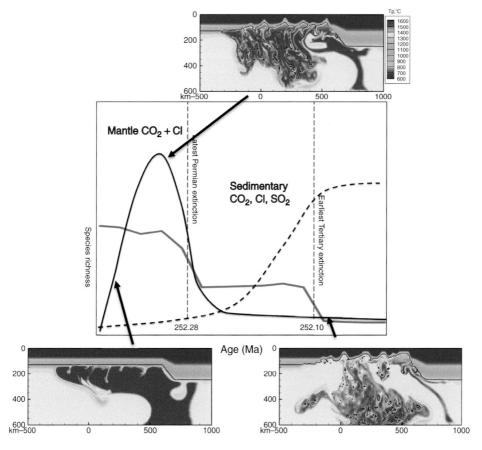

Figure 10.6   For figure caption, see text, p.160.

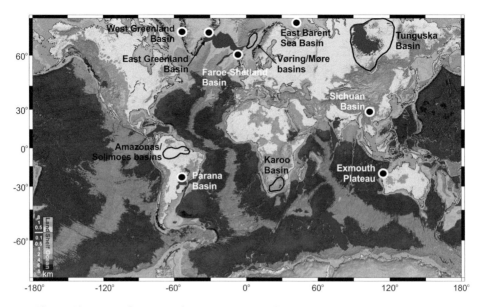

Figure 12.1    For figure caption, see text, p.179.

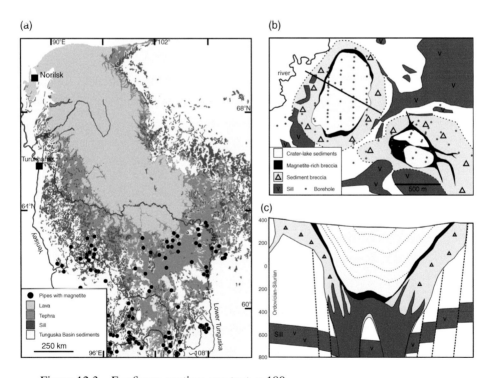

Figure 12.3    For figure caption, see text, p.180.

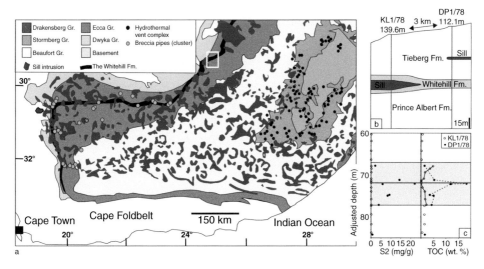

Figure 12.4    For figure caption, see text, p.181.

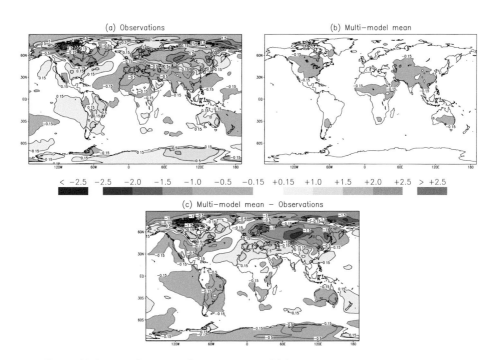

Figure 13.2    For figure caption, see text, p.201.

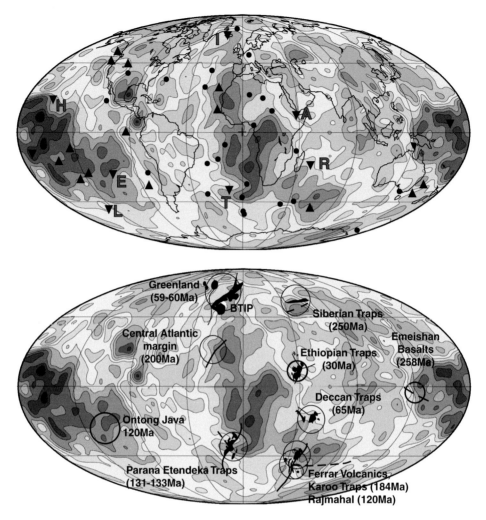

Figure 15.3    For figure caption, see text, p.233.

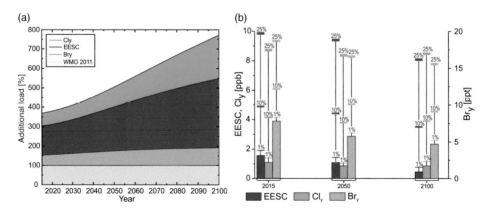

Figure 16.3    For figure caption, see text, p.254.

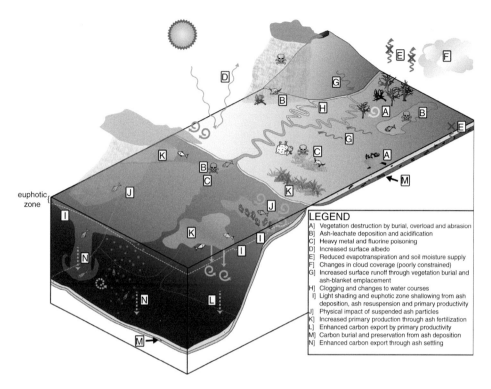

euphotic zone {

LEGEND
A] Vegetation destruction by burial, overload and abrasion
B] Ash-leachate deposition and acidification
C] Heavy metal and fluorine poisoning
D] Increased surface albedo
E] Reduced evapotranspiration and soil moisture supply
F] Changes in cloud coverage (poorly constrained)
G] Increased surface runoff through vegetation burial and ash-blanket emplacement
H] Clogging and changes to water courses
I] Light shading and euphotic zone shallowing from ash deposition, ash resuspension and primary productivity
J] Physical impact of suspended ash particles
K] Increased primary production through ash fertilization
L] Enhanced carbon export by primary productivity
M] Carbon burial and preservation from ash deposition
N] Enhanced carbon export through ash settling

Figure 17.3    For figure caption, see text, p.263.

Figure 17.6    For figure caption, see text, p.267.

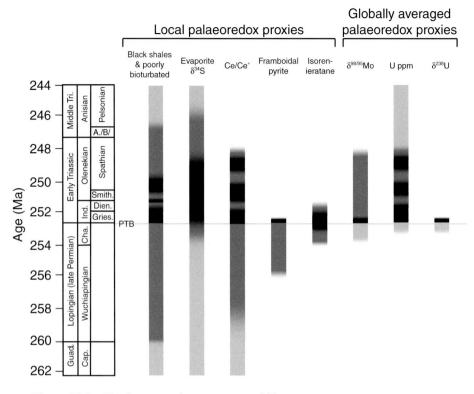

Figure 18.2    For figure caption, see text, p.282.

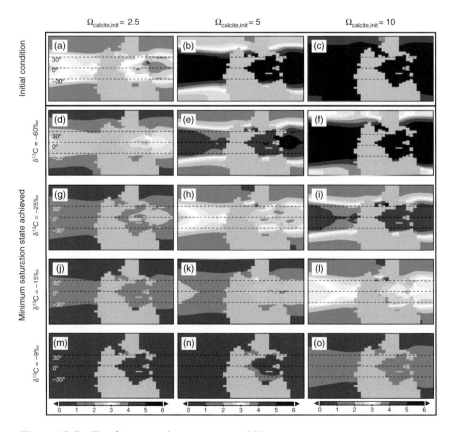

Figure 19.5    For figure caption, see text, p.300.

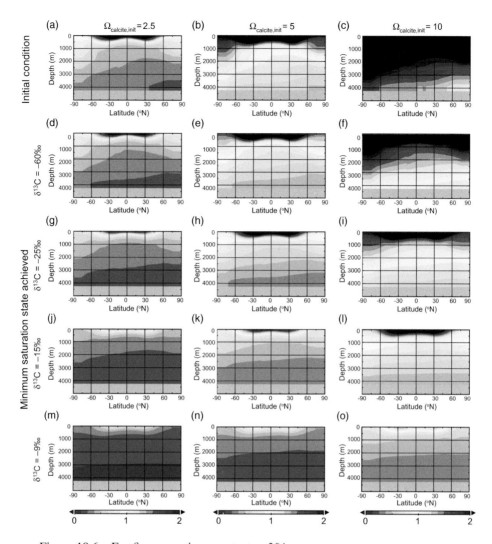

Figure 19.6    For figure caption, see text, p.301.

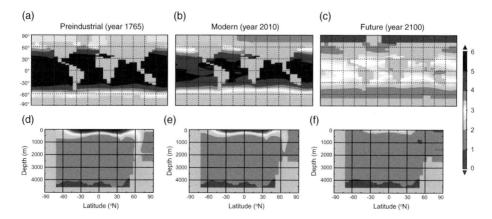

Figure 19.7    For figure caption, see text, p.302.

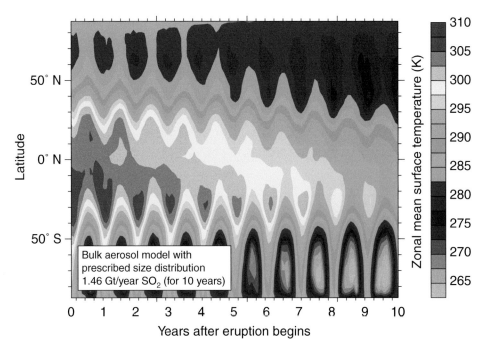

Figure 20.3    For figure caption, see text, p.313.

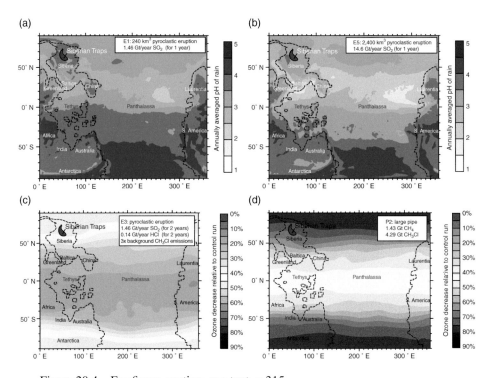

Figure 20.4    For figure caption, see text, p.315.

# 12

# Volatile generation and release from continental large igneous provinces

HENRIK SVENSEN, KIRSTEN E. FRISTAD, ALEXANDER G. POLOZOV AND
SVERRE PLANKE

## 12.1 Introduction

The temporal and causal link between large igneous provinces (LIPs) and rapid changes in the environment and the biosphere has been explored and debated for decades (Wignall, 2001; Stothers, 1993). Early studies suggested that violent sulfur-rich LIP eruptions could have caused global cooling on short annual timescales and that simultaneous release of mantle carbon from lava flows could have initiated global warming on longer centurial timescales, resulting in a potentially chaotically varying global climate (Wignall, 2001). However, upon detailed investigation of sedimentary sections, it was suggested that any carbon release to the atmosphere was characterized by vast quantities of $^{12}$C-enriched gas. Since mantle $CO_2$ is relatively $^{12}$C-depleted, gas hydrates were introduced as a potential source to explain the isotope excursions associated with the Palaeocene–Eocene Thermal Maximum (PETM), the Toarcian event and the end-Permian event (Hesselbo et al., 2000; Dickens et al., 1997). As a consequence, focus turned away from the role of LIP processes.

In the 1990s, regional metamorphic processes were suggested as potential drivers of climate change. Pioneering work by Derrill M. Kerrick and co-workers attempted to link Cenozoic greenhouse events (the PETM and others) to large-scale $CO_2$ degassing generated by prograde metamorphism of carbonate-bearing rocks (Nesbitt et al., 1995; Kerrick et al., 1995). The 5–10-million-year timescale of orogeny or crustal extension can, however, not explain rapid (10–100 kyr) and transient climatic events and thus regional metamorphism is unlikely to affect short-term global climate change. In 2004, Svensen et al. suggested that processes in the sub-volcanic parts of the North Atlantic Igneous Province (NAIP) could have triggered the PETM by greenhouse-gas generation and release from contact aureoles around sills. Similar processes were later suggested to have been involved in

*Volcanism and Global Environmental Change*, eds. Anja Schmidt, Kirsten E. Fristad and Linda T. Elkins-Tanton.
Published by Cambridge University Press. © Cambridge University Press 2015.

the end-Permian, the Toarcian, and the end-Triassic events (Svensen *et al.*, 2009; McElwain *et al.*, 2005; Ruhl and Kürschner, 2011; Svensen *et al.*, 2007). During the last decade, gas hydrate dissociation, impacts, sluggish ocean circulation and anoxia, lava $CO_2$ degassing and contact metamorphic $CH_4$ degassing are among the hypotheses that are discussed in the scientific community as triggers for ancient rapid global warming events (e.g. Wignall, 2001; Knoll *et al.*, 2007; Cohen *et al.*, 2007; Cui *et al.*, 2013). We emphasize that there is a current consensus that Siberian Traps volcanic and metamorphic processes triggered the end-Permian crisis.

Here we show that contact metamorphism in volcanic basins has the potential to generate a mass of $CH_4$, $CO_2$, $SO_2$, and halocarbons in the $>10^3$–$10^4$ Gt range, and that the gases produced in these provinces have been released to the atmosphere via vertical pipe structures and basin seepage processes. The large volumes and compositions of gases produced by contact metamorphism indicate that volcanic and metamorphic processes may have first-order influence on environmental change and mass extinctions.

## 12.2 The anatomy of continental LIPs

During the formation of continental LIPs, magma is commonly emplaced as sills and dikes in the upper crust. When the magma volumes are high and the emplacement takes place within sedimentary basins, these basins are classified as 'volcanic basins'. Volcanic basins are present along rifted continental margins and on lithospheric cratons (Coffin and Eldholm, 1994) (Figure 12.1). Sill thicknesses can exceed 350 m, sandwiched between sedimentary sequences and associated contact metamorphic aureoles. Here we present some of the key aspects on the structure and anatomy of volcanic basins associated with three of the largest LIPs, as summarized in Figure 12.2. We focus on the sediment compositions in the basins, the distribution of sills, the formation of degassing pipes and the composition and volume of the gas formed.

### 12.2.1 The Siberian Traps and the Tunguska Basin

The Siberian Traps covered several million square kilometres of Siberia with kilometre-thick stacks of lava over a period of 0.5–1 Myr (Kamo *et al.*, 2006; Mundil *et al.*, 2004). On the way to the surface, the lavas in Siberia passed through a series of thick sedimentary sequences collectively known as the Tunguska Basin (Figure 12.3Aa). The basin contains enormous volumes of Cambrian evaporites, with up to 2.5-km-thick sequences of halite-rich strata, anhydrite and carbonates across a 2 million $km^2$ area (Petrychenko *et al.*, 2005; Zharkov, 1984).

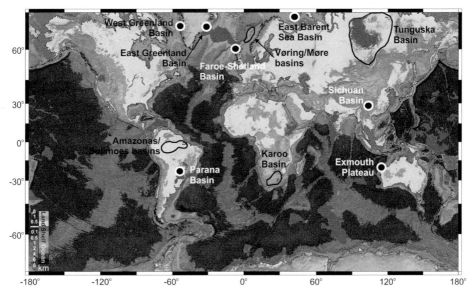

Figure 12.1   Distribution of LIPs and volcanic basins. Basalt flows are shown in red colour. Sedimentary basins injected by sills and dykes (i.e. volcanic basins) mentioned in the text are outlined with heavy lines. Other major volcanic basins are indicated by black filled circles. A black and white version of this figure will appear in some formats. For the colour version, please refer to the plate section.

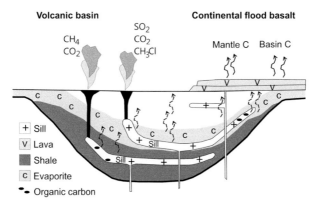

Figure 12.2   Schematic cross section through a volcanic basin, showing sills and dykes, degassing from contact aureoles and subsequent pipe formation, and degassing from the flood basalts. The composition of the gases formed during contact metamorphism depends on the composition of the sediments.

Sills and dykes are abundant throughout the Tunguska Basin, and form sheets up to 450 m thick, locally comprising up to 70% of the basin fill (Fedorenko and Czamanske, 1997; Vasil'ev *et al.*, 2000). Contact metamorphism and dolerite–sediment mixing led to overpressure within salt-dominated lithologies and resulted in the formation of explosion pipes that ejected tephra and a range of gases to the

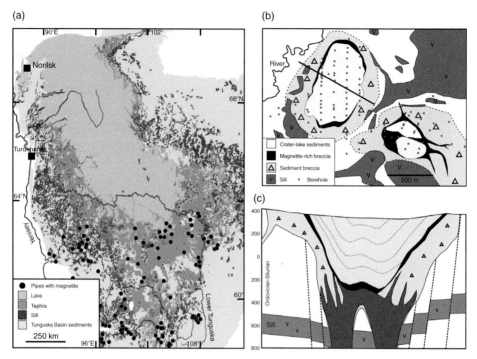

Figure 12.3    (a) Geological map of East Siberia showing exposed flood basalts, sills and explosion pipes that formed during the end-Permian. (b) Map view of two pipes located about 100 km west of Bratsk in the southern parts of the Tunguska Basin (Benedyuk, 1987). The central parts of the pipes are occupied by crater-lake sediments. The eastern pipe is mineralized with strontium carbonate, the eastern pipe (the October pipe) with magnetite. (c) Cross section through the October pipe based on numerous, fully cored boreholes drilled during mineral prospecting in the 1980s. The grey area within the pipe is a magnetite-rich breccia. A black and white version of this figure will appear in some formats. For the colour version, please refer to the plate section.

atmosphere (Von der Flaass and Naumov, 1995; Von der Flaass, 1997; Svensen *et al.*, 2009). Various parts of the Tunguska Basin are characterized by different volume of sills, volcaniclastics and lava flows (Vasil'ev *et al.*, 2000). In the Tunguska Basin, spectacular pipes with up to 1.6-km-wide subaerial explosion craters formed during the end-Permian (Figure 12.3b,c). The pipes are rooted in metasomatic aureole zones 2–4 km down in the basin and are filled by breccias with a compositional range reflecting the underlying country rock, commonly designated as either 'basalt' or 'magnetite' pipes (Von der Flaass, 1997). Magnetite-rich pipes are most common in the southern halite-rich region of the basin where hydrothermal brines rose to the surface through the pipes, leaving massive deposits rich in iron, strontium, copper and other ores. Borehole exploration has revealed twelve pipes containing explosion craters hosting lacustrine deposits (Von der Flaass and Naumov, 1995; Fristad, 2012). The mineral and

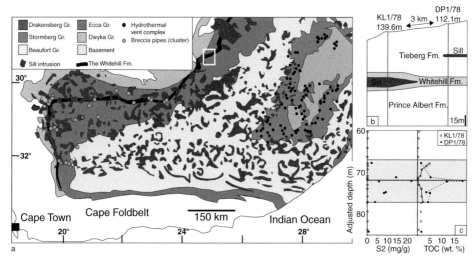

Figure 12.4   (a) Geological map of the Karoo Basin in South Africa. Two different types of degassing structures are present in the basin: breccia pipes in the organic-rich shales in the west, and hydrothermal vent complexes in the upper, sand-dominated levels in the basin. (b) Cross section showing a 15-m-thick sill emplaced in organic-rich shale and two boreholes studied for contact metamorphism (Aarnes *et al.*, 2011). (c) Data from the contact aureole and the unaffected background shale. Note the destruction of the pyrolysis hydrocarbon productivity (the S2 parameter) in the aureole and the lowering of the total organic carbon (TOC) content from about 15 wt.% to zero at the contact with the sill. A black and white version of this figure will appear in some formats. For the colour version, please refer to the plate section.

isotopic record preserved in the breccia pipes and crater deposits reflect mobilization and transport of carbon and sulfur from the Tunguska Basin to the end-Permian environment (Aarnes *et al.*, 2011).

### 12.2.2  *The Karoo province and the Karoo Basin*

The Upper Carboniferous to Jurassic Karoo Supergroup in South Africa has a maximum cumulative thickness of 12 km and a preserved maximum thickness of 5.5 km (Tankard *et al.*, 2009). The current area with outcropping Karoo sediments in South Africa is about 630,000 km$^2$ (Figure 12.4a), but time-equivalent sedimentary rocks are present throughout southern Africa. The depositional environments range from marine to fluvial and finally eolian (Catuneanu *et al.*, 1998). The Karoo Basin is overlain by 1.6 km of volcanic rocks of the Drakensberg Group, consisting mainly of stacked basalt flows erupted in a continental environment. The plumbing system of the flood basalts is a basin-scale complex consisting of sills and dykes (Marsh and Eales, 1984; Chevallier and Woodford, 1999).

(a)                                          (b)

(c)                                          (d)

Figure 12.5   Various outcrop manifestations of hydrothermal vent complexes and breccia pipes in the Karoo Basin. (a) The Witkop I hydrothermal vent complex, a sand-dominated crater-like deposit. (b) Unnamed hydrothermal vent complex representing a former crater in Elliot Formation sandstone filled with Clarens Formation sand. (c) Breccia pipe near Loriesfontein in the western Karoo. (d) Satellite image of breccia pipes (shown by white arrows) north of Loriesfontein. More than 400 pipes are mapped in this area. Dirt road (line) for scale.

The thickest sill in the basin is about 220 m thick, most are in the 10–60 m range. The sills were emplaced at 182.6 Ma, likely within a 400 kyr interval as suggested by Monte Carlo simulations of sill ages (Svensen *et al.*, 2012). The composition of the sills is mainly tholeiitic, with a few evolved intrusions (andesitic) (Marsh and Eales, 1984; Neumann *et al.*, 2011). Hundreds of breccia pipes and hydrothermal vent complexes (HVCs) are rooted in the contact aureoles in the Karoo Basin (Figure 12.4b,c) and formed during pressure build-up related to devolatilization (Jamtveit *et al.*, 2004; Svensen *et al.*, 2007; Aarnes *et al.*, 2012). The HVCs commonly crop out in the uppermost 400–500 m of the basin (Figure 12.4a). Examples of breccia pipes and HVCs on outcrop scales are shown in Figure 12.5.

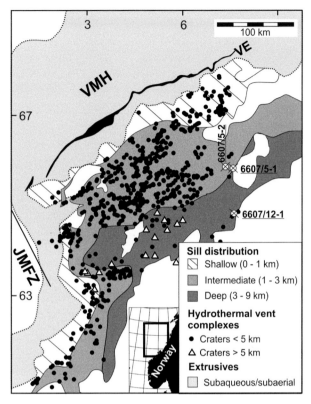

Figure 12.6    Extrusives (basalt flows and hyaloclastites), sills and HVCs in the Vøring Basin, offshore Norway. The sills and pipes were formed during emplacement of the NAIP during the PETM. Abbreviations: VE: Vøring escarpment. VMH: Vøring margin high. JMFZ: Jan Mayen fracture zone.

### 12.2.3  The NAIP and the Vøring and Møre basins

The Vøring and Møre basins offshore Norway contain a voluminous magmatic complex of dominantly subhorizontal sills that intruded Cretaceous sedimentary rocks during the opening of the northeast Atlantic (Planke *et al.*, 2005; Cartwright and Møller Hansen, 2006). The sills cover an area of more than 80,000 km$^2$ offshore mid-Norway, and stacks of two to five sills are commonly present in the basins (Figures 12.6 and 12.7). Sill intrusions are identified as high-amplitude reflections, commonly displaying saucer-shaped geometries (Figure 12.7b).

More than 700 HVCs have been mapped, and several thousand vent complexes are likely present in these basins. In this marine setting, the craters of HVCs can reach a size of more than 10 km in diameter (Planke *et al.*, 2005). The HVCs commonly have an eye-shaped geometry at the top Palaeocene palaeo-surface (Figure 12.7b). The base of the semi-circular eye structures are crater floors, as documented by an industry borehole drilled through an eye structure (Svensen

(a)

(b)

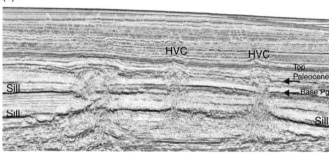

Figure 12.7   (a) A 60-m sill and the associated contact aureole in Horsedal on Northern Jameson Land, East Greenland. The sill was emplaced about 56 Ma just above the Triassic–Jurassic boundary in shales near the base of the Lower Jurassic Kap Stewart Formation. (b) Two-dimensional seismic reflection profile from the Vøring Basin showing sills and HVCs.

*et al.*, 2003). The borehole data show that the infill of the eye is undercompacted, with unusually high porosities. The undercompaction is likely due to high pore pressure in the crater sediments maintained by vertical seepage of fluids in the vent complex since the earliest Eocene. Pipe-like vertical zones of disturbed seismic reflections connect the eye with underlying sills. Commonly, the craters of the HVCs are located directly above the termination of transgressive sills at depth (Planke *et al.*, 2005).

## 12.3  Sill and aureole volumes

Vast volumes of sedimentary rocks are heated in volcanic basins following sill emplacement (Figure 12.2). Sills may extend laterally over several hundred kilometres and commonly have thicknesses of 50–150 m (Figure 12.7). The volume of individual sills is difficult to measure due to limited three-dimensional seismic imaging of stacked sill reflections, and few continuous field exposures and boreholes. Estimating the maximum individual sill volume in a basin is important for understanding the basin devolatilization reactions during contact metamorphism. Using the well-constrained Karoo Basin as an example, 100-m-thick sills commonly cover $100 \times 100$ km$^2$. This gives a volume of 1,000 km$^3$, which is considerable and even comparable in volume to the largest silicic volcanic eruptions and single basaltic flows (Bryan *et al.*, 2010). At the extreme, but still realistic end, single sill intrusions may have volumes up to 18,000 km$^3$ ($300 \times 300$ km $\times$ 200 m).

The thicknesses of metamorphic aureoles commonly extend about the same distance as the sill thickness, both below and above the sill (Figure 12.4c). This means that a single sill emplacement event may heat up to 36,000 km$^3$ of sedimentary rocks in the end-member example given above. The aureole size ultimately depends on both the geological setting and the measuring methods (Aarnes *et al.*, 2010). Sills emplaced at deep levels in a basin will have thicker aureoles compared to a shallower sill because the sill cooling is less efficient. Fluid flow in the aureole may lead to a more widespread aureole and lower maximum temperatures in cases where the host sedimentary rocks have high permeability (Iyer *et al.*, 2013).

The timescale of metamorphism around a sill of that size will depend on the emplacement time of the sill, depth of emplacement, the geothermal gradient, the magma temperature and the host-rock composition and water content (Aarnes *et al.*, 2011). Although metamorphism in the aureole may last for > 1,000 years, metamorphism is a rapid process and will start immediately following emplacement.

## 12.4 Gas production mechanisms

The chemical composition of the sedimentary rocks heated by igneous intrusions has a profound influence on the metamorphic fluid composition (e.g. Ganino and Arndt, 2009; Svensen *et al.*, 2004; Svensen *et al.*, 2009). For instance, organic-rich shale generates $CH_4$ during contact metamorphism, whereas coal generates $CO_2$-dominated fluids. In addition, water is generated. Since many sedimentary basins contained hydrogen-rich kerogen and oil and gas accumulations at the time of sill emplacement, gases like methane ($CH_4$) and ethane ($C_2H_6$) may have dominated the fluid. If limestones or dolostones are heated, the generated fluid will be dominated by $^{13}$C-enriched $CO_2$, or $^{13}$C-depleted $CH_4$ if organic matter or graphite is present. Some of the key metamorphic reactions taking place when carbonates, marls and shales are heated are presented below (Tracy and Frost, 1991):

$$5 \text{ dolomite} + 8 \text{ quartz} + H_2O = \text{tremolite} + 3 \text{ calcite} + 7 \text{ CO}_2 \qquad (12.1)$$
$$\text{tremolite} + 3 \text{ calcite} + 2 \text{ quartz} = 5 \text{ diopside} + 3 \text{ CO}_2 + H_2O \qquad (12.2)$$

$$3 \text{ dolomite} + \text{diopside} = 4 \text{ calcite} + 2 \text{ forsterite} + 2 \text{ CO}_2 \qquad (12.3)$$
$$\text{dolomite} = \text{periclase} + \text{calcite} + \text{CO}_2 \qquad (12.4)$$
$$2\text{CH}_2\text{O} = \text{CH}_4 + \text{CO}_2 \qquad (12.5)$$

Reaction (12.3) will produce olivine (forsterite) when the pore fluid composition is $CO_2$-rich. Reaction (12.5) is a generalized way of describing devolatilization of organic matter.

Evaporites with anhydrite and rock salt may generate $SO_2$ and HCl and, if organic matter or petroleum is present, $CH_4$ and halocarbons such as methyl chloride ($CH_3Cl$) and methyl bromide ($CH_3Br$) may also form. This was recently confirmed by experiments with natural rock salt from the Tunguska Basin in East Siberia (Svensen *et al.*, 2009). The samples were heated to 275 °C in a gas chromatography mass spectrometer (Svensen *et al.*, 2009) without crushing, to simulate contact metamorphism. The results show that at room temperature, following crushing, the released gas contained butane, benzene, and sulfur dioxide. No halocarbons were detected. When the samples were heated, the concentrations of sulfur dioxide had increased significantly (up to 130 times), and the concentration of other sulfur gases and hydrocarbons decreased. Interestingly, halocarbons $CH_3Cl$ and $CH_3Br$ were identified in all heating runs.

Previous experiments have solely focussed on biological samples (e.g. Keppler *et al.*, 2000). Even though the details about halocarbon formation during our experiments remain poorly understood we attribute the formation of these molecules to reactions between hydrocarbon compounds within the fluid inclusions and dissolved chloride from the rock salt.

## 12.5 Constraining gas generation in volcanic basins

Numerous approaches can be used for constraining gas generation in contact aureoles, including thermodynamic calculations based on mineral-water equilibrium, rock heating and pyrolysis, numerical simulations and estimates based on aureole data. Here we present an approach utilizing geochemical analysis of aureole rocks.

Based on metamorphism of carbon-rich sediments, it is possible to estimate carbon production potential in an intruded sedimentary basin. This method requires the contact aureole volume (area $A$ × thickness $h$) and the amount of carbon in wt.% converted to $CH_4$ or $CO_2$, $F_C$. The total mass of carbon, $W_C$, produced in contact aureoles is:

$$W_C = F_C \times A \times h \times \rho,$$

where $\rho$ is the rock density (2,400 kg/m$^3$). The area $A$ is measured from seismic or field data. The thickness $h$ of the metamorphosed sedimentary layer can be estimated from field and seismic observations or the aureole total organic carbon (TOC) and vitrinite reflectivity data. The total mass of carbon produced, $W_C$, can be converted to equivalents of $CH_4$ (= $W_C \times 1.34$) and $CO_2$ (= $W_C \times 3.66$).

If only 1 wt.% of the organic carbon in shale or siltstone is transformed into gaseous carbon compounds, the gas production potential associated with a 5,000–20,000 $km^3$ sill intrusion is 230–920 Gt C (corresponding to 310–1200 Gt $CH_4$). More gas may be generated if the shales have higher TOC contents and if the sills are emplaced within hydrocarbon-bearing lithologies. This means that a single melt batch injected into an organic-bearing sedimentary basin can generate sufficient $CH_4$, rapidly enough, to cause global warming (Aarnes *et al.*, 2010). Since the inner contact aureoles reach peak metamorphic conditions (typically 400–500 °C) shortly after sill emplacement (10–500 years), the metamorphic reactions and associated fluid production are also very fast.

In addition to the heated sedimentary rocks, the igneous sub-volcanic and the volcanic parts of LIPs also release volatiles. Basaltic melts contain $H_2O$, $CO_2$ and $SO_2$ that is released during magma ascent and decompression. The carbon content of un-degassed basaltic magmas is poorly constrained, but may be on the order of 13 Mt $CO_2/km^3$ (Self *et al.*, 2005). Studies from sills in the Siberian Traps have suggested that high-temperature interactions between dolerite and evaporites can explain high halogen concentrations in melt inclusions (Black *et al.*, 2012).

## 12.6 Gas-release mechanisms

When sedimentary host rocks are heated by magma intrusion, the resulting over-pressure may lead to hydrofracturing and the formation of vertical pipes. Both dehydration reactions and pore-fluid boiling are important processes for a rapid increase in the local fluid pressure in contact aureoles (Jamtveit *et al.*, 2004). Additional processes that can increase the local fluid pressure and lead to venting include volatile exsolution from the melt and interactions between near-surface water and the melt. Vent structures and breccia pipes are characteristic features of the three volcanic basins presented here, although they vary greatly in formation, size and degassing style. The dehydration-related hydrofracturing was recently explored by Aarnes *et al.* (2012), showing that the organic content in a shale is important for determining if the aureole fractures during heating. Increasing the organic carbon content from 1 to 10 wt.% has an effect comparable to a three orders of magnitude decrease in permeability. In cases where the permeability of the sedimentary rocks is relatively high, metamorphic and igneous gases may be released pervasively and seep out at the surface without any vent being formed. A modern analogy of such seep activity can be found in the Salton Sea area in California (Mazzini *et al.*, 2011).

## 12.7 Environmental implications

As sedimentary basins are among the main crustal reservoirs of organic and inorganic carbon (e.g. IPCC, 2007), there is a huge potential for $CH_4$ and $CO_2$ generation in volcanic basins. A link between contact metamorphism and the PETM was suggested in 2004, based on detailed seismic imaging and borehole studies in the Vøring and Møre basins, offshore Norway (Svensen *et al.*, 2004), and this provided a new framework for investigating LIP processes and the link to environmental changes. The cornerstone of this hypothesis is rapid gas generation from contact metamorphism of organic material and subsequent release of greenhouse gases to the atmosphere.

Recent work has provided evidence that degassing from volcanic basins can be linked to other climate events, including the Toarcian (Lower Jurassic), the Triassic–Jurassic, the end-Permian, and the end-Guadalupian (e.g. Svensen *et al.*, 2007; Svensen *et al.*, 2009; Ganino and Arndt, 2009; McElwain *et al.*, 2005; Beerling *et al.*, 2007). This hypothesis is supported by geological constraints (e.g. observed contact aureoles, sill intrusions, and vertical pipe structures) as well as the fact that metamorphism of organic carbon leads to the generation of $^{12}$C-enriched $CH_4$. This can better explain the available geochemical data than the emission of $^{12}$C-depleted mantle $CO_2$ due to LIP lava degassing.

In eastern Siberia, the sill emplacement event took place at the end of the Permian, and thick sills are present throughout the basin. Mass balance calculations suggest that 10,000–30,000 Gt C could have been generated during contact metamorphism of organic matter, accompanied by 4,500–13,000 Gt $CH_3Cl$ (Svensen *et al.*, 2009). The presence of a large number of pipe structures rooted in evaporite lithologies suggests that the gases were released to the atmosphere. Violent degassing of sulfur-, carbon-, and halogen-rich gases, as suggested by the pipes in Siberia, makes it likely that the gases could have passed through the otherwise rather undisturbed tropopause and been injected into the stratosphere. This may explain some of the terrestrial consequences of the end-Permian mass extinctions, as mutated pollen grains are ascribed to ozone-layer breakdown and damaging ultraviolet-B radiation (Visscher *et al.*, 2004; Foster and Afonin, 2005). This hypothesis is supported by the heating experiments mentioned here and atmospheric chemistry modelling of end-Permian ozone stability (Black *et al.*, 2014). Contact metamorphism of coal and other carbonaceous sediments in the Tunguska Basin in eastern Siberia generated carbon gases that induced global warming (Svensen *et al.*, 2004; Retallack and Krull, 2006; Visscher *et al.*, 2004; Svensen *et al.*, 2009).

## 12.8 Summary

- Volatile generation in the sub-volcanic part of LIPs occurs during sill intrusions into the sedimentary strata of volcanic basins. The style of emplacement (deep-seated intrusion versus eruption) and type of sedimentary cover (coals, shale, carbonates, evaporites) determine the amount and the composition of generated gases.
- Thermogenic gases release mainly through pipes, fissures and small vent structures, offering a fast and efficient mechanism for inducing environmental change or even mass extinction.
- More work is needed to explore the halocarbon degassing hypothesis as presented here. New quantitative experiments on rock salt must be conducted to improve composition and volume estimates of pipe degassing scenarios for eastern Siberia.
- The role of gas release from cooling magma and from recrystallizing country rock following magma–sediment interaction and sediment assimilation must be investigated to improve atmospheric gas release estimates.
- We suggest that the main reason why the Siberian Traps is associated with a major extinction event is the chemical composition of the heated sedimentary rocks. A key fact is that one of the world's biggest salt deposits was intruded by 1,200 °C igneous sills in eastern Siberia at the same time as the end-Permian extinction.

## Acknowledgements

This study was supported by a Centre of Excellence grant to CEED and by a grant to H. Svensen, both from the Norwegian Research Council. We thank TGS for permission to use seismic data, and Yuri Podladchikov, Seth Burgess, Dougal Jerram, Stephane Polteau and Clement Ganino for the company and scientific input during fieldwork in Siberia.

## References

Aarnes, I., Fristad, K., Planke, S. & Svensen, H. 2011. The impact of host-rock composition on devolatilization of sedimentary rocks during contact metamorphism around mafic sheet intrusions. *Geochemistry Geophysics Geosystems*, **12**, Q10019.

Aarnes, I., Podladchikov, Y. & Svensen, H. 2012. Devolatilization-induced pressure build-up: Implications for reaction front movement and breccia pipe formation. *Geofluids*, **12**, 265–279.

Aarnes, I., Svensen, H., Connolly, J. A. D. & Podladchikov, Y. Y. 2010. How contact metamorphism can trigger global climate changes: modeling gas generation around igneous sills in sedimentary basins. *Geochimica et Cosmochimica Acta*, **74**, 7179–7195.

Beerling, D. J., Harfoot, M., Lomax, B. & Pyle, J. A. 2007. The stability of the stratospheric ozone layer during the end-Permain eruption of the Siberian Traps. *Philosophical Transactions of the Royal Society A –Mathematical Physical and Engineering Sciences*, **365**, 1843–1866.

Benedyuk, P. F. 1987. *Report of the results of prospecting on the flanks of the October ore-field for the 1985–1987 years. Graphics applications* (in Russian).

Black, B. A., Elkins-Tanton, L. T., Rowe, M. C. & Peate, I. U. 2012. Magnitude and consequences of volatile release from the Siberian Traps. *Earth and Planetary Science Letters*, **317–318**, 363–373.

Black, B. A., Lamarque, J.-F., Shields, C. A., Elkins-Tanton, L. T. & Kiehl, J. T. 2014. Acid rain and ozone depletion from pulsed Siberian Traps magmatism. *Geology*, **42**, 67–70.

Bryan, S. E., Peate, I. U., Peate, D. W. *et al.* 2010. The largest volcanic eruptions on Earth. *Earth-Science Reviews*, **102**, 207–229.

Cartwright, J. & Møller Hansen, D. 2006. Magma transport through the crust via interconnected sill complexes. *Geology*, **34**, 929–932.

Catuneanu, O., Hancox, P. J. & Rubidge, B. S. 1998. Reciprocal flexural behaviour and contrasting stratigraphies: a new basin development model for the Karoo retroarc foreland system, South Africa. *Basin Research*, **10**, 417–439.

Chevallier, L. & Woodford, A. 1999. Morpho-tectonics and mechanism of emplacement of the dolerite rings and sills of the western Karoo, South Africa. *South African Journal of Geology*, **102**, 43–54.

Coffin, M. F. & Eldholm, O. 1994. Large igneous provinces: crustal structure, dimensions, and external consequences. *Reviews of Geophysics*, **32**, 1–36.

Cohen, A. S., Coe, A. L. & Kemp, D. B. 2007. The late Palaeocene–early Eocene and Toarcian (Early Jurassic) carbon isotope excursions: a comparison of their time scales, associated environmental changes, causes and consequences. *Journal of the Geological Society*, **164**, 1093–1108.

Cui, Y., Kump, L. R. & Ridgwell, A. 2013. Initial assessment of the carbon emission rate and climatic consequences during the end-Permian mass extinction. *Palaeogeography, Palaeoclimatology, Palaeoecology*, **389**, 128–136.

Dickens, G. R., Castillo, M. M. & Walker, J. C. G. 1997. A blast of gas in the latest Paleocene: simulating first-order effects of massive dissociation on oceanic methane hydrate. *Geology*, **25**, 259–262.

Fedorenko, V. & Czamanske, G. K. 1997. Results of new field and geochemical studies of the volcanic and intrusive rocks of the Maymecha–Kotuy area, Siberian Flood-Basalt Province, Russia. *International Geology Review*, **39**, 479–531.

Foster, C. B. & Afonin, S. A. 2005. Abnormal pollen grains: an outcome of deteriorating atmospheric conditions around the Permian–Triassic boundary. *Journal of the Geological Society*, **162**, 653–659.

Fristad, K. E. 2012. The Oktyabr'sk volcanic crater sediments: petrography, geochemistry, and environmental implications of Siberian Trap diatreme eruption products. Dissertation, University of Oslo.

Ganino, C. & Arndt, N. T. 2009. Climate changes caused by degassing of sediments during the emplacement of large igneous provinces. *Geology*, **37**, 323–326.

Hesselbo, S. P., Grocke, D. R., Jenkyns, H. C. *et al.* 2000. Massive dissociation of gas hydrate during a Jurassic oceanic anoxic event. *Nature*, **406**, 392–395.

IPCC 2007. *Climate Change 2007. Fourth Assessment Report (AR4).* Intergovernmental Panel on Climate Change, Geneva, 2007.

Iyer, K., Rüpke, L. & Galerne, C. Y. 2013. Modeling fluid flow in sedimentary basins with sill intrusions: implications for hydrothermal venting and climate change. *Geochemistry, Geophysics, Geosystems*, **14**, 5244–5262.

Jamtveit, B., Svensen, H., Podladchikov, Y. & Planke, S. 2004. Hydrothermal vent complexes associated with sill intrusions in sedimentary basins. *Geological Society of London Special Publication*, **234**, 233–241.

Kamo, S. L., Crowley, J. & Bowring, S. A. 2006. The Permian–Triassic boundary event and eruption of the Siberian flood basalts: an inter-laboratory U–Pb dating study. *Geochimica et Cosmochimica Acta*, **70**, A303–A303.

Keppler, F., Eiden, R., Niedan, V., Pracht, J. & Scholer, H. F. 2000. Halocarbons produced by natural oxidation processes during degradation of organic matter. *Nature*, **403**, 298–301.

Kerrick, D. M., Mckibben, M. A., Seward, T. M. & Caldeira, K. 1995. Convective hydrothermal $CO_2$ emission from high heat flow regions. *Chemical Geology*, **121**, 285–293.

Knoll, A. H., Bambach, R. K., Payne, J. L., Pruss, S. & Fischer, W. W. 2007. Paleophysiology and end-Permian mass extinction. *Earth and Planetary Science Letters*, **256**, 295–313.

Marsh, J. S. & Eales, H. V. 1984. The chemistry and petrogenesis of igneous rocks of the Karoo central area, Southern Africa. *Special Publication of the Geological Society of South Africa*, **13**, 27–67.

Mazzini, A., Svensen, H., Etiope, G., Onderdonk, N. & Banks, D. 2011. Fluid origin, gas fluxes and plumbing system in the sediment-hosted Salton Sea Geothermal System (California, USA). *Journal of Volcanology and Geothermal Research*, **205**, 67–83.

McElwain, J. C., Wade-Murphy, J. & Hesselbo, S. P. 2005. Changes in carbon dioxide during an oceanic anoxic event linked to intrusion into Gondwana coals. *Nature*, **435**, 479–482.

Mundil, R., Ludwig, K. R., Metcalfe, I. & Renne, P. R. 2004. Age and timing of the Permian mass extinctions: U/Pb dating of closed-system zircons. *Science*, **305**, 1760–1763.

Nesbitt, B. E., Mendoza, C. A. & Kerrick, D. M. 1995. Surface fluid convection during Cordilleran extension and the generation of metamorphic $CO_2$ contributions to Cenozoic atmospheres. *Geology*, **23**, 99–101.

Neumann, E.-R., Svensen, H., Galerne, C. Y. & Planke, S. 2011. Multistage evolution of dolerites in the Karoo Large Igneous Province, central South Africa. *Journal of Petrology*, **52**, 959–984.

Petrychenko, O. Y., Peryt, T. M. & Chechel, E. I. 2005. Early Cambrian seawater chemistry from fluid inclusions in halite from Siberian evaporites. *Chemical Geology*, **219**, 149–161.

Planke, S., Rassmussen, T., Rey, S. S. & Myklebust, R. (eds.) 2005. Seismic characteristics and distribution of volcanic intrusions and hydrothermal vent complexes in the Vøring and Møre basins. In Doré, A. G. & Vining, B. A. (eds.), *Petroleum Geology: North-West Europe and Global Perspectives – Proceedings of the 6th Petroleum Geology Conference*. London: Geological Society, pp. 833–844.

Retallack, G. J. & Krull, E. S. 2006. Carbon isotopic evidence for terminal-Permian methane outbursts and their role in extinctions of animals, plants, coral reefs, and peat swamps. *Geological Society of America Special Paper*, **399**, 249–268.

Ruhl, M. & Kürschner, W. M. 2011. Multiple phases of carbon cycle disturbance from large igneous province formation at the Triassic–Jurassic transition. *Geology*, **39**, 431–434.

Self, S., Thordarson, T. & Widdowson, M. 2005. Gas fluxes from flood basalt eruptions. *Elements*, **1**, 283–287.

Stothers, R. B. 1993. Flood basalts and extinction events. *Geophysical Research Letters*, **20**, 1399–1402.

Svensen, H., Corfu, F., Polteau, S., Hammer, O. & Planke, S. 2012. Rapid magma emplacement in the Karoo Large Igneous Province. *Earth and Planetary Science Letters*, **325**, 1–9.

Svensen, H., Planke, S., Chevallier, L. *et al.* 2007. Hydrothermal venting of greenhouse gases triggering Early Jurassic global warming. *Earth and Planetary Science Letters*, **256**, 554–566.

Svensen, H., Planke, S., Jamtveit, B. & Pedersen, T. 2003. Seep carbonate formation controlled by hydrothermal vent complexes: a case study from the Vøring Basin, the Norwegian Sea. *Geo-Marine Letters*, **23**, 351–358.

Svensen, H., Planke, S., Malthe-Sorenssen, A. *et al.* 2004. Release of methane from a volcanic basin as a mechanism for initial Eocene global warming. *Nature*, **429**, 542–545.

Svensen, H., Planke, S., Polozov, A. G. *et al.* 2009. Siberian gas venting and the end-Permian environmental crisis. *Earth and Planetary Science Letters*, **277**, 490–500.

Tankard, A., Welsink, H., Aukes, P., Newton, R. & Stettler, E. 2009. Tectonic evolution of the Cape and Karoo basins of South Africa. *Marine and Petroleum Geology*, **26**, 1379–1412.

Tracy, R. J. & Frost, B. R. 1991. Phase-equilibria and thermobarometry of calcareous, ultramafic and mafic rocks, and iron formations. *Reviews in Mineralogy*, **26**, 207–289.

Vasil'ev, Y. R., Zolotukhin, V. V., Feoktistov, G. D. & Prusskava, S. N. 2000. Evaluation of the volume and genesis of Permo-Triassic Trap magmatism on the Siberian Platform. *Russian Geology and Geophysics*, **41**, 1696–1705.

Visscher, H., Looy, C. V., Collinson, M. E. *et al.* 2004. Environmental mutagenesis during the end-Permian ecological crisis. *Proceedings of the National Academy of Sciences of the United States of America*, **101**, 12952–12956.

Von Der Flaass, G. S. 1997. Structural and genetic model of an ore field of the Angaro-Ilim type (Siberian Platform). *Geology of Ore Deposits*, **39**, 461–473.

Von Der Flaass, G. S. & Naumov, V. A. 1995. Cup-shaped structures of iron ore deposits in the south of the Siberian Platform (Russia). *Geology of Ore Deposits*, **37**, 340–350.

Wignall, P. B. 2001. Large igneous provinces and mass extinctions. *Earth-Science Reviews*, **53**, 1–33.

Zharkov, M. A. 1984. *Paleozoic Salt Bearing Formations of the World*. Berlin: Springer-Verlag.

# Part Three

## Modes of volcanically induced global environmental change

# 13

# Volcanism, the atmosphere and climate through time

ANJA SCHMIDT AND ALAN ROBOCK

## 13.1 Introduction

Volcanism can affect the environment on timescales of weather (days to weeks) and climate (months to years). In the past 250 years, the atmospheric and climatic effects of several volcanic eruptions have been witnessed and documented by scientist and non-scientist alike. For example, the 1783–1784 Laki eruption in Iceland, which was followed by hot summer temperatures and cold winter temperatures in central Europe (Thordarson and Self, 2003; Oman *et al.*, 2006a; Schmidt *et al.*, 2012b), was described in great detail by an Icelandic priest (Steingrímsson, 1788). The 1815 Tambora eruption in Indonesia caused the 'Year Without a Summer' in 1816, which inspired Mary Shelley's *Frankenstein*.

The scientific understanding of the climatic effects caused by short-lived explosive volcanic eruptions has advanced greatly in the past five decades, mainly due to theoretical work and observations following the eruptions of Agung in 1963, Mount St Helens in 1980, El Chichón in 1982 and, in particular, Mount Pinatubo in 1991 (for reviews see Robock, 2000; Timmreck, 2012).

In the past two decades, the climate relevance of effusive volcanic activity injecting gases mainly into the troposphere has become more recognized (e.g. Graf *et al.*, 1997; Schmidt *et al.*, 2012a), and also the atmospheric and climatic effects of small- to moderate-sized explosive eruptions (e.g. Solomon *et al.*, 2011; Neely *et al.*, 2013). However, we are only beginning to understand how large-volume flood basalt eruptions such as the Deccan Traps in India at around 65 Ma may have affected the environment.

This chapter provides an overview of the impact of various eruption types and styles on Earth's atmosphere and climate through time.

*Volcanism and Global Environmental Change*, eds. Anja Schmidt, Kirsten E. Fristad and Linda T. Elkins-Tanton. Published by Cambridge University Press. © Cambridge University Press 2015.

## 13.2 Volcanism on Earth

The size of a volcanic eruption can be expressed in terms of the volume (km$^3$) or mass (kg) of magma produced. The Volcanic Explosivity Index (VEI) uses volcanological data such as eruption column height and ejecta volume to measure the relative magnitude of an eruption's explosivity using an open-ended logarithmic scale (except for VEI between 0 and 2) (Newhall and Self, 1982). The fragmentation of volatile-rich magma during explosive volcanic eruptions produces large quantities of volcanic tephra, and generally the likelihood that ejecta reach stratospheric altitudes increases with increasing VEI. There are, however, examples of VEI 5 eruptions, such as the 1980 Mount St Helens eruption, which injected little sulfur into the stratosphere, and thus had no impact on climate (Robock, 1981).

One of the largest eruptions in the twentieth century, the 1991 Mount Pinatubo eruption, had a VEI of 6 and produced a bulk magma volume of about 10 km$^3$. For context, the 1815 Tambora eruption (VEI 7) had a bulk volume $>$ 100 km$^3$. The 1783–1784 Laki eruption in Iceland greatly exceeded twentieth-century lava volumes, yielding about 15 km$^3$ of extrusives in eight months (Thordarson and Self, 2003).

Past episodes of continental flood basalt (CFB) volcanism produced huge lava volumes on the order of tens of millions of cubic kilometres. Examples include the emplacement of the Deccan Traps, 65 Ma (Late Cretaceous; $> 10^6$ km$^3$); or the Columbia River Basalt Group (Western United States), 17–10 Ma (Middle Miocene; $\sim$ 210,000 km$^3$), including the Roza flow, 14.7 Ma ($\sim$ 1,300 km$^3$). As also discussed in Chapters 5 and 11, CFB volcanism is typified by numerous, recurring large-volume eruptive phases, each lasting decades or longer. On geological timescales, the climactic pulses that build up an entire CFB province are short-lived – typically about 1 Ma or less. For the assessment of the environmental perturbations it is, however, important to recognize the pulsed nature of CFB volcanism, with the duration of non-eruptive phases (usually centuries to millennia) outlasting the duration of individual eruptive phases (years to decades). An individual decade-long eruptive phase would have yielded between $10^3$ and $10^4$ km$^3$ of extrusives building up pāhoehoe-dominated lava flow fields (see Chapter 11 for details; Self *et al.*, 2006; 2008). Despite emplacing huge magma volumes, individual eruptions would rarely have exceeded a VEI of 4 and, typically, volcanic gases would have been injected into the upper troposphere and lower stratosphere (see Chapter 11). There is, however, evidence that some CFB provinces were associated with more violent silicic eruptions (e.g. Bryan and Ferrari, 2013; see also Chapter 1].

## 13.3 Volcanic gas emissions

Water vapour ($H_2O$) and carbon dioxide ($CO_2$) are the most abundant volatile species released during a volcanic eruption, but in the short term their effect on atmospheric composition is negligible because of the insignificant relative contribution to the high atmospheric background concentrations of $H_2O$ and $CO_2$. However, on the timescale of the age of Earth, volcanic 'outgassing' has been the source of our current atmosphere (see also Chapter 14).

Compared to 35,000 Teragrams (Tg) of anthropogenic $CO_2$ emissions per year, estimates of the $CO_2$ flux from present-day subaerial and submarine volcanism range between 130 and 440 Tg (Gerlach, 2011, and references therein). Self *et al.* (2008) estimated that the Deccan Traps released up to 14 Tg of $CO_2$ per $km^3$ of lava erupted (assuming a degassing efficiency of 80%). For a typical decade-long CFB eruptive phase producing a total lava volume of 1,000 $km^3$ this equates to 1400 Tg of $CO_2$ per year, which is one order of magnitude smaller than the current anthropogenic flux.

Volcanic eruptions also release halogen species (mainly bromine oxide, hydrogen chloride, hydrogen bromide and hydrogen fluoride), which play an important role in volcanic plume chemistry (von Glasow, 2010). Over the industrial era, anthropogenic chlorofluorocarbon (CFC) emissions have resulted in high stratospheric chlorine and bromine concentrations. These halogens destroy stratospheric $O_3$ efficiently, leading to phenomena such as the Antarctic 'ozone hole'. Stratospheric $O_3$ depletion was also observed after the eruptions of El Chichón in 1982 and Mount Pinatubo in 1991 because volcanic aerosol particles serve as surfaces for heterogeneous reactions promoting conversion of less reactive (anthropogenic) chlorine/bromine species into more reactive forms (Solomon *et al.*, 1998; Solomon, 1999 for a review). Therefore, in general, the chemical impact of volcanic eruptions is to exacerbate anthropogenic-driven $O_3$ destruction. Anthropogenic chlorine concentrations are steadily declining; hence, future eruptions will not deplete stratospheric $O_3$ except if an eruption itself injects sufficient chlorine (in the form of hydrochloric acid) into the stratosphere; this is a matter of debate (e.g. Tabazadeh and Turco, 1993; Kutterolf *et al.*, 2013; see also Chapter 16). Chapter 20 discusses how halogen-bearing species emitted by flood basalt eruptions could have resulted in global-scale $O_3$ depletion during the end-Permian (252 Ma) emplacement of the Siberian Trap province.

To date, sulfur dioxide ($SO_2$) is the sole volcanic volatile species that has been observed to alter the radiative balance of the atmosphere, because of its conversion to volcanic sulfuric acid aerosol. Sulfur species contribute between 2% and 35% by volume of the gas phase, and $SO_2$ and hydrogen sulfide ($H_2S$) are most abundant. In the troposphere, $H_2S$ rapidly oxidizes to $SO_2$, which commonly has a chemical lifetime of hours to days. If released into the stratosphere, the lifetime of $SO_2$ increases to about 3 weeks due to slower removal processes than in the troposphere. $SO_2$ is removed from the atmosphere via dry and wet deposition ('acid rain/snow') or oxidization to form sulfuric acid aerosol. Gas-phase oxidation of $SO_2$ by the hydroxyl radical (OH·) forms sulfuric acid ($H_2SO_4$) vapour, which is a low volatility compound that forms new particles ('nucleation') and/or rapidly condenses to grow existing particles to larger sizes. Within clouds, $SO_2$ undergoes aqueous-phase oxidation via reactions with dissolved hydrogen peroxide ($H_2O_2$) or ozone ($O_3$). Volcanic sulfuric acid particles are typically composed of 75% by weight of $H_2SO_4$ and 25% by weight of $H_2O$ (Bekki, 1995).

There has not been a large stratospheric injection of volcanic $SO_2$ since the 1991 Mount Pinatubo eruption, which released about 20 Tg of $SO_2$ into the stratosphere. For comparison, the current anthropogenic $SO_2$ flux is about 116 Tg per year. The 1982 El Chichón eruption injected ~ 7 Tg of $SO_2$ into the stratosphere. Between the years 2000 and 2012 there has been a series of small- to moderate-sized eruptions (VEI 3–4) injecting at the most ~ 1.3 Tg of $SO_2$ during an individual eruption (Nabro, 2011, in Eritrea). In comparison, individual decade-long CFB eruptive phases would have released up to 5,000 Tg of $SO_2$ (Self *et al.*, 2008). The Roza flow, the best-studied CFB eruptive phase, released about 1,200 Tg of $SO_2$ per year for a decade or longer into altitudes between 7 and 13 km (Thordarson and Self, 1996). For context, Laki was one of the largest historic flood basalt events and injected about 120 Tg of $SO_2$ in 8 months (Thordarson and Self, 2003, and references therein).

The annual sulfur flux from continuously degassing and sporadically erupting volcanoes is about 13 Tg, based on measurements at 49 volcanoes 'only' (Andres and Kasgnoc, 1998). The uncertainties on this estimate are large (about ± 50%); hence, there is also a large uncertainty on the magnitude of the radiative effects induced by these emissions (Graf *et al.*, 1997; Schmidt *et al.*, 2012a). With the advent of the satellite era, more numerous measurements of the volcanic sulfur flux have become available (e.g. Carn *et al.*, 2013) but the picture is still far from complete.

### 13.4 Atmospheric and climatic impacts of volcanic aerosol

Figure 13.1 illustrates how volcanic gases and particles affect atmospheric composition, chemistry and Earth's climate. Tephra is a major constituent of a volcanic cloud associated with explosive eruptions. Airborne tephra particles rapidly fall out of the atmosphere within minutes to days, hence their climatic effects are mostly negligible (Robock, 1981; Niemeier *et al.*, 2009). Continent-sized ash deposits produced by super-eruptions (VEI $\geq$ 8), however, may induce short-term climate

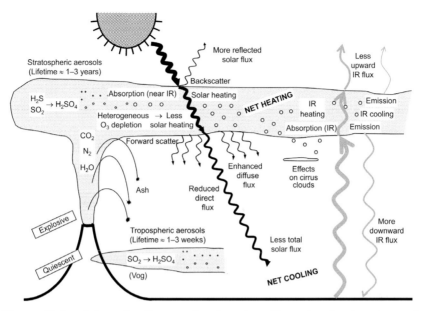

Figure 13.1. Schematic diagram showing the effect of volcanic gases and particles on the atmospheric composition and Earth's climate. Adapted from Plate 1 of Robock (2000).

change (Jones *et al.*, 2007), and Chapter 17 discusses the environmental impacts associated with ash deposition.

In contrast to airborne volcanic tephra, volcanic sulfuric acid aerosol particles can alter the radiative balance of Earth on timescales relevant for climate change due to their ability to scatter and absorb solar radiation (aerosol direct effect). Sulfuric acid aerosol scatters radiation across the entire solar spectrum due to its optical properties and typical particle radius of 0.5 μm (although there is a small degree of absorption in the near-infrared spectrum). Once in the stratosphere, volcanic sulfuric acid aerosol has a typical e-folding time of 9–12 months for tropical eruptions and 2–4 months for high-latitude eruptions (Kravitz and Robock, 2011). The top of a stratospheric aerosol cloud absorbs solar radiation in the near-infrared, whereas the lower stratosphere is heated by absorption of upward thermal infrared radiation (Stenchikov *et al.*, 1998). An aerosol cloud also forward scatters some of the incoming solar radiation, resulting in enhanced downward diffuse radiation, thereby increasing the sky's brightness. The net radiative effect of explosive volcanic eruptions is, however, a reduction in surface temperatures because the scattering exceeds the absorption efficiency.

In the case of the 1991 eruption of Mount Pinatubo, stratospheric aerosol concentrations remained elevated for about 2 years, and globally averaged stratospheric temperatures increased by around 2–3 K 3 months after the eruption

(Labitzke and McCormick, 1992). About 18 months following the June 1991 Mount Pinatubo eruption, the peak reduction in globally averaged lower-tropospheric temperatures reached 0.5 K, with climate models predicting a cooling of the same magnitude when including water-vapour feedbacks and removing the El Niño–Southern Oscillation signal (Soden *et al.*, 2002, and references therein).

Kirchner *et al.* (1999) showed that the aerosol formed after tropical eruptions results in an enhanced pole-to-equator heating gradient in the lower stratosphere that creates a stronger polar vortex and associated positive mode of the Arctic Oscillation in tropospheric circulation. This results in a warming over northern America, northern Europe and Russia, and a cooling over the Middle East, during the first winter, and sometimes the second winter, following a tropical eruption (Robock and Mao, 1992; Graf *et al.*, 1993). This indirect advective effect on temperature is stronger than the direct radiative effect that dominates at lower latitudes and during summer. Figure 13.2 highlights that while observations confirm this 'winter warming' response the majority of current climate models fail to reproduce the magnitude of this effect (Driscoll *et al.*, 2012).

Research on tropical and high-latitude eruptions revealed that feedbacks between the additional aerosol loading and atmospheric dynamics can weaken the Asian and African summer monsoon systems (Oman *et al.*, 2006b). For high-latitude eruptions in particular, the season of eruption is important for assessing the magnitude of the climatic response (Schmidt *et al.*, 2010; Kravitz and Robock, 2011). Trenberth and Dai (2007) showed that after the 1991 Mount Pinatubo eruption precipitation over land and continental freshwater discharge decreased significantly between October 1991 and September 1992. Following Robock and Liu (1994), Haywood *et al.* (2013) showed that the asymmetric aerosol loading between the northern and southern hemispheres after eruptions can influence the sea surface temperature gradient in the Atlantic, which in turn affects Sahelian precipitation rates (three of the four driest Sahelian summers between 1900 to 2010 were preceded by an eruption in the northern hemisphere).

Figure 13.1 also shows that aside from the aerosol direct effect, aerosol particles can also alter cloud amount and albedo by acting as cloud condensation nuclei or ice nuclei, thus modifying the optical properties and lifetime of clouds ('aerosol indirect effects'). Whether volcanic particles alter cirrus cloud properties has been the subject of several studies with contradicting results (e.g. Sassen, 1992; Luo *et al.*, 2002; see Robock *et al.*, 2013 for a review). In contrast, the ability of volcanic sulfates to act as cloud condensation nuclei and their indirect effect on low-level warm clouds has been confirmed by means of measurements

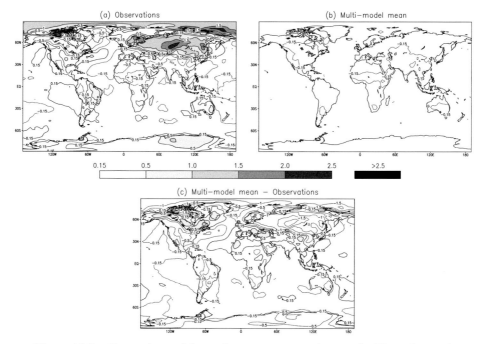

Figure 13.2    Comparison of the surface temperature changes (in K) as observed (a) and simulated by current climate models (b) averaged for two winter seasons after a major eruption in the tropics. The difference between modelled and observed surface temperature changes is shown in (c) with dashed contours (blue colours in colour version) indicating that modelled temperatures are too cold and grey-shaded contours (yellow colours in colour version) indicating that modelled temperatures are too warm compared to the observations. Figures reproduced based on data analysed and presented in Driscoll *et al.* (2012). A black and white version of this figure will appear in some formats. For the colour version, please refer to the plate section.

(e.g. Mather *et al.*, 2003), satellite retrievals (e.g. Gassó, 2008), and modelling (e.g. Graf *et al.*, 1997; Schmidt *et al.*, 2010, 2012a).

## 13.5  Volcano-climate interactions through time

Rampino and Self (1982) noted that eruption size does not correlate with the magnitude of the climatic effect. Pinto *et al.* (1989) used a one-dimensional microphysics model to demonstrate that the volcanic impact on climate becomes self-limited with increasing $SO_2$ release because microphysical processes such as coagulation cause particles to grow to large sizes, which have a lower optical depth per unit mass and fall out of the stratosphere faster. Timmreck *et al.* (2010) and English *et al.* (2013) simulated an eruption thought to be representative of the Youngest Toba Tuff eruption (74 ka) using three-dimensional microphysics models. These model simulations suggested a global mean temperature response about three times weaker (change of about −3.5 K) than previously estimated using climate models that did not consider microphysical processes. Pinto *et al.* (1989) and Bekki (1995) noted that oxidant depletion of OH· could prolong the climate response producing sulfuric acid aerosol more slowly and increasing the lifetime of $SO_2$, but Robock *et al.* (2009) suggested that this would be a small effect.

A reasonable rule of thumb might be that a minimum amount of 5 Tg of $SO_2$ injected into the stratosphere is likely to significantly alter Earth's radiative balance on timescales of months to years; however, there are complicating factors such as the season and latitude of an eruption as well as what temperature change we consider significant. In any case, simply extrapolating the magnitude of the climatic effects of present-day explosive volcanism to, for example, CFB volcanism is bound to be flawed due to the non-linear relationship between eruption magnitude and climatic effect, as well as the differences in eruption style.

### 13.5.1 *Ancient large-scale CFB eruptions*

Five Phanerozoic periods of CFB volcanism in particular have been the subject of a long-standing debate about their association with major environmental changes as evident from the proxy record of, for example, the abundances and diversity of planktonic foraminifera. However, the mechanisms by which CFB volcanism may have triggered these environmental changes remain enigmatic (for reviews see Officer *et al.*, 1987 and Wignall, 2001; see also Chapter 11).

Most studies on CFB volcanism suggested, by analogy to present-day volcanism, a short-term cooling effect (lasting years to decades) from $SO_2$ and the formation of sulfuric acid aerosol; together with a long-term warming effect (lasting tens to thousands of years) from the seemingly 'high' volcanic $CO_2$ emissions. However, the annual volcanic $CO_2$ flux during an eruptive phase equates to about 4% of the current annual anthropogenic flux (Section 13.3). To date, there has been only one carbon-cycle modelling study, and it found a negligible effect of volcanic $CO_2$ emissions from the Deccan Traps on Late Cretaceous temperatures (Caldeira and Rampino, 1990). Whether volcanic $CO_2$ affected long-term weathering rates and/or ocean acidity could be addressed in future using carbon-cycle models. Alternative explanations for the long-term warming trend observed at the time of, for example, the Siberian Traps revolve around the dissociation of gas hydrates (see also Chapters 10, 12 and 20). Release of isotopically light carbon from gas hydrates may explain the negative

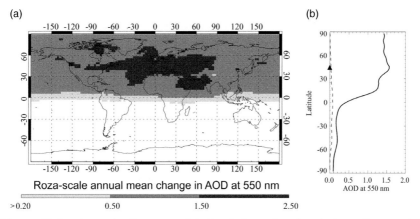

Figure 13.3    Aerosol–climate model simulation of a decade-long Roza-scale eruption (releasing 1200 Tg of $SO_2$ per year) with (a) showing the annual mean change in aerosol optical depth (AOD) at 550 nm for the tenth year of the eruption with respect to a pre-industrial control simulation, and (b) showing the latitudinal mean AOD for a pre-industrial control simulation (dashed) and the volcanically perturbed simulation (solid). The latitude of the eruption is indicated by the black triangle.

carbon isotope excursions at the end-Permian, rather than these excursions being due to the release of volcanic carbon (Wignall, 2001, and references therein).

For the 14.7 Ma decade-long Roza flow, Thordarson and Self (1996) estimated a change in aerosol optical depth (AOD) at 550 nm of between 7 and 13, which is at least 2.5 times larger than that for Toba (peak AOD of 2.6; *English et al.*, 2013). However, Thordarson and Self (1996) assumed that all $SO_2$ released is converted to sulfuric acid aerosol of a certain size. Accounting for oxidant availability and microphysical processes in a global model simulation of a Roza-scale eruption, A. Schmidt (unpublished data) found an annual mean AOD change of 1.2 when averaged over the northern hemisphere (Figure 13.3), which is significantly lower than previously estimated but nonetheless would have resulted in a substantial perturbation of climate if maintained for a decade. Since the climatic perturbations are relatively short-lived on geological timescales, A. Schmidt (unpublished data) suggested that better constraints on the frequency and duration of individual eruptions as well as non-eruptive phases are needed to fully quantify the magnitude and duration of the climatic and environmental effects (see also Chapter 11).

### 13.5.2 Post-2000 small- to moderate-sized eruptions (VEI 3–4)

The rate of global warming slowed between the years 2005 and 2012. Among other contributing factors, VEI 3–4 volcanic eruptions during that period contributed a small negative radiative forcing by increasing stratospheric aerosol concentrations (Solomon *et al.*, 2011; Vernier *et al.*, 2011; Neely *et al.*, 2013). Fyfe *et al.* (2013) estimated a global surface temperature response of at least –0.07°C.

The June 2011 eruption of Nabro (VEI 3–4) injected about 1.3 Tg of $SO_2$ into the upper troposphere (9–14 km), where seemingly the climate impacts would have been more muted than for an injection into the stratosphere. Bourassa *et al.* (2012), however, suggested that deep convection and interaction with the Asian summer monsoon lofted Nabro's $SO_2$ into the stratosphere, hence prolonging the aerosol radiative forcing.

These recent findings highlight the necessity for detailed observations of small-magnitude eruptions (VEI of 3 to 4) to allow for a better understanding of present-day climate change. Such findings might also change our view of what eruption magnitude and mass of $SO_2$ released into the atmosphere we need to consider relevant for climate change on decadal timescales, and also provide information of relevance for stratospheric geoengineering proposals with sulfuric acid aerosol.

## 13.6 Summary

Observations have confirmed that volcanic eruptions can have numerous impacts on our environment, mainly via the sulfuric acid aerosol direct effects, resulting in a reduction of surface temperatures, and changes in water-vapour concentrations and stratospheric heating rates. The latter results in pole-to-equator gradients that induce an indirect advective effect on surface temperatures referred to as winter warming. Transient and asymmetric volcanic aerosol forcings have been shown to weaken the Asian and African summer monsoons and to cause precipitation deficits in the Sahel. There is no observational evidence that eruptions force certain circulation modes such as El Niño but the issue remains a matter of debate. There is increasing evidence for an aerosol indirect effect on liquid water clouds induced by volcanic sulfuric acid particles but whether, and to what extent, explosive volcanic eruptions affect ice clouds remains debated.

Future eruptions will present an opportunity to test our current understanding and fill gaps in our ability to monitor volcanic clouds and to obtain detailed measurements. Climate models and observations have greatly enhanced our understanding of the climatic impacts of eruptions. However, we have not yet obtained sufficient observations of, for example, the temporal evolution of the aerosol size distribution in both the stratosphere and troposphere for different eruption sizes, which will be an acid test for climate and aerosol models.

## Acknowledgements

We thank Steve Self for a review that helped to improve this chapter. Anja Schmidt is funded by an Academic Fellowship from the School of Earth and Environment, University of Leeds. Alan Robock is supported by US National Science Foundation grants AGS-1157525 and CBET-1240507. We are very grateful to Simon Driscoll for providing the data for Figure 13.2.

## References

Andres, R. J. and A. D. Kasgnoc (1998), A time-averaged inventory of subaerial volcanic sulfur emissions, *J. Geophys. Res.*, **103**, 25251–25262.

Bekki, S. (1995), Oxidation of volcanic $SO_2$ – a sink for stratospheric OH and $H_2O$, *Geophys. Res. Lett.*, **22**, 913–916.

Bourassa, A. E., A. Robock, W. J. Randel *et al.* (2012), Large volcanic aerosol load in the stratosphere linked to Asian monsoon transport, *Science*, **337**, 78–81.

Bryan, S. E. and Ferrari, L. (2013), Large igneous provinces and silicic large igneous provinces: progress in our understanding over the last 25 years. *Geol. Soc. Amer. Bull.*, **125**, 1053–1078.

Caldeira, K., and M. R. Rampino (1990), Carbon dioxide emissions from deccan volcanism and a K/T boundary greenhouse effect, *Geophys. Res. Lett.*, **17**, 1299–1302.

Carn, S. A., N. A. Krotkov, K. Yang *et al.* (2013), Measuring global volcanic degassing with the Ozone Monitoring Instrument (OMI), *Geol. Soc., London, Spec. Publ*, **380**, 229–257.

Driscoll, S., A. Bozzo, L. J. Gray *et al.* (2012), Coupled Model Intercomparison Project 5 (CMIP5) simulations of climate following volcanic eruptions, *J. Geophys. Res.*, **117**, D17105, doi: 10.1029/2012JD017607.

English, J. M., O. B. Toon, and M. J. Mills (2013), Microphysical simulations of large volcanic eruptions: Pinatubo and Toba, *J. Geophys. Res.*, **118**, 1880–1895, doi: 10.1002/jgrd.50196.

Fyfe, J. C., von Salzen, K., Cole, J. N. S. *et al.* (2013), Surface response to stratospheric aerosol changes in a coupled atmosphere–ocean model, *Geophys. Res. Lett.*, **40**, 584–588.

Gassó, S. (2008), Satellite observations of the impact of weak volcanic activity on marine clouds, *J. Geophys. Res.*, **113**, D14S19, doi: 10.1029/2007JD009106.

Gerlach, T. (2011), Volcanic versus anthropogenic carbon dioxide, *Eos, Trans. Amer. Geophys. Union*, **92**, 201–202.

Graf, H.-F., I. Kirchner, A. Robock, and I. Schulte (1993), Pinatubo eruption winter climate effects: model versus observations, *Climate Dynam.*, **9**, 81–93.

Graf, H.-F., J. Feichter, and B. Langmann (1997), Volcanic sulfur emissions: estimates of source strength and its contribution to the global sulfate distribution, *J. Geophys. Res.*, **102**, 727–738.

Haywood, J. M., A. Jones, N. Bellouin *et al.* (2013), Asymmetric forcing from stratospheric aerosols impacts Sahelian rainfall, *Nature Clim. Change*, **3**, 660–665.

Jones, M., R. Sparks and P. Valdes (2007), The climatic impact of supervolcanic ash blankets, *Climate Dynam.*, **29**, 553–564.

Kirchner, I., G. L. Stenchikov, H.-F. Graf *et al.* (1999), Climate model simulation of winter warming and summer cooling following the 1991 Mount Pinatubo volcanic eruption, *J. Geophys. Res.*, **104**, 19,039–19,055.

Kravitz, B. and A. Robock (2011), The climate effects of high latitude volcanic eruptions: the role of the time of year. *J. Geophys. Res.*, **116**, D01105, doi: 10.1029/2010JD014448.

Kutterolf, S., T. H. Hansteen, K. Appel *et al.* (2013), Combined bromine and chlorine release from large explosive volcanic eruptions: a threat to stratospheric ozone?, *Geology*, **41**(6), 707–710.

Labitzke, K. and M. P. McCormick (1992), Stratospheric temperature increases due to Pinatubo aerosols, *Geophys. Res. Lett.*, **19**, 207–210.

Luo, Z., W. B. Rossow, T. Inoue *et al.* (2002), Did the eruption of the Mt. Pinatubo volcano affect cirrus properties?, *J. Climate*, **15**, 2806–2820.

Mather, T. A., D. M. Pyle, and C. Oppenheimer (2003), Tropospheric volcanic aerosol. In *Volcanism and The Earth's Atmosphere*, A. Robock and C. Oppenheimer, Eds., Geophysical Monograph 139 (American Geophysical Union, Washington, DC), 189–212.

Neely, R. R., O. B. Toon, S. Solomon *et al.* (2013), Recent anthropogenic increases in $SO_2$ from Asia have minimal impact on stratospheric aerosol, *Geophys. Res. Lett.*, **40**, 999–1004.

Newhall, C. G. and S. Self (1982), The Volcanic Explosivity Index (VEI): An estimate of explosive magnitude for historical volcanism, *J. Geophys. Res.*, **87**(C2), 1231–1238.

Niemeier, U., C. Timmreck, H. F. Graf *et al.* (2009), Initial fate of fine ash and sulfur from large volcanic eruptions, *Atmos. Chem. Phys.*, **9**, 9043–9057.

Officer, C.B., A Hallam, C.L. Drake *et al.* (1987), Late Cretaceous and paroxysmal Cretaceous–Tertiary extinctions, *Nature*, **326**, 143–149.

Oman, L., A. Robock, G. L. Stenchikov *et al.* (2006a), Modeling the distribution of the volcanic aerosol cloud from the 1783–1784 Laki eruption, *J. Geophys. Res.*, **111**, D12209, doi: 10.1029/2005JD006899.

Oman, L., A. Robock, G. L. Stenchikov *et al.* (2006b), High-latitude eruptions cast shadow over the African monsoon and the flow of the Nile, *Geophys. Res. Lett.*, **33**, L18711.

Pinto, J. P., R. P. Turco and O. B. Toon (1989), Self-limiting physical and chemical effects in volcanic eruption clouds, *J. Geophys. Res.*, **94**(D8), 11165–11174.

Rampino, M. R. and S. Self (1982), Historic eruptions of Tambora (1815), Krakatau (1883), and Agung (1963), their stratospheric aerosols, and climatic impact, *Quatern. Res.*, **18**, 127–143.

Robock, A. (1981), The Mount St. Helens volcanic eruption of 18 May 1980: minimal climatic effect, *Science*, **212**, 1383–1384.

Robock, A. (2000), Volcanic eruptions and climate, *Rev. Geophys.*, **38**, 191–219.

Robock, A. and J. Mao (1992), Winter warming from large volcanic eruptions, *Geophys. Res. Lett.*, **19**, 2405–2408.

Robock, A. and Y. Liu (1994), The volcanic signal in Goddard Institute for Space Studies three-dimensional model simulations. *J. Climate*, **7**, 44–55.

Robock, A., C. M. Ammann, L. Oman *et al.* (2009), Did the Toba volcanic eruption of ~ 74ka B.P. produce widespread glaciation? *J. Geophys. Res.*, **114**, D10107, doi: 10.1029/2008JD011652.

Robock, A., D. G. MacMartin, R. Duren *et al.* (2013), Studying geoengineering with natural and anthropogenic analogs. *Clim. Change*, **121**, 445–458.

Sassen, K. (1992), Evidence for liquid-phase cirrus cloud formation from volcanic aerosols: climatic implications, *Science*, **257**, 516–519.

Schmidt, A., K. S. Carslaw, G. W. Mann *et al.* (2010), The impact of the 1783–1784 AD Laki eruption on global aerosol formation processes and cloud condensation nuclei, *Atmos. Chem. Phys.*, **10**, 6025–6041, doi: 10.1029/2012JD018414.

Schmidt, A., K. S. Carslaw, G. W. Mann *et al.* (2012a), Importance of tropospheric volcanic aerosol for indirect radiative forcing of climate, *Atmos. Chem. Phys.*, **12**, 7321–7339.

Schmidt, A., T. Thordarson, L. D. Oman *et al.* (2012b), Climatic impact of the long-lasting 1783 Laki eruption: inapplicability of mass-independent sulfur isotopic composition measurements, *J. Geophys. Res.*, **117**, D23116.

Self, S., S. Blake, K. Sharma *et al.* (2008), Sulfur and chlorine in Late Cretaceous Deccan magmas and eruptive gas release, *Science*, **319**, 1654–1657.

Self, S., M. Widdowson, T. Thordarson *et al.* (2006), Volatile fluxes during flood basalt eruptions and potential effects on the global environment: a Deccan perspective, *Earth Planet. Sci. Lett.*, **248**, 518–532.

Soden, B. J., R. T. Wetherald, G. L. Stenchikov *et al.* (2002), Global cooling following the eruption of Mt. Pinatubo: a test of climate feedback by water vapor. *Science*, **296**, 727–730.

Solomon, S. (1999), Stratospheric ozone depletion: a review of concepts and history, *Rev. Geophys.*, **37**, 275–316.

Solomon, S., J. S. Daniel, R. R. Neely *et al.* (2011), The persistently variable "background" stratospheric aerosol layer and global climate change, *Science*, **333**, 866–870.

Solomon, S., R. W. Portmann, R. R. Garcia *et al.* (1998), Ozone depletion at mid-latitudes: coupling of volcanic aerosols and temperature variability to anthropogenic chlorine, *Geophys. Res. Lett.*, **25**, 1871–1874.

Steingrímsson, J. (1788), Fullkomid skrif um Sídueld (A complete description on the Sída volcanic fire), *Safn til Sögu Íslands*, Copenhagen, 1907–1915, pp. 1– 57.

Stenchikov, G. L., I. Kirchner, A. Robock *et al.* (1998), Radiative forcing from the 1991 Mount Pinatubo volcanic eruption. *J. Geophys. Res.*, **103**, 13,837–13,857.

Tabazadeh, A. and R. P. Turco (1993), Stratospheric chlorine injection by volcanic eruptions: HCl scavenging and implications for ozone, *Science*, **260**, 1082–1086.

Thordarson, T. and S. Self (1996), Sulfur, chlorine and fluorine degassing and atmospheric loading by the Roza eruption, Columbia River Basalt Group, Washington, USA, *J. Volcanol. Geotherm. Res.*, **74**, 49–73.

Thordarson, T. and S. Self (2003), Atmospheric and environmental effects of the 1783–1784 Laki eruption: a review and reassessment, *J. Geophys. Res.*, **108**(D1), 4011, doi: 10.1029/2001JD002042.

Timmreck, C. (2012), Modeling the climatic effects of large explosive volcanic eruptions, *Wiley Interdisc. Rev: Clim. Change*, **3**, 545–564.

Timmreck, C., H.-F. Graf, S. J. Lorenz *et al.* (2010), Aerosol size confines climate response to volcanic super-eruptions, *Geophys. Res. Lett.*, **37**, L24705, doi: 10.1029/2010GL045464.

Trenberth, K. E. and A. Dai (2007), Effects of Mount Pinatubo volcanic eruption on the hydrological cycle as an analog of geoengineering, *Geophys. Res. Lett.*, **34**, L15702, doi: 10.1029/2007GL030524.

Vernier, J. P., L. W. Thomason, J. P. Pommereau *et al.* (2011), Major influence of tropical volcanic eruptions on the stratospheric aerosol layer during the last decade, *Geophys. Res. Lett.*, **38**, L12807, doi: 10.1029/2011GL047563.

von Glasow, R. (2010), Atmospheric chemistry in volcanic plumes, *Proc. Natl. Acad. Sci. USA*, **107**, 6594–6599.

Wignall, P. B. (2001), Large igneous provinces and mass extinctions, *Earth-Sci. Rev.*, **53**, 1–33.

# 14

# Volcanic emissions: short-term perturbations, long-term consequences and global environmental change

TAMSIN A. MATHER AND DAVID M. PYLE

## 14.1 Introduction

Volcanism plays a vital role in the outgassing and cycling of major gases and other volatiles between the Earth's mantle and its surface reservoirs. As explored elsewhere in this book, volatile-cycling by volcanism is one of the key mechanisms by which processes *within* our planet drive change at the surface. Earth's surface response to volcanism has been suggested to be catastrophic in nature over both shorter ($10^1$–$10^2$ years, e.g. following Toba-scale explosive eruptions) and longer timescales ($10^5$–$10^6$ years e.g. large igneous provinces). However, volatile outgassing and recycling have also played a major role in the development and maintenance of Earth's atmosphere, contributing to the persistence of conditions on the planet required for life to emerge and evolve (e.g. Schaefer and Fegley, 2007). For example, on geological timescales it is hypothesised that atmospheric carbon dioxide ($CO_2$) concentration is regulated by a dynamic equilibrium between volcanic outgassing and consumption by chemical weathering of silicate minerals and deposition of carbonates (since ~ 200 Ma, e.g. Berner *et al.*, 1983). This stabilising mechanism has been invoked as one of the key attributes that allowed Earth to become and remain habitable (Kasting and Catling, 2003). However, proxy records suggest that $CO_2$ concentrations have varied significantly over the last 500 million years (e.g. Wallmann, 2001). Many explanations are proposed for this variability but significant fluctuations in the 'background' carbon flux from the world's volcanoes is one likely piece of the puzzle.

Magmatic gases are generally dominated by water and $CO_2$. However, in much of this chapter we focus on sulfur dioxide ($SO_2$): (a) due to its importance in Earth's radiative balance following conversion to sulfate aerosol (see also Chapter 13) and (b) because it is generally the best quantified volcanic gas species due to its low concentration in the background atmosphere, and the availability

*Volcanism and Global Environmental Change*, eds. Anja Schmidt, Kirsten E. Fristad and Linda T. Elkins-Tanton.
Published by Cambridge University Press. © Cambridge University Press 2015.

of straightforward spectroscopic (for plume gases; see also Chapters 6 and 8) and microanalytical (for rock samples) measurement techniques (see also Chapter 11). Where data are available we also discuss volcanic $CO_2$ because of its climatic importance. We explore the contrasting roles of volcanoes in maintaining and perturbing the atmosphere. First (Section 14.2), we discuss the 'background' flux of magma and volcanic gas to Earth's surface and atmosphere, and the significance of the different scales of activity that contribute to the long-term planetary degassing budget. We then discuss ways in which deviations from long-term average degassing patterns have been implicated in sustained periods of planetary change (Section 14.3). We draw on evidence both from present-day studies of active volcanism, and the geological record.

## 14.2 'Background' activity, short-term perturbations and time-averaging

Like many geological phenomena, volcanic activity spans a vast range of scales (Figure 14.1), both in terms of the typical return-interval for characteristic events (up to $\sim 10^5$–$10^6$ years for individual 'supereruptions' and up to $10^7$–$10^8$ years for large igneous provinces (LIPs)), and in terms of the size of individual eruptions (e.g. up to $1$–$2 \times 10^{16}$ kg for individual silicic 'supereruptions', or lava flows from a LIP; Mason *et al.*, 2004a; Bryan *et al.*, 2010). Since the largest events also tend to be the rarest, individual examples of these events may be significant short-term perturbations, but contribute relatively little to long-term average fluxes of magma and gas.

Long-term rates of magmatism and associated gas fluxes may be estimated using a range of different approaches, including direct observations, geological evidence (reconstruction of past eruptions), proxy records (e.g. volcanic deposition to ice cores), and experimental and modelling approaches. Table 14.1 summarises recent 'best-guess' estimates for current rates of terrestrial magmatism. The present-day long-term-averaged eruptive flux associated with subaerial volcanoes is of the order of $0.5$–$3$ km$^3$/yr, or $1,000$–$8,000$ Tg/yr of magma. One major area of uncertainty in these estimates is the balance between intrusive and extrusive fluxes, which is very poorly known.

Long-term average fluxes of gas could in principle be estimated from these magmatic fluxes, by using solubility and speciation models for volcanic gases in magmas that are calibrated against experimental data (e.g. Moretti *et al.*, 2003; Scaillet and Pichavant, 2003). However, these models are sensitive to assumptions about magma compositions and magma storage conditions prior to eruption, making it hard to generalise to a global volatile budget. Instead, efforts have focussed on using petrological evidence and experimental approaches to constrain emissions for individual eruptions (e.g. Chesner and Luhr, 2010). Rates of volcanic sulfur emission have also been inferred from analysis of ice-core records, and present-day remote-sensing measurements (e.g. Pyle *et al.*, 1996).

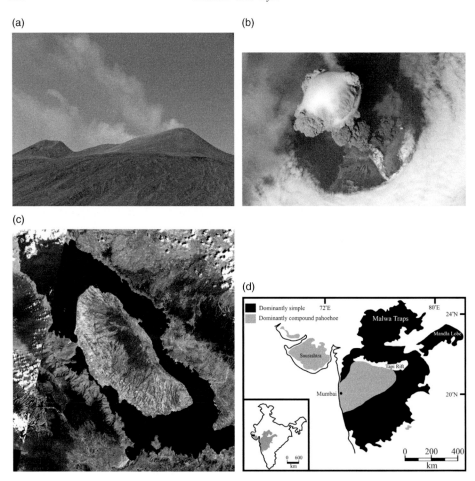

Figure 14.1   Photos illustrating different modes of volcanic emissions to our atmosphere. (a) Persistent 'background' emissions: the non-explosive plumes emanating from Mt Etna (Italy)'s North East Crater (cone on the right), Voragine and Bocca Nova (not visible) in August 2004 taken from the North East Flank. Etna is currently one of the major persistent volcanic gas sources both during and between eruptive episodes (photo: David Pyle). (b) Sporadic explosive eruptions: an early stage of the 12 June 2009 eruption of Sarychev Peak (Kuril Islands, northeast of Japan) taken from the International Space Station (photo: the ISS Crew Earth Observations experiment and Image Science & Analysis Laboratory, NASA Johnson Space Center). (c) The largest-scale explosive eruptions on Earth (M8): the Toba caldera (lake 30 by 100 km) in Sumatra, Indonesia, produced the Young Toba Tuff eruption 75,000 years ago, with an estimated volume of 2,800 km$^3$, the largest volcanic eruption in the last 2 million years. The caldera probably formed in stages (photo: ASTER image, 28 January 2006, NASA/ GSFC/METI/Japan Space Systems, and US/Japan ASTER Science Team). (d) Large igneous provinces: map showing the present-day extent of the subaerial Deccan Traps, India (inset shows their scale compared to India as a whole). Adapted from Brown *et al.* (2011).

Table 14.1 Estimates of global, time-averaged fluxes of magma in different tectonic settings.

(a) Estimates of global time-averaged magmatic fluxes (from Crisp, 1984)

| Setting | km³/yr, intruded | km³/yr, extruded | Tg/yr magma erupted subaerially[a] |
|---|---|---|---|
| Arc magmatism | 2.5–8 | 0.4–0.6 | 1,000–1,500 |
| Continental intraplate magmatism | 0.1–1.5 | 0.03–0.1 | 75–250 |
| Oceanic intraplate magmatism | 1.5–2 | 0.3–0.4 | 750–1,000 |
| Ocean ridge magmatism (submarine) | 18 | 3 | 0 |
| Total | 22–30 | 3.7–4.1 | 1,800–2,800 |

(b) Global magmatic fluxes in volcanic arcs

| Method | km³/yr | Tg/yr[a] |
|---|---|---|
| Erupted tephra flux | 0.4[b] | 1000 |
| Flux based on arc length and productivity | 0.3–1.7[c] | 750–4300 |
| Flux based on volcano productivity | 1.7–3.4[d] | 4,300–8,500 |

(c) Time-averaged volcanic output rates, by field or volcano (from White *et al.*, 2006)

| | km³/yr | Tg/yr[a] |
|---|---|---|
| Flood basalts (during eruptive periods) | $9 (\pm 2) \times 10^{-1}$ | $2,200 \pm 500$ |
| Oceanic volcanoes, other than flood basalts | $2.8 (\pm 0.4) \times 10^{-2}$ | $70 \pm 10$ |
| Continental volcanoes, other than flood basalts | $4.4 (\pm 0.8) \times 10^{-3}$ | $11 \pm 2$ |

[a] Assuming magma density of 2,500 kg/m³.
[b] Based on explosive eruption rates over past 2,000 years (Pyle, 1995).
[c] For a global arc length of 33,000 km and a time-averaged magmatic flux of (10–50) km³ per Myr per km of arc, based on compilations of volcano size (southern volcanic zone, Chile, past 500 kyr; Völker *et al.*, 2011; Central America, past 600 kyr; Carr *et al.*, 2007; accounting for cumulate masses and tephra, Kutterolf *et al.*, 2008), and studies of emplacement rates in the Sierra Nevada Batholith (Paterson *et al.*, 2011).
[d] Assuming 839 active arc volcanoes, and a magmatic flux per volcano of $(2–4) \times 10–3$ km³ per yr (Syracuse and Abers, 2006; White *et al.*, 2006).

Here we focus on emissions direct to the atmosphere rather than those from submarine activity. Table 14.2 summarises some of the characteristics of how different subaerial types of volcanism inject gas into the atmosphere. While the final environmental impacts of volcanic activity depend on many factors, here we

Table 14.2 *Examples of gas injections from some different types of volcanic emissions to our atmosphere. A similar analysis could have been done for volcanic products such as ash, but these are more difficult to quantify for flood basalt volcanism and have a shorter atmospheric lifetime than gaseous species so we use gas emissions to illustrate the differences in modes of emission.*

| Mode of volcanism/ volcanic event | Age of activity | Duration total | Estimate of total subaerial volatiles released (Tg except where indicated) | | Height of injection (km) | References |
|---|---|---|---|---|---|---|
| | | | $SO_2$ | $CO_2$ | | |
| Present-day persistent emissions | Present day | Continuous | 13–18 (Tg/yr) | 65–540 (Tg/yr) | In terms of $SO_2$: ~ 1–4 Tg/yr on average to stratosphere from 'large' eruptions, ~ 9 Tg/yr from passive degassing to the troposphere and rest from sporadic 'small' eruptions to troposphere. Passive emissions vary with volcano summit height (most > 1 km). | Stoiber et al. (1987); Andres and Kasgnoc (1998); Halmer et al. (2002); Pyle and Mather (2003); Mather et al. (2003); Burton et al. (2013). |
| Pinatubo (largest eruption of the satellite era) | 1991 | 9 h (climactic phase) | 12–26 (20 usually accepted value) | 42 | > 30 | Krueger et al. (1995); McCormick et al. (1995); Westrich and Gerlach (1992); Gerlach et al. (1996) |

| Name | Age/date | Duration | Volume (km³) | SO₂ mass (Mt) | Notes | References |
|---|---|---|---|---|---|---|
| Tambora (largest historic eruption) | 1815 | 5–6 d | 20–200; 50–58 (petrological) | 43 | | Oppenheimer (2003); Self et al. (2004). |
| Young Toba Tuff (largest quaternary eruption) | ~74 ka | 9–14 d | 35–3,300 | 32–40 | | Oppenheimer (2002); Chesner and Luhr (2010) |
| **Flood basalt volcanism** | | | | | | |
| Columbia River Basalt Group | ~17–6 Ma | | | | | |
| Grande Ronde Basalts | ~17–15.6 Ma | 0.4 Myr | $^a 1 \times 10^6$ | Not estimated | | Blake et al. (2010) |
| Roza Member | ~14.7 Ma | ~10 yr | ~12,420 | Not estimated | ~9,620 Mt SO₂ released at vents and lofted by the eruption columns to 7–13 km altitude | Thordarson and Self (1996) |
| Afro-Arabia province | ~29.5 Ma | 1–2 Ma | $^b 8.6–14.6 \times 10^5$ (just silicic products) | Not estimated | | Scaillet and Macdonald (2006) |
| Deccan Traps | ~68–60 Ma | Focussed in 4 Myr period with most activity within a 1 Myr period. Pulses of 10–100 yr with hiatuses of $10^3–10^4$ yr | $^c 3.5–6.5 \times 10^6$ | $^d 1.4 \times 10^7$ | Gas distributed throughout the atmosphere from the near surface to the tropopause and above | Self et al. (2006); Self et al. (2008) |

Table 14.2 (cont.)

| Mode of volcanism/ volcanic event | Age of activity | Duration total | Estimate of total subaerial volatiles released (Tg except where indicated) | | Height of injection (km) | References |
|---|---|---|---|---|---|---|
| | | | $SO_2$ | $CO_2$ | | |
| Paraná–Etendeka province | ~ 132 Ma | ~ 1 Myr | [e]$6.2$–$10.8 \times 10^6$ | Not estimated | | Marks et al. (2014) |
| Siberian Traps | ~ 252–250 Ma | Bulk of the eruption occurred during <1 Myr | ~ $12.6$–$68 \times 10^6$ | $8.5 \times 10^7$ | Uncertain but likely with at least some injection into the stratosphere | Black et al. (2012); Tang et al. (2013) |

[a] Delivered in intermittent bursts of < 1 to 30 Gt separated by long non-eruptive intervals.

[b] Just for the peralkaline silicic magmas associated with the flood basalts.

[c] This is based on an estimate of $10^6$ $km^3$ of lava for the Deccan. Evidence suggests that fluxes of $10^2$–$10^3$ Tg of $SO_2$ per year might have been sustained during pulses over a decade. Scaillet and Macdonald (2006) suggest a further $7.2$–$12.2 \times 10^3$ Tg $SO_2$ (based on observed silicic rocks associated with the Deccan) or $1.8$–$3.0 \times 10^6$ Tg $SO_2$ (assuming Deccan basalts were able to produce peralkaline derivatives at the same yield as for the Afro-Arabian province) may have been released from peralkaline rhyolites associated with the Deccan.

[d] Based on the Self et al. (2006) estimates for $10^3$ $km^3$ scaled up to $10^6$ $km^3$. They estimate 220–1,110 Tg/yr for 10–50 yr pulses of $10^3$ $km^3$.

[e] Scaillet and Macdonald (2006) suggest a further $9.2 \times 10^3$ Tg $SO_2$ for the metaluminous silicic magmas associated with the flood basalts.

use the amounts of gas released by different types of volcanic activity as an indication of the scale of these effects.

'Present-day' emissions of $SO_2$, measured at the world's volcanoes since the advent of ground- and satellite-based ultraviolet (UV) spectrometers in the late 1970s yield a time-averaged $SO_2$ flux of 13–18 Tg/yr (Table 14.2). This includes 'passive' emissions to the atmosphere, such as those from the ongoing 30-year-long eruption of Kīlauea (Hawai`i), where most degassing is from lava lakes or lava flows; and from sporadic explosive or non-explosive eruptions of varying sizes (e.g. Pinatubo and Hudson in 1991 and Nyiragongo in 2002). Previous studies estimating continuous $SO_2$ emissions (summarised in Table 14.2) all compile the contributions from these different types of volcanism over the modern period of observations and time-average in some way. For example, Pyle and Mather (2003) estimated that ~ 40% of the time-averaged flux was from continuous 'passive' degassing, ~ 40% from sporadic emissions from smaller explosive eruptions ($< 10^{13}$ kg tephra, 20–30 eruptions/yr), ~ 10% from sporadic emissions from non-explosive eruptions (1–2 eruptions/yr) and ~ 10% from sporadic emissions to the stratosphere from larger explosive eruptions ($> 10^{13}$ kg tephra, 1–2 eruptions/century).

The environmental consequences of these time-averaged emissions may range from local impacts on farmland and animal and human health from downwind fumigation by persistent, non-explosive degassing volcanoes (e.g. Masaya in Nicaragua; Delmelle *et al.*, 2002), to the global effects recorded in terms of temperature decrease and ozone depletion after sporadic but relatively large eruptions into the stratosphere (e.g. the 1991 eruption of Pinatubo). Table 14.2 summarises $SO_2$ emissions estimated from the Pinatubo (1991, largest eruption of the satellite era), Tambora (1815, largest historic eruption) and the Young Toba Tuff (~ 74 ka, largest Quaternary eruption) eruptions (see also Chapter 2). There is convincing historical evidence for the impacts of Tambora (Oppenheimer, 2003) and, although their degree is debated, Toba's environmental impacts were certainly widespread; unsurprising, given the short timescales of release and their wide dispersal (e.g. Petraglia *et al.*, 2012; Timmreck *et al.*, 2012).

While the short-term injection of $SO_2$ was significant in each of these cases, we can illustrate the contributions of eruptions of these sizes to 'background' planetary outgassing by considering the size–recurrence interval relationships. The Pinatubo eruption injected 12–26 Tg of $SO_2$ into the atmosphere, with 20 Tg often taken as the accepted value (McCormick *et al.*, 1995; Krueger *et al.*, 1995). Based on the geological record, a Pinatubo-sized eruption occurs on average about every 100 years (Pyle, 1995). Thus, the time-averaged $SO_2$ flux from Pinatubo-scale eruptions is about 0.2 Tg/yr. This is almost two orders of magnitude smaller than the measured present-day global volcanic $SO_2$ flux and shows that while Pinatubo was an important global perturbation on a $\sim$ 3-year timescale, on geological timescales, such eruptions are just a small part of the background outgassing of the planet. Sulfur emissions from eruptions such as Tambora and the Young Toba Tuff remain uncertain but have been estimated using petrological methods, or from proxy measurements of sulfate in ice cores (e.g. Zielinski, 2000). Tambora is estimated to have released 50–58 Tg of $SO_2$ over 5–6 days, while estimates for Toba range from 35–3,300 Tg of $SO_2$ over 9–14 days (Oppenheimer, 2003; Self *et al.*, 2004), with petrological estimates suggesting an upper limit of 200–400 Tg $SO_2$ (Chesner and Luhr, 2010). Based on the geological record, a Tambora-scale eruption occurs about every 1,000 years (Pyle, 1995) yielding a time-averaged $SO_2$ flux of $\sim$ 0.05–0.06 Tg/yr; insignificant over geological time compared to the flux from 'background' emissions activity. For Toba-scale eruptions (M8–M9) the geological record suggests that the recurrence rate drops sharply, to approximately one every 1 Myr (based on statistics for the last 13.5 Myr; Mason *et al.*, 2004a). Thus, $SO_2$ emissions from Toba-scale eruptions account for less than $\sim 3 \times 10^{-3}$ Tg/yr. In summary, while large-scale short-lived eruptions are undoubtedly important global perturbations on timescales of years to decades, over geological time (e.g. time-averaging over $> 10^5$–$10^6$ years) sporadic, individual short-lived eruptions are a relatively insignificant part of the background outgassing budget of the planet (see also Chapter 11).

Short-lived explosive eruptions, however, dominate the time-averaged flux of volcanic $SO_2$ emissions to the stratosphere and ash into the environment. The time-averaged fluxes of ash and gas from explosive eruptions of different magnitudes are illustrated in Figure 14.2. The interplay between magma composition and plume height (more basaltic magmas are able to carry more sulfur but tend to have lower plumes) means that medium-sized (M4) eruptions dominate the flux of volcanic S to the stratosphere; whilst tephra flux scales with eruption magnitude until the drop off in recurrence rate for M8 and 9 eruptions.

The present-day volcanic $CO_2$ flux is not yet particularly well constrained. Marty and Tolstikhin (1998) suggested a preferred range of 180–440 Tg/yr (Gerlach, 2011). More recently Burton *et al.* (2013) compiled estimates in the range 65–540 Tg/yr, with the larger estimates including emissions from volcanic lakes and tectonically and hydrothermally active areas (Burton *et al.*, 2013). Some of this flux might be supplied from thermal and metamorphic reactions of basement rocks (e.g. limestones) with magma (e.g. Parks *et al.*, 2013), but the size of this contribution is not known. The submarine volcanic $CO_2$ flux is likely to be substantial ($\sim$ 100 Tg/yr), but the current view is that alteration of ocean crust currently consumes more C (precipitated as carbonates) than is released at submarine vents (e.g. Burton *et al.*, 2013). Estimates of $CO_2$ emissions from sporadic large eruptions are sparse due to the low solubility of $CO_2$ in

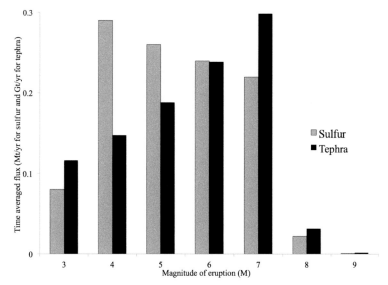

Figure 14.2   The time-averaged volcanic $SO_2$ flux to the stratosphere (Mt/yr) and erupted tephra flux (Gt/yr) from explosive volcanic eruptions of different size (magnitude, M; Pyle *et al.*, 1996). The tephra flux is calculated from the rates of explosive volcanism over the past 2,000 years (Pyle, 1995), while the $SO_2$ flux is calculated from the relationship between eruption size (M) and $SO_2$ release determined for eruptions with M3–6, adjusted for the amount that reaches the stratosphere (Pyle *et al.*, 1996). Uncertainty in the sulfur emission rate is likely to be of the order of a factor of 2. Gas emissions decline slightly from M4–7, because smaller (M3, 4) eruptions tend to be more S-rich per unit mass of magma (basalts, andesites) than (M6, 7) eruptions (dacites, rhyolites). Both fluxes drop off for large magnitude eruptions (M8, 9), reflecting their much lower recurrence rates (Mason *et al.*, 2004a).

magma, making petrological methods challenging to apply; and high concentrations in the background atmosphere hinder reliable spectroscopic quantification. Gerlach *et al.* (1996) estimated that the 1991 Pinatubo eruption released 42 Tg $CO_2$, about an order of magnitude lower than the current best estimates for the global annual flux of $CO_2$ from 'background' volcanism. Time-averaging over the $\sim$ 100-year recurrence rate of a Pinatubo-scale eruption, this flux is insignificant ($\sim$ 0.4 Tg/yr).

Although there is clear evidence for global effects following eruptions of the scale of Tambora and Toba, recent modelling studies also suggest that global temperature perturbations following this type of volcanism recover on decadal timescales, without tipping the Earth systems into any different mode of behaviour (Jones *et al.*, 2005; Robock *et al.*, 2009; Timmreck *et al.*, 2010). In the next section we consider the modes of volcanism that may have the potential to cause prolonged and significant perturbation to the Earth system.

## 14.3 Potential prolonged volcanic perturbations to the Earth system

As described above there is no evidence as yet that individual large explosive eruptions cause significant prolonged perturbations to the Earth system when considered on typical geological timescales. However, some periods of geological history are characterised by significant deviations from 'background' volcanic behaviour. These may result from some configuration of factors within the planet's interior causing a fundamental change in the rate or nature of volcanism, or may be a response to feedbacks with surface processes. We shall consider four ways that deviations from 'background' volcanic activity might significantly affect the Earth system on geological timescales: (i) temporal clustering of individual explosive eruptions; (ii) long-term persistently elevated rates of volcanism originating from deep in the planet's interior, namely LIP volcanism; (iii) changes in rates and/or style of volcanism in response to direct feedbacks from the surface environment; and (iv) systematic and significant perturbations to the magma volatile load and composition of the gases emitted by volcanism.

### (i) Temporal clustering of individual explosive eruptions

Although we can estimate recurrence times from the geological record for explosive eruptions of different magnitudes these values are just averages, and the stochastic nature of volcanic activity means that explosive eruptions and stratospheric $SO_2$ inputs may cluster purely by chance. For example, Miller *et al.* (2012) suggest that decadally-paced explosive volcanism over a $\sim$ 50-year period and sea-ice/ocean feedbacks might be linked to the onset of the Little Ice Age ($\sim$ AD 1350–1850).

In some tectonic settings, there is also evidence for clusters of large-magnitude explosive eruptions of silicic magma that together form large-scale silicic 'flare ups'. Prominent examples include the Altiplano–Puna complex of the central Andes (10–1 Ma; de Silva and Grosnold, 2007), and the Southern Rocky Mountain volcanic field of Colorado and New Mexico (38–23 Ma; Lipman, 2007). These 'flare-ups' can be highly productive, with total volumes of erupted silicic magma exceeding $10^4$ km$^3$ (Altiplano–Puna), erupted at a time-averaged rate typically $\sim 10^{-3}$–$10^{-2}$ km$^3$/yr (de Silva and Gosnold, 2007). From the records of known large explosive eruptions, Mason *et al.* (2004a) identified two pulses of 'super-eruptions': from 36 to 25 Ma, and 13.5 Ma to present. During these periods, the mean interval between M 8 and larger eruptions may have been dropped to $< 10^5$ years. Even at this higher recurrence rate, the time-averaged $SO_2$ emissions from Toba-scale events would be insignificant relative to background fluxes. To date, only limited evidence has been found linking such 'flare-ups' to prolonged and significant perturbations to the Earth system (Cather *et al.*, 2009).

## (ii) Continental LIPs and flood volcanism

Flood basalt volcanism releases large quantities of gas over prolonged periods of time, as well as being associated with huge outpourings of lava and ash onto the surface of the planet. The link between this activity and evidence of global change, the style of eruption and its potential impacts are explored in detail elsewhere in this book (e.g. Chapter 11) but here we consider the significance of the estimated gas emissions compared to other modes of volcanism.

Studies of continental flood basalt provinces show that while the emplacement of the provinces as a whole may span a million years or so, during this period there will be pulses of heightened activity associated with the emplacement of individual flow units (see also Chapter 11). As illustrated in Table 14.2, estimated $SO_2$ emissions for entire individual flood basalt provinces range from $\sim 3 \times 10^6$ to $7 \times 10^7$ Tg $SO_2$. Averaged over a timescale of $\sim 1$ Myr, which reflects their broad duration, this yields time-averaged $SO_2$ fluxes of 3.5–6.5 Tg/yr for the Deccan Traps, 6.2–10.8 Tg/yr for the Paraná–Etendeka and 12.6–68 Tg/yr for the Siberian Traps. These fluxes represent a significant enhancement (20–200%) compared to the 'present-day background' continuous volcanic flux. The pulses of heightened activity thought to characterise the emplacement of these provinces means that these broad enhancements include phases of highly elevated flux separated by hiatuses. For example, the Roza flow of the Columbia River Basalt Group may have been emplaced in about a decade (Thordarson and Self, 1996) yielding an average $SO_2$ flux of 1,242 Tg/yr during this period. Similarly, fluxes of 100–1,000 Tg/yr of $SO_2$ probably characterised decadal-scale pulses of Deccan volcanism (Self *et al.*, 2006). These fluxes would overwhelm global persistent emissions similar to those seen today, increasing global volcanic sulfur emissions by 10- to 100-fold during these pulses of activity.

Clusters of large-magnitude silicic eruptions may also be associated with continental LIP provinces (e.g. Paraná–Etendeka and Afro-Arabian LIPs, Bryan *et al.*, 2010). Like the 'flare-ups', these provinces can be highly productive, with total volumes of erupted silicic magma up to $\sim 5 \times 10^4$ km$^3$ (Paraná–Etendeka), and a time-averaged rate of silicic magmatism exceeding 0.05 km$^3$/yr (Bryan *et al.*, 2010). The potential emissions from these silicic provinces may well have been significant. Scaillet and Macdonald (2006) used experiments to investigate the solubility of sulfur in peralkaline and metaluminous silicic magmas representative of this type of activity, and found that the origin of the silicic magma is critical to its capacity to deliver large quantities of volatiles to the atmosphere. In particular, peralkaline rhyolites that are the products of extensive fractional crystallisation of basic magma in extensional rift settings like Ethiopia can carry significant amounts of dissolved sulfur, and yield substantial emissions. Scaillet and Macdonald (2006) estimate that the 60,000 km$^3$ rhyolites from the Afro-Arabian province may have released as much as 9–15 $\times$ 10$^5$ Tg SO$_2$. In contrast, the metaluminous silicic magmas from the Paraná–Etendeka province, likely the products of crustal melting, may not have been volatile saturated at depth, and had much less capacity to carry dissolved sulfur (estimated total release of up to 1 $\times$ 10$^4$ Tg SO$_2$ from the 20,000 km$^3$ rhyolites from the Paraná province).

There are limited estimates of carbon dioxide emissions for continental flood basalt volcanism (Table 14.2). Self *et al.* (2006) estimate emissions during 1,000 km$^3$ pulses of Deccan activity lasting 10–50 years at 1.4–11 Pg CO$_2$ in total, or ~ 220–1,100 Tg/yr. In contrast to SO$_2$, this CO$_2$ injection is trivial compared to the atmospheric reservoir (currently ~ 3 $\times$ 10$^6$ Tg CO$_2$ and likely higher at the end of the Cretaceous), and hundreds to thousands of years between eruptive episodes would allow the atmosphere sufficient time to re-equilibrate (Self *et al.*, 2006).

### (iii) Feedbacks from the Earth's surface modulating the rate/style of volcanism

While flood basalt volcanism may be the most spectacular symbol of the potential link between volcanism and global change, more subtle long-term changes in the rates of global volcanism have also been proposed to perturb the global environment over Earth history. The major feedbacks that have been explored are those between the hydrological cycle and volcanism. Although short (annual) timescale feedbacks have been proposed (Mason *et al.*, 2004b), feedbacks on longer time-scales such as glacial cycles have received more attention and have more relevance in terms of the links between volcanism and global change.

Many studies have discussed the possible coupling between volcanism and glaciation. In the 1970s, Bray (1977) postulated that volcanism might have driven Pleistocene glaciations. The reverse hypothesis, that rapid climate change leading to ice-retreat, glacial unloading, and sea-level change might trigger volcanic activity, was proposed by Rampino *et al.* (1979). Subsequent work has revealed a number of more- or less-compelling examples where this hypothesis might hold. There is strong evidence for a burst of subaerial volcanic eruptions in Iceland shortly after the last glacial maximum (LGM) with time-averaged eruption rates several times higher than at present (Sigvaldason *et al.*, 1992; Maclennan *et al.*, 2002). There is also weak evidence for increased continental eruption rates in both northern Europe and eastern California following deglaciation, over several glacial cycles (Glazner *et al.*, 1999; Nowell *et al.*, 2006). Each of these settings, however, features volcanism that results from decompression melting of the mantle, and where decompression following ice removal may lead to increased mantle melt production (e.g. Jull and McKenzie, 1996; Pagli and Sigmundsson, 2008).

In contrast, evidence from the post-glacial explosive eruption histories of southern Chile, Kamchatka and the Cascades shows no statistically significant enhancement in rates of volcanism following deglaciation at arc volcanoes (Watt *et al.*, 2013). There is though a suggestion that some of the largest post-glacial eruptions in southern Chile occurred shortly after ice retreat, consistent with magma ponding beneath glaciated volcanoes and released once the ice has been removed (Watt *et al.*, 2013).

Modulation of volcanic activity, and degassing, by glacial cycles may mean that non-arc volcanoes accounted for a relatively higher proportion of global volcanic emissions in the early post-glacial period than at present-day (Watt *et al.*, 2013). However, the global significance of these changes is still under debate. Huybers and Langmuir (2009) suggested that global volcanic activity increased, perhaps as much as six-fold, from 12 to 7 ka, due to enhanced eruption rates in formerly glaciated regions. They postulated that associated increases in volcanic emissions may have contributed significantly to the post-glacial increase in atmospheric $CO_2$. However, modelling studies of the oceanic carbon cycle (e.g. Roth and Joos, 2012) find only a small role for volcanic forcing in post-glacial $CO_2$, while a more detailed analysis of volcanic eruption records (Watt *et al.*, 2013) shows sparser evidence for a post-glacial volcanic 'pulse', with a maximum two-fold increase in global eruption rates, relative to the present day, between 13 and 7 ka. This suggests that although volcanism may have been an important source of $CO_2$ in the early Holocene, it is unlikely to have been a dominant control on changes in atmospheric $CO_2$ after the LGM.

Other studies have argued that rapid post-glacial sea-level rise (Nakada and Yokose, 1992) and fall (Rampino *et al.*, 1979; Wallmann *et al.*, 1988) may have led to enhanced volcanic eruption rates (McGuire *et al.*, 1997). However, the data needed to prove these hypotheses remain incomplete, and the evidence for an association is still weak. More recently, Lund and Asimow (2011) have suggested that since mantle decompression rates induced by changes in sea level during glacial–interglacial cycles may approach the same order of magnitude as those due to plate spreading, there may be significant sea-level driven variations in submarine magma flux. This remains to be tested.

### *(iv) Perturbations to magma volatile load and the composition of the volcanic gases emitted*

Other chapters in this book deal with major perturbations to the volatile load or composition of flood basalt magmatism (Chapters 10–12). These chapters discuss potentially enhanced fluxes of volatiles, for example, from the recycled oceanic crustal component of the mantle plume (Sobolev *et al.*, 2011), and release of volatiles from sedimentary rocks by magmatic heating (Ganino and Arndt, 2009). Analogous long-term consequences of variability in the composition of subduction-zone or other non-LIP magmatic gases due to variations in the arc crustal compositions over geological time have yet to be explored. Other links between volcanic gas composition and factors such as mantle oxidation state, volatile cycling and average degassing pressure (e.g. Holland, 2002; 2009; Gaillard *et al.*, 2011) have been linked to the oxygenation of Earth's atmosphere; however, our understanding of the exact nature of such controls on gas composition remains incomplete.

It has been suggested that subduction of carbonate-rich sediments (analogous to Central America today) could be linked to increased volcanic $CO_2$ outgassing and periods of global warming. The deep sources of magmatic gases at subduction zones include both volatiles from the mantle wedge, and components brought into the mantle with the downgoing slab. For example, carbon may be carried into the mantle in sediments (carbonates, or organic C) or in the hydrothermally altered basaltic crust, and serpentinised peridotite. A poorly constrained portion of this C is released by progressive metamorphism during subduction (Gorman *et al.*, 2006). Therefore, the closure of the Tethys Ocean and subduction of large volumes of marine carbonate has been postulated as a major factor contributing to late Cretaceous to early Cenozoic elevated atmospheric $CO_2$ concentrations (Edmond and Huh, 2003; Kent and Muttoni, 2008; Johnston *et al.*, 2011). Without a better understanding of the factors controlling the efficiency of C recycling through subduction zones, however, these ideas remain poorly understood (Gorman *et al.*, 2006).

## 14.4 Summary

This chapter is intended to put the discussions of specific modes of volcanic impacts on Earth's atmosphere into context. Even the largest short-lived explosive eruptions contribute little to atmospheric composition on geological timescales, even though they represent significant perturbations on decadal timescales. Due to their scale and prolonged nature, LIPs are one manifestation of volcanism that has the potential to impact the Earth's atmosphere on geological timescales ($10^4$–$10^5$ yr), comparable to the carbon-cycle perturbations such as ocean anoxic events. There is still much to be done to understand how these significant and prolonged LIP events may affect the global environment. Equally, much remains to be done to understand more subtle feedbacks between volcanism and the environment, such as changes in glacial loading; and the consequences of spatially and temporally varying fluxes of volatile elements (including carbon and halogens) through the world's subduction zones. Improving our global geological eruption records as well as our understanding of volcanic environmental impacts, and the factors controlling both the composition of volcanic emissions and volatile recycling between the surface and mantle on the present-day Earth, are key to furthering our understanding of volcanism's role in shaping our and other planets.

# References

Andres, R.J. and Kasgnoc, A.D. (1998). A time-averaged inventory of subaerial volcanic sulfur emissions. *Journal of Geophysical Research*, **103**, 25251–25261.

Berner, R.A., Lasaga, A.C. and Garrels, R.M. (1983). The carbonate–silicate geochemical cycle and its effect on atmospheric carbon dioxide over the past 100 million years. *American Journal of Science*, **283**, 641–683.

Black, B.A., Elkins-Tanton, L.T., Rowe, M.C. and Peate I.U. (2012). Magnitude and consequences of volatile release from the Siberian Traps. *Earth and Planetary Science Letters*, **317–318**, 363–373.

Blake, S., Self, S., Sharma, K. and Sephton, S. (2010). Sulfur release from the Columbia River Basalts and other flood lava eruptions constrained by a model of sulfide saturation. *Earth and Planetary Science Letters*, **299**, 328–338.

Bray, J.R. (1977). Pleistocene volcanism and glacial initiation. *Science*, **197**, 251–254.

Brown, R.J., Blake, S., Dondre, N.R., Phadnis, V.M. and Self, S. (2011). 'A'ā lava flows in the Deccan Volcanic Province, India, and their significance for the nature of continental flood basalt eruptions. *Bulletin of Volcanology*, **73**, 737–752.

Bryan, S.E., Peate, I.U., Peate, D.W. *et al.* (2010). The largest volcanic eruptions on Earth. *Earth-Science Reviews*, **102**, 207–229

Burton, M.R., Sawyer, G.M. and Granieri, D. (2013). Deep carbon emissions from volcanoes. *Reviews in Mineralogy & Geochemistry*, **75**, 323–354.

Carr, M.J., Saginor, I., Alvarado, G.E. *et al.* (2007). Element fluxes from the volcanic front of Nicaragua and Costa Rica. *Geochemistry Geophysics Geosystems*, **8**, Q06001.

Cather, S.M., Dunbar, N.W., McDowell, F.W., McIntosh, W.C. and Scholle, P.A. (2009). Climate forcing by iron fertilization from repeated ignimbrite eruptions: the icehouse–silicic large igneous province (SLIP) hypothesis. *Geosphere*, **5**, 315–324.

Chesner, C.A. and Luhr, J.F. (2010). A melt inclusion study of the Toba Tuffs, Sumatra, Indonesia. *Journal of Volcanology and Geothermal Research*, **197**, 259–278.

Crisp, J.A. (1984). Rates of magma emplacement and volcanic output. *Journal of Volcanology and Geothermal Research*, **20**, 177–211.

Delmelle, P., Stix, J., Baxter, P.J., Garcia-Alvarez, J. and Barquero, J. (2002). Atmospheric dispersion, environmental effects and potential health hazard associated with the low-altitude gas plume of Masaya volcano, Nicaragua. *Bulletin of Volcanology*, **64**, 423–434.

de Silva, S.L. and Gosnold, W.D. (2007). Episodic construction of batholiths: insights from the spatiotemporal development of an ignimbrite flare-up. *Journal of Volcanology and Geothermal Research*, **167**, 320–335.

Edmond, J.M. and Huh, Y. (2003). Non-steady state carbonate recycling and implications for the evolution of atmospheric P-$CO_2$. *Earth and Planetary Science Letters*, **216**, 125–139.

Gaillard, F., Scaillet, B. and Arndt, N.T. (2011). Atmospheric oxygenation caused by a change in volcanic degassing pressure. *Nature*, **478**, 229–232.

Ganino, C. and Arndt, N.T. (2009). Climate changes caused by degassing of sediments during the emplacement of large igneous provinces. *Geology*, **37**, 323–326.

Gerlach, T.M. (2011). Volcanic versus anthropogenic carbon dioxide. *EOS, Transactions AGU*, **92**, 201–202.

Gerlach, T.M., Westrich, H.R. and Symonds, R.B. (1996). Preeruption vapor in magma of the climactic Mount Pinatubo eruption: source of the giant stratospheric sulfur dioxide cloud. In *Fire and Mud: Eruptions and Lahars of Mount Pinatubo, Philippines*, ed. C.G. Newhall and R.S. Punongbayan, University of Washington Press, 415–433.

Glazner, A.F., Manley, C.R., Marron, J.S. and Rojstaczer, S. (1999). Fire or ice: anti-correlation of volcanism and glaciation in California over the past 800,000 years. *Geophysical Research Letters*, **26**, 1759–1762.

Gorman, P.J., Kerrick, D.M. and Connolly, J.A.D. (2006). Modeling open system metamorphic decarbonation of subducting slabs. *Geochemistry Geophysics Geosystems*, **7**, Q04007.

Halmer, M.M., Schmincke, H.U. and Graf, H.F. (2002). The annual volcanic gas input into the atmosphere, in particular into the stratosphere: a global data set for the past 100 years. *Journal of Volcanology and Geothermal Research*, **115**, 511–528.

Holland, H.D. (2002). Volcanic gases, black smokers, and the Great Oxidation Event. *Geochimica et Cosmochimica Acta*, **66**, 3811–3826.

Holland, H.D. (2009). Why the atmosphere became oxygenated: a proposal. *Geochimica et Cosmochimica Acta*, **73**, 5241–5255.

Huybers, P. and Langmuir, C. (2009). Feedback between deglaciation, volcanism, and atmospheric $CO_2$. *Earth and Planetary Science Letters*, **286**, 479–491.

Johnston, F.K.B., Turchyn, A.V. and Edmonds, M. (2011). Decarbonation efficiency in subduction zones: implications for warm Cretaceous climates. *Earth and Planetary Science Letters*, **303**, 143–152.

Jones, G.S., Gregory, J.M., Stott, P.A., Tett, S.F.B. and Thorpe, R.B. (2005). An AOGCM simulation of the climate response to a volcanic super-eruption. *Climate Dynamics*, **25**, 725–738.

Jull, M. and McKenzie, D. (1996). The effect of deglaciation on mantle melting beneath Iceland. *Journal of Geophysical Research*, **101**, 21815–21828.

Kasting, J.F. and Catling, D. (2003). Evolution of a habitable planet. *Annual Review of Astronomy and Astrophysics*, **41**, 429–463.

Kent, D.V. and Muttoni, G. (2008). Equatorial convergence of India and early Cenozoic climate trends. *Proceedings of the National Academy of Sciences*, **105**, 16065–16070.

Krueger, A.J., Walter, L.S., Bhartia, P.K. *et al.* (1995). Volcanic sulfur dioxide measurements from the total ozone mapping spectrometer instruments. *Journal of Geophysical Research*, **100**, 14,057–14,076.

Kutterolf, S., Freundt, A. and Perez, W. (2008). Pacific offshore record of Plinian arc volcanism in Central America: 2. Tephra volumes and erupted masses. *Geochemistry Geophysics Geosystems,* **9**, Q02S02.

Lipman, P.W. (2007). Incremental assembly and prolonged consolidation of Cordilleran magma chambers: evidence from the Southern Rocky Mountain volcanic field. *Geosphere*, **3**, 42–70.

Lund, D.C. and Asimow, P.D. (2011). Does sea level influence mid-ocean ridge magmatism on Milankovitch timescales? *Geochemistry Geophysics Geosystems*, **12**, Q12009.

Maclennan, J., Jull, M., McKenzie, D., Slater, L. and Gronvold, K. (2002). The link between volcanism and deglaciation in Iceland. *Geochemistry Geophysics Geosystems*, **3**, 1062–1087.

Marks, L., Keiding, J., Wenzel, T. *et al.* (2014). F, Cl, and S concentrations in olivine-hosted melt inclusions from mafic dikes in NW Namibia and implications for the environmental impact of the Paraná–Etendeka Large Igneous Province. *Earth and Planetary Science Letters*, **392**, 39–49.

Marty, B. and Tolstikhin, I.N. (1998). $CO_2$ fluxes from mid-ocean ridges, arcs and plumes. *Chemical Geology*, **145**, 233–248.

Mason, B.G., Pyle, D.M. and Oppenheimer, C. (2004a). The size and frequency of the largest eruptions on Earth. *Bulletin of Volcanology*, **66**, 735–748.

Mason, B.G., Pyle, D.M., Dade, W.B. and Jupp, T. (2004b). Seasonality of volcanic eruptions. *Journal of Geophysical Research*, **109**, B04206.

Mather, T.A., Pyle, D.M. and Oppenheimer, C. (2003). Tropospheric volcanic aerosol. *AGU Geophysical Monograph*, **139**, 189–212.

McCormick M.P., Thomason, L.W. and Trepte, C.R. (1995). Atmospheric effects of the Mt. Pinatubo eruption. *Nature*, **373**, 399–404.

McGuire, W.J., Howarth, R.J., Firth, C.R. *et al.* (1997). Correlation between rate of sea-level change and frequency of explosive volcanism in the Mediterranean. *Nature*, **389**, 473–476.

Miller, G.H., Geirsdóttir, Á., Zhong, Y.F. *et al.* (2012). Abrupt onset of the Little Ice Age triggered by volcanism and sustained by sea-ice/ocean feedbacks. *Geophysical Research Letters*, **39**, L02708.

Moretti, R., Papale, P. and Ottonello, G. (2003). A model for the saturation of C–O–H–S fluids in silicate melts. *Geological Society of London Special Publication*, **213**, 81–101.

Nakada, M. and Yokose, H. (1992). Ice-age as a trigger of active quaternary volcanism and tectonism. *Tectonophysics*, **212**, 321–329.

Nowell, D.A.G., Jones, M.C. and Pyle, D.M. (2006). Episodic Quaternary volcanism in France and Germany. *Journal of Quaternary Science*, **21**, 645–675.

Oppenheimer, C. (2002). Limited global change due to the largest known Quaternary eruption, Toba ~ 74 kyr BP? *Quaternary Science Reviews*, **21**, 1593–1609.

Oppenheimer, C. (2003). Climatic, environmental and human consequences of the largest known historic eruption: Tambora volcano (Indonesia) 1815. *Progress in Physical Geography*, **27**, 230–259.

Pagli, C. and Sigmundsson, F. (2008). Will present day glacier retreat increase volcanic activity? Stress induced by recent glacier retreat and its effect on magmatism at the Vatnajokull ice cap, Iceland. *Geophysical Research Letters*, **35**, L09304.

Parks, M.M., Caliro, S., Chiodini, G. *et al.* (2013). Distinguishing contributions to diffuse $CO_2$ emissions in volcanic areas from magmatic degassing and thermal decarbonation using soil gas $^{222}$Rn–$\delta^{13}$C systematics: application to Santorini volcano, Greece. *Earth and Planetary Science Letters*, **377–378**, 180–190.

Paterson, S.R., Okaya, D., Memeti, V., Economos, R. and Miller, R.B. (2011). Magma addition and flux calculations of incrementally constructed magma chambers in continental margin arcs: combined field, geochronologic, and thermal modeling studies. *Geosphere*, **7**, 1439–1468.

Petraglia, M., Ditchfield, P., Jones, S., Korisettar, R. and Pal, J.N. (2012). The Toba volcanic super-eruption, environmental change and hominin occupation history in India over the last 140,000 years. *Quaternary International*, **258**, 119–134.

Pyle, D.M. (1995). Mass and energy budgets of explosive volcanic eruptions. *Geophysical Research Letters*, **22**, 563–566.

Pyle, D.M., Beattie, P.D. and Bluth, G.J.S. (1996). Sulphur emissions to the stratosphere from explosive volcanic eruptions. *Bulletin of Volcanology*, **57**, 663–671.

Pyle, D.M. and Mather, T.A. (2003). The importance of volcanic emissions in the global atmospheric mercury cycle. *Atmospheric Environment*, **37**, 5115–5124.

Rampino, M.R., Self, S. and Fairbridge, R.W. (1979). Can rapid climatic change cause volcanic eruptions? *Science*, **206**, 826–829.

Robock, A., Ammann, C.M., Oman, L. *et al.* (2009). Did the Toba volcanic eruption of similar to 74 ka BP produce widespread glaciation? *Journal of Geophysical Research*, **114**, D10107.

Robock, A. (2000). Volcanic eruptions and climate. *Reviews of Geophysics*, **38**, 191–219.

Roth, R. and Joos, F. (2012). Model limits on the role of volcanic carbon emissions in regulating glacial–interglacial $CO_2$ variations. *Earth and Planetary Science Letters*, **329**, 141–149.

Scaillet, B. and Macdonald, R. (2006). Experimental and thermodynamic constraints on the sulphur yield of peralkaline and metaluminous silicic flood eruptions. *Journal of Petrology*, **47**, 1413–1437.

Scaillet, B. and Pichavant, M. (2003). Experimental constraints on volatile abundances in arc magmas and their implications for degassing processes. *Geological Society of London Special Publication*, **213**, 23–52.

Schaefer, L. and Fegley, B. (2007). Outgassing of ordinary chondritic material and some of its implications for the chemistry of asteroids, planets, and satellites. *Icarus*, **186**, 462–483.

Self, S., Gertisser, R., Thordarson, T., Rampino, M.R. and Wolff, J.A. (2004). Magma volume, volatile emissions, and stratospheric aerosols from the 1815 eruption of Tambora. *Geophysical Research Letters*, **31**, L20608.

Self, S., Widdowson, M., Thordarson, T. and Jay, A.E. (2006). Volatile fluxes during flood basalt eruptions and potential effects on the global environment: a Deccan perspective. *Earth and Planetary Science Letters*, **248**, 518–532.

Self, S., Blake, S., Sharma, K., Widdowson, M. and Sephton, S. (2008). Sulfur and chlorine in Late Cretaceous Deccan magmas and eruptive gas release. *Science*, **319**, 1654–1657.

Sigvaldason, G.E., Annertz, K. and Nilsson, M. (1992). Effect of glacier loading/deloading on volcanism: postglacial volcanic production rates of the Dyngjufjöll area, central Iceland. *Bulletin of Volcanology*, **54**, 385–392.

Sobolev, S.V., Sobolev, A.V., Kuzmin, D.V. *et al.* (2011). Linking mantle plumes, large igneous provinces and environmental catastrophes. *Nature*, **477**, 312–316.

Stoiber, R.E., Williams, S.N. and Huebert, B. (1987). Annual contribution of sulfur dioxide to the atmosphere by volcanoes. *Journal of Volcanology and Geothermal Research*, **33**, 1–8.

Syracuse, E.M. and Abers, G.A. (2006). Global compilation of variations in slab depth beneath arc volcanoes and implications. *Geochemistry Geophysics Geosystems*, **7**, Q05017.

Tang, Q., Zhang, M., Li, C., Yu, M. and Li, L. (2013). The chemical compositions and abundances of volatiles in the Siberian large igneous province: constraints on magmatic $CO_2$ and $SO_2$ emissions into the atmosphere. *Chemical Geology*, **339**, 84–91.

Thordarson, T. and Self, S. (1996). Sulfur, chlorine and fluorine degassing and atmospheric loading by the Roza eruption, Columbia River Basalt Group, Washington, USA. *Journal of Volcanology and Geothermal Research*, **74**, 49–73.

Timmreck, C., Graf, H-F., Lorenz, S.J. *et al.* (2010). Aerosol size confines climate response to volcanic super-eruptions. *Geophysical Research Letters*, **37**, L24705.

Timmreck, C., Graf, H.F., Zanchettin, D. *et al.* (2012). Climate response to the Toba super-eruption: regional changes. *Quaternary International*, **258**, 30–44.

Völker, D., Kutterolf, S. and Wehrmann, H. (2011). Comparative mass balance of volcanic edifices at the southern volcanic zone of the Andes between 33 °S and 46 °S. *Journal of Volcanology and Geothermal Research*, **205**, 114–129.

Wallmann, P.C., Mahood, G.A. and Pollard, D.D. (1988). Mechanical models for correlation of ring-fracture eruptions at Pantelleria, Strait of Sicily, with glacial sea-level drawdown. *Bulletin of Volcanology*, **50**, 327–339.

Wallmann, K. (2001). Controls on the Cretaceous and Cenozoic evolution of seawater composition, atmospheric $CO_2$ and climate. *Geochimica et Cosmochimica Acta*, **65**, 3005–3025.

Watt, S.F.L., Pyle, D.M. and Mather, T.A. (2013). The volcanic response to deglaciation: evidence from glaciated arcs and a reassessment of global eruption records. *Earth-Science Reviews*, **122**, 77–102.

Westrich, H.R. and Gerlach, T.M. (1992). Magmatic gas source for the stratospheric $SO_2$ cloud from the June 15, 1991, eruption of Mount Pinatubo. *Geology*, **20**, 867–870.

White, S.M., Crisp, J.A. and Spera, F.J. (2006). Long-term volumetric eruption rates and magma budgets. *Geochemistry Geophysics Geosystems*, **7**, Q03010.

Zielinski, G.A. (2000). Use of paleo-records in determining variability within the volcanism–climate system. *Quaternary Science Reviews*, **19**, 417–438.

# 15

# Evidence for volcanism triggering extinctions: a short history of IPGP contributions with emphasis on paleomagnetism

VINCENT COURTILLOT, FRÉDÉRIC FLUTEAU AND JEAN BESSE

## 15.1 Introduction

The question of mass extinctions came to the forefront with the discovery of anomalous iridium concentrations in a thin clay layer at the Cretaceous–Tertiary boundary (KTB) near Gubbio (Italy) and the proposal by Alvarez *et al.* (1980) that an impact of an asteroid was the likely cause of anomalies observed at that boundary. This came less than two decades after the plate tectonics revolution; many scientists who were involved in the latter became interested in the former. One of us (V.C.), who had assisted Xavier Le Pichon in teaching the first plate tectonics class in France in 1972, became a convert. In 1982, during the first joint field trip to Tibet between Chinese and Western geoscientists, Claude Allègre, Paul Tapponnier and V.C. discussed at length what seemed to be the solution to the KTB extinction. We taught our students the "Alvarez impact hypothesis" in the following years.

This was also the time when we started forming the paleomagnetic laboratory at IPG (Paris). Having first analysed Mesozoic and Cenozoic red sandstones and andesitic lavas from Tibet, we wanted to see if paleomagnetism could better constrain the convergence history of the Indian and Asian plates. Next, we aimed at India and focused on the thick pile of Deccan lavas that, according to older papers, might have erupted over tens of millions of years from the Cretaceous to the Miocene. But only one (in general) and at most two (on rare occasions) reversals were found in thousands of meters of lava outcrops: the duration of volcanism had been quite short. The K–Ar method indicated an age of 65 Ma, with an uncertainty of 1 to 2 Myr; a tooth of a fossil ray (found in sediments sandwiched between the lower lava flows) implied the Maestrichtian. Put together, these three lines of evidence led to the conclusion that volcanism had lasted only about 1 Myr, straddling the KTB (Courtillot *et al.*, 1986a, b; Figure 15.1). The Deccan Traps ranked "as one of the

*Volcanism and Global Environmental Change*, eds. Anja Schmidt, Kirsten E. Fristad and Linda T. Elkins-Tanton.
Published by Cambridge University Press. © Cambridge University Press 2015.

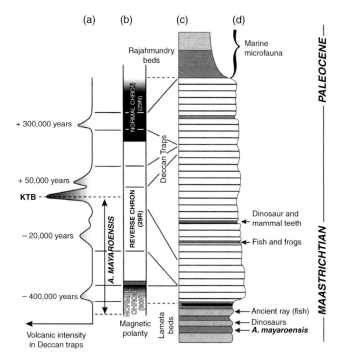

Figure 15.1   A synthesis of paleomagnetic, paleontologic and geochronologic data from the Deccan Traps (as updated in Courtillot, 1999). (a) Proposed changes in volcanic intensity; (b) magnetic polarity; (c) geological section of the Deccan; (d) key fossils.

largest volcanic catastrophes in the last 200 Ma", leading us to conclude, "our hypothesis that the Deccan eruptions should be linked to the KTB and must be accounted for by KTB scenarii may eventually allow tests of 'internal' vs. 'external' models for this major geological event". Recalling the effects of the 1783 eruption of Laki in Iceland, Courtillot *et al.* (1986b) suggested that "It would be interesting to test whether the Deccan lavas were unusually rich in sulfur, and to model the injection of S in the atmosphere related to more than $10^6 \, km^3$ of lava erupted over less than 1 Ma". This was the start of our involvement in the volcanism/extinction connection.

McElhinny (1968) and Kono *et al.* (1972) had already used paleomagnetic reversals to argue for an upper value of 3 to 5 Myr for the duration of volcanism. McLean (1981) proposed the first climatic scenario of the consequences of volcanism based on a massive injection of $CO_2$. Officer and Drake (1985) argued that the KTB events must have lasted 10 to 100 kyr rather than the mere instant of the impact and linked this to Deccan volcanism. For several years, proponents of the impact scenario were pitted against proponents of a volcanic scenario.

We complemented our work on the Deccan and produced a short summary for the general public (Courtillot, 1990). A more technical review with new results was published by Vandamme *et al.* (1991). Surprisingly, to this day, many textbooks aimed at undergraduate geosciences students in the USA mention only the impact hypothesis; in Europe, some undergraduate textbooks have mentioned the two hypotheses on equal footing for two decades.

## 15.2  The trap–extinction correlation

What about other flood basalts and extinction events? Could one uncover common features and mechanisms? Rampino and Stothers (1988) showed that there was a remarkable correlation between 11 flood basalt provinces and mass extinctions going back to 250 Ma. They observed that these events occurred quasi-periodically every ~ 30 Myr and suggested that "showers of impacting comets may be the cause". In an update going back to 300 Ma, Courtillot (1994) argued that the flood basalt–extinction age correlation had further improved and that continental flood basalt volcanism "appears as the main candidate for causing most extinction events". The case for an impact–extinction correlation remained strong for the KTB, but there was no other occurrence and there were cases of large impacts dated at a time without global extinction events. A further update by Courtillot and Renne (2003) strengthened these conclusions (Figure 15.2), extending them back to 360 Ma (Frasnian–Fammenian extinction; Ricci *et al.*, 2013). In addition, it was realized that large sub-oceanic igneous provinces were the main source of global anoxia events (e.g. Wignall, 2001).

In the Phanerozoic, one is left with only end-Ordovician without an associated magmatic counterpart. Buggisch *et al.* (2010) suggested that volcanism could have triggered the first Late Ordovician icehouse. They reconstructed Late Ordovician paleotemperatures and found evidence of a short-lived cooling, that could have initiated a glacial episode, associated with the remains of huge ash eruptions. The extinction comprised two phases/pulses at the base and in the middle of the Hirnantian stage, i.e. separated by ~ 1 Myr at ~ 446 Ma; the two pulses have been linked to intense glaciation (e.g. Finnegan *et al.*, 2012; Armstrong and Harper, 2013). Courtillot and Renne (2003) suggested that a trap might be related to these events. It remains to be found, should any part of it have been preserved from subduction, collision or erosion.

This left open important questions, such as the mechanisms that lead to continental flood basalt eruption (a problem for deep-Earth geoscientists) and of environmental change triggered by flood basalt eruption (a problem for climate modelers and paleontologists).

## 15.3  Traps, continental breakup, plumes and the deep Earth

Based on ideas that can be traced back to Vogt (1972) and Morgan (1981), Richards *et al.* (1989) concluded that, "the largest flood basalt events mark the

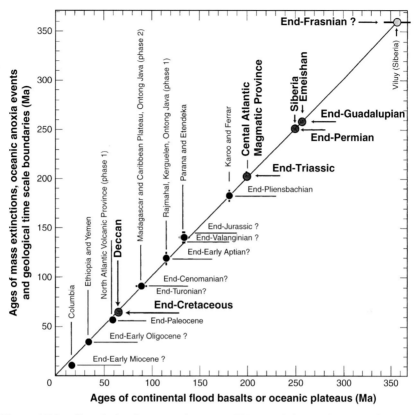

Figure 15.2   Correlation between the ages of LIPs and those of mass extinctions and oceanic anoxia events (from Courtillot and Renne, 2003). The larger symbols are the major mass extinctions.

earliest volcanic activity of many major hot spots, which are thought to result from deep mantle plumes. … flood basalts represent plume 'heads' and hot spots represent continuing magmatism associated with the remaining plume conduit or 'tail'." This started another debate on the reality and depth of origin of plumes, as intense as the impact vs. volcanism debate. The main flood basalt provinces that correlate with the main mass extinctions originate from the deepest mantle (Courtillot *et al.*, 2003) and there is often a close connection between flood basalts and continental breakup (Courtillot *et al.*, 1999). A link between magnetic superchrons and "killer plumes" has been proposed (Courtillot and Olson, 2007). As foreseen by Morgan (1981), hotspots are major geodynamical players: they link core and mantle convection, flood basalt eruption and major changes in the fluid envelopes and biosphere. There are a dozen such punctuated events in the Phanerozoic, with a random distribution in time (Courtillot *et al.*, 2004): the Earth's heat engine, trying to get rid of its excess heat.

A strong spherical harmonic degree two in geoid and seismic tomography suggests that the Earth's mantle is divided into two boxes by subducted plates. This can be traced back at least 400 Myr (Storey, 1995; Greff-Lefftz and Besse, 2012). The two boxes extend from the transition zone to the core–mantle boundary (CMB). Present-day convection in the lower mantle is dominated by a quadrupolar mode, in which superplumes rise from roughly antipodal equatorial regions under Africa and the central Pacific. Surface hotspots may rise from distinct depths (Courtillot *et al.*, 2003): seven from the deepest part of the lower mantle, probably anchored on chemical heterogeneities deposited in the $D''$ layer. Six out of seven of these primary hotspots are found at the margins of the superplumes (Figure 15.3). The original locations of flood basalts at the time of eruption are also along the edges of the present-day seismic low-velocity anomalies imaged above the CMB (Burke and Torsvik, 2004; Davaille *et al.*, 2005). Torsvik *et al.* (2008) have proposed that the edge of the Indo-African superswell at the CMB has not moved significantly with respect to the spin axis of the Earth in the past 300 Myr. Greff-Lefftz and Besse (2012) have proposed an extension of this analysis back to 410 Ma.

## 15.4 The tempo of flood basalt volcanism

A nagging problem remains the limits of time resolution imposed by geochrono-logical methods. Seen from a general perspective, most continental flood basalts appear similar: the volumes of extruded lava are on the order of 1 $Mkm^3$ (spread less than a factor 10) and their durations are on the order of 1 Myr. Yet, their environmental consequences are diverse. The factors that could explain this diversity include geographical location of eruptions, height of volcanic plumes, composition, amounts and fluxes of volatiles or the presence of sediments intruded by the rising magma. We have focused on (and attempted to test) the hypothesis that the tempo of volcanism is the culprit: eruption would take place as a short series of intense, brief volcanic pulses, and the separation in time between these pulses, compared to the time required by the ocean to return to equilibrium (a few thousand years), could be the main factor.

We returned to the Deccan (Chenet *et al.*, 2007, 2008, 2009), and then to the Karoo in Lesotho and South Africa (Moulin *et al.*, 2011, unpublished work). Other groups have documented the Columbia (Mankinen *et al.*, 1985, 1987; Jarboe *et al.*, 2008), North Atlantic Igneous Province (NAIP) (Greenland) (Riisager *et al.*, 2002, 2003), Central Atlantic Magmatic Province (CAMP; Knight *et al.*, 2004) and the Siberian Traps (Pavlov *et al.*, 2011; also see Chapter 5). This has recently been reviewed by Courtillot and Fluteau (2014). We are not aware of similarly advanced studies for the other Phanerozoic continental flood basalts. A key piece of

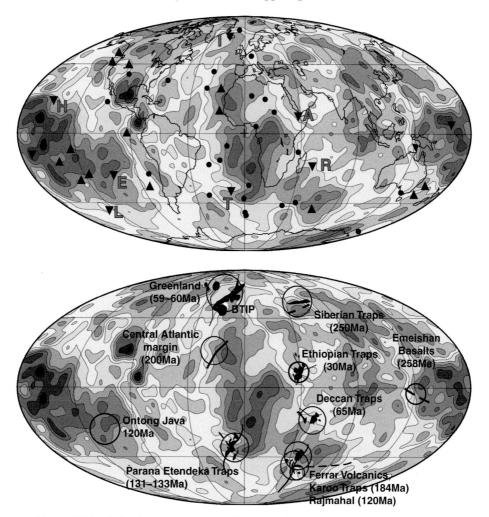

Figure 15.3   Seismic velocity anomalies near the core–mantle boundary (color scale from +3% (fast) to –3% (slow) anomalies in shear velocity; Ritsema *et al.*, 2011). (Top) Distribution of 49 hotspots from the catalog used in Courtillot *et al.* (2003). The seven "primary" hotspots are shown as downward-pointing triangles (H = Hawaii, E = Easter Island, L = Louisville, T = Tristan da Cunha, I = Iceland, A = Afar, R = Reunion). Upward-pointing triangles are hotspots with possible deep origin. Dots indicate other hotspots. (Bottom) Same distribution with locations of main flood basalt provinces at appropriate locations at time of eruption (Davaille *et al.*, 2005). When possible, a reconstruction with present day outcrops of lava (in black) is indicated, with LIP name and age. See stability of plume head and tail location in time near edges of slow quadrupolar anomalies. A black and white version of this figure will appear in some formats. For the colour version, please refer to the plate section.

information has come from detailed flow-by-flow *magnetic stratigraphies* of thick sections of lava flows, in which sequences of superimposed lava flows with the same paleomagnetic direction have been identified. These directional groups (DGs) appear to have cooled in a time too short to have recorded secular variation. These DGs correspond to single eruptive events (SEEs), with a volume often larger than 1000 km$^3$. Their emplacement could have lasted less than a decade (Thordarson and Self, 2003).

There are now six flood basalt provinces for which sufficient data are available to propose a common model of embedded timescales of volcanism: large igneous province (LIP) volcanism appears to occur in a highly discontinuous way, with embedded timescales on the order of 1 to a few million years (full trap emplacement), 100 to 10 kyr (volcanic phase) and 100 to 10 years (SEEs) (Figure 15.4). This model draws in particular from the work of Chenet *et al.* (2008, 2009) on the Deccan, supported by similar observations from the Columbia, Brito-Arctic, Karoo–Ferrar, CAMP and Siberian Traps. It is not always easy to distinguish the time corresponding to impingement of the plume head from early rifting to drifting stages in the cases when opening of an ocean basin occurred. In most cases, when sufficient geochronologic data are available, volcanism is concentrated in a few major phases, each on the order of 1 Myr in duration or less. Finer analysis (based for instance on identification of $\delta^{18}O$ or $\delta^{13}C$ isotopic anomalies in correlative marine sections) shows that these major phases lasted on the order of 100 kyr in duration (e.g. Keller *et al.*, 2012). These phases contain a number of SEEs, some larger than 10,000 km$^3$! Failure to record significant geomagnetic secular variation implies cooling times of these SEEs (flow fields) on the order of a decade to at most a century, whereas recognition of small secular variation loops or structure in the path of a reversing field implies time constants on the order of centuries to a few millennia at most (Chenet *et al.*, 2008, 2009).

The recognition of DGs/SEEs allows one to propose that the total time during which a continental flood basalt was erupted (i.e. without intervening quiescence periods) was on the order of only several thousand years; this is about 1% of the total duration. Many authors note a lack of sediment layers, weathered crusts or paleosols between flows. The DG method is unfortunately "irregular," in that it allows one to identify DGs at the decadal timescale, but not the time between two successive flows with distinctly different paleomagnetic directions. The study of inter-trappean sediments would seem more promising.

Interpretation of these patterns is important for plume dynamics, melt generation and eruption, and eruptive tectonics could be a key element in evaluating the "killing potential" of a continental flood basalt. Our working hypothesis is that the same numbers $N$, volumes $V$, and $CO_2$ and $SO_2$ contents of two separate continental flood basalts could lead to vastly different biotic and environmental

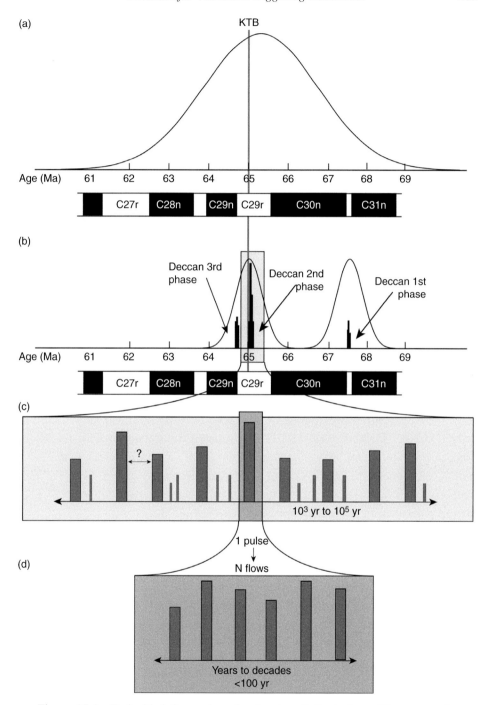

Figure 15.4  Embedded timescales of volcanic activity in traps (Deccan case).
(a) Reversal timescale near the KTB (indicated by the vertical line) and schematic
spread of the total database of age determinations indicating activity over several

consequences, depending on whether the $N$ flow fields (SEEs) were erupted thousands of years apart, allowing for the ocean to re-equilibrate the ocean–atmosphere system, or only decades or centuries apart, possibly leading to a runaway effect and a crisis.

## 15.5 Climate scenarios

The causal link between trap emplacement and mass extinctions remains to be fully understood and the killing mechanisms quantified. LIP emplacement is associated with massive release of gases to the atmosphere–ocean system. Gases such as $CO_2$, $SO_2$ and HCl are known to be potential causes of severe environmental perturbations. $CO_2$ can lead to global warming and ocean acidification on a 1 Myr-long timescale; sulfate aerosols resulting from the oxidation of $SO_2$ by hydroxyl radicals cause surface cooling on short timescales (years to decades); chlorine may cause ozone depletion (Wignall, 2001). Release of these gases may lead to warming and/or cooling episodes, poisoned environments, marine anoxia, mutagenesis and hypercapnia (Knoll *et al.*, 2007; Keller *et al.*, 2012).

Caldeira and Rampino (1990) were among the first to quantify the consequences of the release of $CO_2$ during Deccan trap emplacement, using a biogeochemical carbon-cycle model. They estimated that up to 8,800 Gt of $CO_2$ could have been emitted (original lava volume 2.6 $Mkm^3$). Released over 100 kyr, this would have increased partial pressure of $CO_2$ ($pCO_2$) by ~ 65 ppm and temperature by only ~ 0.7 °C. Dessert *et al.* (2001) re-evaluated the impact of $CO_2$ release by Deccan trap volcanism; assuming that 70,000 Gt of $CO_2$ were released, the $pCO_2$ would have increased by 1000 ppmv and then decreased in ~ 1 Myr, reaching steady state at a lower than pre-Deccan level (due to consumption of $CO_2$ by chemical weathering of basalts; see also Self *et al.*, 2006).

The early work assumed that magma degassing occurred in steady state over the duration of trap emplacement. But the existence and distribution of pulses must be taken into account (Paris *et al.*, 2012). Emplacement of the CAMP at the Triassic–Jurassic boundary is associated with one of the five largest mass extinctions. It is correlated with major carbon-cycle perturbations marked by a carbonate production crisis and two negative isotopic excursions of sedimentary organic carbon (Hesselbo *et al.*, 2002; Galli *et al.*, 2005). Paris *et al.* (2012) have

Figure 15.4 (*cont.*) Myr; (b) As for (a), with outline of three peak phases of activity over some $10^5$ years (from Chenet *et al.*, 2007, 2008, 2009); (c) hypothetical (random) distribution of pulses within the largest, second phase; (d) individual flows within a single eruptive event or pulse (duration less than $10^2$ years based on paleomagnetism).

attempted to model the climate response to CAMP pulsed volcanism, using a geochemical model coupled with a three-dimensional climate model. Eruptive history is modelled as a series of ten identical volcanic pulses, each lasting 100 yr and equally spaced in time over 500 kyr. The total amount of $CO_2$ released is 77,700 Gt. The model simulates a decrease by 30% of carbonate production and reproduces the negative carbon isotopic excursions (but only by assuming a – 20‰ isotopic composition for $CO_2$ released by CAMP or by adding a source of highly depleted $^{13}C$).

$SO_2$ emissions during continental flood basalt emplacement have also been proposed as a major agent of environmental perturbations (Wignall, 2001; Courtillot *et al.*, 2006b; Self *et al.*, 2006). $SO_2$ is rapidly oxidized in the atmosphere, leading to the formation of sulfate aerosols (Robock, 2000). In the case of stratospheric injection, the lifetime of sulfate aerosols is about 1–2 yr, with global cooling at the surface as a response. The amount of $SO_2$ released in a LIP eruption is estimated using the relationship between the $TiO_2/FeO$ ratio and S established for recent and historical tholeiitic lava flows (Self *et al.*, 2006, 2008). The amount of $SO_2$ is 3.6–6.5 $Tg/km^3$ of erupted magma. Chenet *et al.* (2009) proposed that volcanic pulses had emitted up to 1 Gt/yr of $SO_2$ over 100 yr and between 6,800 and 17,000 Gt for the whole traps. Mussard *et al.* (2014) have simulated the impact of the joint release of $SO_2$ and $CO_2$ using an approach similar to Paris *et al.* (2012). A series of 280 volcanic pulses lasting 10 yr each and equally spaced in time (1,500 yr) over 400 kyr was considered. The release of 28,000 Gt of $CO_2$ (alone) increases the $pCO_2$ by 150 ppmv and sea surface temperature by 1.5 °C at mid-to-low latitudes after 400 kyr. When $CO_2$ (28,000 Gt) and $SO_2$ (6,500 Gt) are released simultaneously, the simulated $pCO_2$ variation is larger, reaching 200 ppmv, and sea surface temperatures warm by 2.5 °C, due to $SO_2$ (which first leads to abrupt cooling events, decreasing the efficiency of silicate weathering and amplifying the rise of $pCO_2$). The injection of $SO_2$ during trap emplacement and its conversion to sulfate aerosols lead to climate changes on a short timescale, shorter than the temporal resolution of most records.

Ganino and Arndt (2009) proposed that the amount of gas released by a continental flood basalt is the sum of direct magma degassing and degassing of sediments intruded by the lava. Contact metamorphism driven by intrusions feeding LIPs may have released large amounts of volatiles (Svensen *et al.*, 2007; Ganino and Arndt, 2009; Retallack and Jahren, 2008; Black *et al.*, 2012). The nature of these volatiles depends on the type of sediment intruded. Svensen *et al.* (2004) estimate that the intrusion of basaltic sills belonging to the NAIP in organic-rich layers in the Vøring and Møre basins would have injected 300 to 3,000 Gt of methane with a highly depleted carbon isotope signature. Such massive release would have occurred at the start of the Initial Eocene Thermal Maximum

(IETM), explaining both the subsequent global warming and a negative excursion in carbon isotopes. Svensen *et al.* (2007) proposed that complexes intruded into the Karoo sedimentary basin would have emitted 27,400 Gt of $CO_2$, and contributed to the global warming observed during the Toarcian (disputed by Gröcke *et al.*, 2009). Ganino and Arndt (2009) estimated that the release of sediment-derived $CO_2$ would have been from 3.6 to 8.6 times larger than that due to magma degassing in the Emeishan Traps (~ 258 Ma). Finally, heating of organic-rich shale and petroleum bearing evaporites by the Siberian Traps would have generated 100,000 Gt of $CO_2$, explaining global warming at the Permo-Triassic boundary and carbon isotopic excursions. The amounts of thermogenic gases are generally extrapolated from the analyses of a small number of outcropping sections of sedimentary formations intruded by dikes and sills. This remains difficult, due to the size of the intrusive systems and potential lateral variability of intruded formations. Also, the precise tempo of emission of gases needs to be evaluated. With the exception of the NAIP (relatively well constrained by palynozones – Svensen *et al.*, 2004), the ages of thermogenic events are known only through isotopic dating of volcanic intrusive complexes. Stratigraphic relationships within dike and sill complexes are not as clear as in a sequence of lava flows.

## 15.6 Summary

Thermal/compositional plumes, a component of convective motions within the Earth's mantle, are considered to be responsible for exceptional volcanism, and also to play a role in localizing continental stretching and breakup. The large mushroom-shaped heads of plumes would generate LIPs. The past ~ 35 years have seen tremendous advances in our understanding of the links between this LIP volcanism and mass extinctions.

The KTB extinction has been the main focus of attention, following the 1980 paper suggesting the impact of an asteroid and the mid-1980s papers suggesting that Deccan volcanism could have been a major player. In order to establish a more general interpretation, other flood basalts, impact craters and mass extinction events had to be studied. Successive reviews of available age determinations of the three kinds of events strengthened the idea that impacts were the exception and flood basalt volcanism the general rule. The correlation is now such that the association of LIPs with mass extinctions is unavoidable. Only the KTB has convincing evidence for co-eval impact and volcanism at the time of extinction.

Successive advances have helped in constraining the timing of volcanism, based on bringing together results from many disciplines. Our group has focused on paleomagnetism, paleontology and geochronology. This has led to a picture of several embedded timescales. In many LIPs, the total duration between the first and last manifestations of volcanism may span on the order of 5 to 10 Myr. But it was found in the late 1980s that the main volumes of volcanism had been emplaced over only $\sim 1$ Myr. It was next found that a small number of climactic phases, having lasted from 10 to 100 kyr, were separated by several hundreds of thousands of years of quiescence. Finally, the analysis of paleomagnetic secular variation, coupled with field observations and petrologic analyses, has shown that single eruptive events with volumes in excess of 1000 km$^3$ and up to 10,000 km$^3$ were emplaced in remarkably short durations of 100 down to possibly 10 years and less! During these climaxes, LIPs emplaced basaltic fluxes up to 40 times (and more?) the global production of all mid-ocean ridges (25 km$^3$/yr).

These gigantic individual flows must have injected enormous amounts of gases, and notably $CO_2$ and $SO_2$, into the atmosphere. One should add to these possibly even larger gas emissions, which are due to contact metamorphism of sediments intruded by dikes and sills. The injection sequence would have led to a compli-cated thermal atmospheric response for each SEE: short-term (months to years) cooling due to sulfate aerosols, longer-term (decades to centuries) warming due to a greenhouse effect, and finally much longer-term (100 kyr) cooling due to $CO_2$ removal and alteration of fresh basaltic lava. It is estimated that 1 km$^3$ of magma could have released $\sim 10$ Tg of $CO_2$, 5 Tg of $SO_2$ and 1 Tg of HCl. The role of $SO_2$ is particularly important in producing early cooling phases and decreasing the efficiency of silicate weathering, leading to an amplification of the rise of $pCO_2$. The actual climate changes would have involved the superimposed responses of the sequence of SEEs erupted by the flood basalt. Our working hypothesis remains that the details of the SEE sequences are the key factor of the severity of the extinctions caused by these climate fluctuations. With SEEs separated in time by less than a few thousand years, a runaway effect would have occurred, leading to a phase of mass extinction.

It is unclear how we can hope to attain the kind of temporal resolution required to test this hypothesis further. Some hope comes from climate modeling of pulsed volcanism, of the type summarized in this paper. But the observational database would greatly benefit from concerted efforts to achieve the kind of analyses now available for the Columbia, NAIP, Deccan, Karoo, CAMP and Siberian traps to all other continental flood basalts.

## References

Alvarez, L., Alvarez, W., Asaro, F., and Michel, H., 1980. Extraterrestrial cause for the Cretaceous–Tertiary extinction – experimental results and theoretical interpretation. *Science*, **208**, 1095–1108.

Armstrong, H.A. and Harper, D.A.T., 2013. End-Ordovician mass extinction (abstract). In *Proceedings of* Volcanism, Impacts and Mass Extinctions: Causes and Effects, March 27–29, 2013, London, The Natural History Museum, p. 23.

Black, B.A., Elkins-Tanton, L.T., Rowe, M.C. and Ukstins Peate, I., 2012. Magnitude and consequences of volatile release from the Siberian Traps. *Earth Planet. Sci. Lett.* **317–318**, 363–373.

Buggisch, W., Joachimski, M.M., Lehnert, O. *et al.*, 2010. Did intense volcanism trigger the first Late Ordovician icehouse? *Geology* **38**, 327–330.

Burke, K. and Torsvik, T.H., 2004. Derivation of large igneous provinces of the past 200 Myr from long-term heterogeneities in the deep mantle. *Earth Planet. Sci. Lett.* **227**, 531– 538.

Caldeira, K. and Rampino, M.R., 1990. Carbon-dioxide emissions from Deccan volcanism and a K/T-boundary greenhouse effect. *Geophys. Res. Lett.* **17**, 1299–1302.

Chenet, A-L., Quidelleur, X., Fluteau, F. and Courtillot, V., 2007. $^{40}K$–$^{40}Ar$ dating of the main Deccan large igneous province: further evidence of KTB age and short duration. *Earth Planet. Sci. Lett.* **263**, 1–15.

Chenet, A-L., Fluteau, F., Courtillot, V., Gérard, M. and Subbarao, K.V., 2008. Determination of rapid Deccan eruptions across the Cretaceous–Tertiary boundary using paleomagnetic secular variation: results from a 1200-m-thick section in the Mahabaleshwar escarpment. *J. Geophys. Res.* **113**, B04101.

Chenet, A-L., Courtillot, V., Fluteau, F. *et al.*, 2009. Determination of rapid Deccan eruptions across the Cretaceous–Tertiary boundary using paleomagnetic secular variation: 2. Constraints from analysis of eight new sections and synthesis for a 3500-m-thick composite section. *J. Geophys. Res.* **114**, B06103.

Courtillot, V., 1990. A volcanic eruption. *Scientific American* **263**, 85–92.

Courtillot, V., 1994. Mass extinctions in the last 300 million years: one impact and seven flood basalts. *Israeli J. Earth Sci.* **43**, 255–266.

Courtillot, V., 1999. *Evolutionary Catastrophes: the Science of Mass Extinction*, Cambridge, Cambridge University Press.

Courtillot, V.E. and Renne, P.R., 2003. On the ages of flood basalt events. *C. R. Geosci.* **335**, 113–140.

Courtillot, V., Davaille, A., Besse, J. and Stock, J., 2003. Three distinct types of hotspots in the Earth's mantle. *Earth Planet. Sci. Lett.* **205**, 295–308.

Courtillot, V. and Olson, P., 2007. Mantle plumes link magnetic superchrons to Phanerozoic mass depletion events. *Earth Planet. Sci. Lett.* **260**, 495–504.

Courtillot, V. and Fluteau, F., 2014. A review of the embedded time scales of flood basalt volcanism, with special emphasis on dramatically short magmatic pulses. *Geol. Soc. Am. Spec. Pap.* **505**, doi: 10.1130/2014.2505 (15).

Courtillot, V., Besse, J., Vandamme, D., Jaeger, J.-J. and Montigny, R., 1986a. Deccan trap volcanism as a cause of biologic extinctions at the Cretaceous–Tertiary boundary? *C. R. Acad. Sci. Paris, Ser.* II **303**, 863–868.

Courtillot, V., Besse, J., Vandamme, D. *et al.*, 1986b. Deccan flood basalts at the Cretaceous/Tertiary boundary? *Earth Planet. Sci. Lett.* **80**, 361–374.

Courtillot, V., Jaupart, C., Manighetti, I., Tapponnier, P., Besse, J., 1999. On causal links between flood basalts and continental breakup. *Earth Planet. Sci. Lett.* **166**, 177–195.

Courtillot, V., Besse, J., Cogné, J.P., Davaille, A. and Renne, P., 2004. No long term changes in LIP production rates: traps as a random sequence of rather similar events, abstract, European Union of Geosciences N 04-A-03797, Nice.

Davaille, A., Stutzmann, E., Silveira, G., Besse, J. and Courtillot, V. 2005. Convective patterns under the Indo-Atlantic "box". *Earth Planet. Sci. Lett.* **239**, 233– 252.

Dessert, C., Dupré, B., François, L. *et al.*, 2001. Erosion of Deccan traps determined by river geochemistry: impact on the global climate and oceanic $^{87}Sr$/$^{86}Sr$. *Earth Planet. Sci. Lett.* **188**, 459–474.

Finnegan, S., Heim, N.A., Peters, S.A. and Fisher, W.W., 2012. Climate change and the selective signature of the Late Ordovician mass extinction. *Proc. Nat. Acad. Sci.* **109**, 6829–6834.

Galli, M.T., Jadoul, F., Bernasconi, S.M. and Weissert, H., 2005. Anomalies in global carbon cycling and extinction at the Triassic/Jurassic boundary: evidence from a marine C-isotope record. *Palaeogeogr. Palaeoclim.* **216**, 203–214.

Ganino, C. and Arndt, N.T., 2009. Climate changes caused by degassing of sediments during the emplacement of large igneous provinces. *Geology* **37**, 323–326.

Greff-Lefftz, M. and Besse, J., 2012. Paleo movement of continents since 300 Ma, mantle dynamics and large wander of the rotational pole. *Earth Planet. Sci. Lett.* **345–348**, 151–158.

Gröcke, D.R., Rimmer, S.M., Yoksoulian, L.E. *et al.*, 2009. No evidence for thermogenic methane release in coal from the Karoo–Ferrar large igneous province. *Earth Planet. Sc. Lett.*, **277**, 204–212.

Hesselbo, S., Robinson, S., Surlyk, F. and Piasecki, S., 2002. Terrestrial and marine extinction at the Triassic–Jurassic boundary synchronized with major carbon-cycle perturbation: a link to initiation of massive volcanism? *Geology* **30**, 251–254.

Jarboe, N.A., Coe, R.S., Renne, P.R., Glen, J.M.G. and Mankinen, E.A., 2008. Quickly erupted volcanic sections of the Steens Basalt, Columbia River Basalt Group: secular variation, tectonic rotation, and the Steens Mountain reversal. *Geochemistry Geophysics Geosystems* **9**, Q11010.

Knight, K.B., Nomade, S., Renne, P.R. *et al.*, 2004. The Central Atlantic Magmatic Province at the Triassic–Jurassic boundary: paleomagnetic and $^{40}$Ar/$^{39}$Ar evidence from Morocco for brief, episodic volcanism. *Earth Planet. Sci. Lett.* **228**, 143–160.

Keller, G., Adatte, T., Bhowmick, P.K. *et al.*, 2012. Nature and timing of extinctions in Cretaceous–Tertiary planktic foraminifera preserved in Deccan in tertrappean sediments of the Krishna–Godavari Basin, India. *Earth Planet. Sci. Lett.* **341–344**, 211–221.

Knoll, A., Bambach, R.K., Payne, J.L., Pruss, S., Fisher, W.W., 2007. Paleophysiology and end-Permian mass extinction. *Earth Planet. Sci. Lett.* **256**, 295–313.

Kono, M., Aoki, Y., Kinoshit, H., 1972. Paleomagnetism of Deccan trap basalts in India. *J. Geomag. Geoelec.* **24**, 49–53

McLean, D., 1981. A test of terminal Mesozoic catastrophe. *Earth Planet. Sci. Lett.* **53**, 103–108.

McElhinny, M.W., 1968. Northward drift of India – examination of recent palaeomagnetic results. *Nature* **217**, 342–344.

Mankinen, E.A., Prévot, M., Grommé, C.S. and Coe, R.S., 1985. The Steen's Mountain (Oregon) geomagnetic polarity transition: 1. Directional history, duration of episodes, and rock magnetism. *J. Geophys. Res.* **90B**, 10,393–10,416.

Mankinen, E.A., Larson, E.L., Grommé, C.S., Prévot, M. and Coe, R.S., 1987. The Steens Mountain (Oregon) geomagnetic polarity transition: 3. Its regional significance. *J. Geophys. Res.* **92B**, 8057–8076.

Morgan, W.J., 1981. Hot spot tracks and the opening of the Atlantic and Indian oceans. In Emiliani, C. (ed.), *The Sea*, New York, Wiley Interscience, vol. 7, pp. 443–487.

Moulin, M., Fluteau, F., Courtillot, V. *et al.*, 2011. An attempt to constrain the age, duration, and eruptive history of the Karoo flood basalt: Naude's Nek section (South Africa). *J. Geophys. Res.* **116B**, B07403.

Mussard, M., Le Hir, G. Fluteau, F., Lefebvre, V. and Goddéris, Y., 2014. Modeling the carbon-sulphate interplays in climate changes related to the emplacement of

Continental Flood Basalts. *Geol. Soc. Am. Spec. Pap.* 505, doi: 10.1130/2014.2505 (15)

Officer, C. and Drake, C., 1985. Terminal Cretaceous environmental events. *Science* **227**, 1161–1167.

Paris, G., Donnadieu, Y., Beaumont, V., Fluteau, F. and Goddéris, Y., 2012. Modeling the consequences on Late Triassic environment of intense pulse-like degassing during the Central Atlantic Magmatic Province using the GEOCLIM model. *Clim. Past Discuss.* **8**, 2075–2110.

Pavlov, V.E., Fluteau, F., Veselovskiy, R.V., Fetisova, A.M. and Latyshev, A.V., 2011. Secular geomagnetic variations and volcanic pulses in the Permian–Triassic Traps of the Norilsk and Maimecha–Kotui provinces. *Izvestiya Physics of the Solid Earth* **47**, 402–417.

Rampino, M.R. and Stothers, R.B., 1988. Flood basalt volcanism during the past 250 million years. *Science* **241**, 663–668.

Retallack, G.J. and Jahren, A.H., 2008. Methane release from igneous intrusion of coal during Late Permian extinction events. *J. Geol.* **116**, 1–20.

Ricci, J., Quidelleur, X., Pavlov, V. *et al.*, 2013. New Ar-40/Ar-39 and K–Ar ages of the Viluy traps (Eastern Siberia): further evidence for a relationship with the Frasnian-Famennian mass extinction. *Palaeogeogr. Palaeoclim. Palaeoecol.*, **386**, 531–540.

Richards, M.A., Duncan, R.A. and Courtillot, V. 1989. Flood basalts and hot spot tracks: plume heads and tails. *Science* **246**, 103–107.

Riisager, P., Riisager, J., Abrahamsen, N. and Waagstein, R., 2002. New paleomagnetic pole and magnetostratigraphy of Faroe Islands flood volcanics, North Atlantic Igneous Province. *Earth Planet. Sci. Lett.* **201**, 261–276.

Riisager, J., Riisager, P. and Pedersen, A.K., 2003. Paleomagnetism of large igneous provinces: case study from west Greenland, North Atlantic Igneous Province. *Earth Planet. Sci. Lett.* **214**, 409–425.

Ritsema, J., Deuss, A., van Heijst,, H.J. and Woodhouse, J.H., 2011. S40RTS: a degree-40 shear-velocity model for the mantle from new Rayleigh wave dispersion, teleseismic traveltime and normal-mode splitting function measurements. *Geophys. J. Int.* **184**, 1223–1236.

Robock, A., 2000. Volcanic eruptions and climate. *Rev. Geophys.* **38**, 191–219.

Self, S., Widdowson, M., Thordarson, T. and Jay, A.E., 2006. Volatile fluxes during flood basalt eruptions and potential effects on the global environment: a Deccan perspective. *Earth Planet. Sci. Lett.* **248**, 518–532.

Self, S., Jay, A.E., Widdowson, M., Keszthelyi, L.P., 2008. Correlation of the Deccan and Rajahmundry Trap lavas: are these the longest and largest lava flows on Earth? *J. Volcanol. Geotherm. Res.* **172**, 3–19.

Storey, B.C., 1995. The role of mantle plumes in continental break-up: case histories from Gondwanaland. *Nature* **377**, 301– 308.

Svensen, H., Planke, S., Malthe-Sørenssen, A. *et al.*, 2004. Release of methane from a volcanic basin as a mechanism for initial Eocene global warming. *Nature* **429**, 542–545.

Svensen, H., Planke, S., Chevallier, L., *et al.*, 2007. Hydrothermal venting of greenhouse gases triggering Early Jurassic global warming. *Earth Planet. Sci. Lett.* **256**, 554–566.

Thordarson, T. and Self, S., 2003. Atmospheric and environmental effects of the 1783–1784 Laki eruption: a review and reassessment. *J. Geophys. Res.* **108**(D1), 4011.

Torsvik, T.H., Smethurst, M.A., Burke, K., Steinberger, B. 2007. Long term stability in deep mantle structure: evidence from the ~ 300 Ma Skagerrak-Centered Large Igneous Province (the SCLIP). *Earth Planet. Sci. Lett.* **267**, 444–452.

Vandamme, D., Courtillot, V., Besse, J., Montigny, R., 1991. Paleomagnetism and age determinations of the Deccan traps (India): results of a Nagpur-Bombay traverse and review of earlier work. *Rev. Geophys.*, **29**, 159–190.

Vogt, P.R., 1972. Evidence for global synchronism in mantle plume convection, and possible significance for geology. *Nature* **240**, 338–342.

Wignall, P.B., 2001. Large igneous provinces and mass extinctions. *Earth Sci. Rev.* **53**, 1–33.

# 16

# Halogen release from Plinian eruptions and depletion of stratospheric ozone

KIRSTIN KRÜGER, STEFFEN KUTTEROLF AND THOR H. HANSTEEN

## 16.1 Introduction

Plinian eruptions are characterized by large mass eruption rates and high eruption columns. They inject gases (water vapour, carbon dioxide, sulfur dioxide and halogen compounds) and solid particles (e.g. ash) directly into the stratosphere (Walker, 1973; Textor *et al.*, 2003a; von Glasow, 2009), where they influence the atmospheric composition and chemistry, in particular sulfuric acid aerosol and ozone concentrations as well as the radiation budget (Robock, 2000; Timmreck, 2012; see also Chapter 13). If a Plinian eruption takes place in the tropics, the climate influence is global, due to the spread of volcanic relevant gases by the Brewer–Dobson circulation in the stratosphere (Toohey *et al.*, 2011). Based on stratospheric circulation timescales sulfur and subsequently formed particles have a residence time of 3–6 years. Therefore they have the ability to cool the surface climate and affect the ozone layer for a few years. If large amounts of volcanic halogens reach the stratosphere, they have the potential to further enhance the depletion of the ozone layer (Stolarski and Butler, 1978; Mankin and Coffey, 1984). Ozone in the stratosphere protects the Earth's surface from the harmful ultraviolet (UV) radiation. Recent model studies revealed substantial ozone depletion and UV amplification due to the Siberian Traps volcanism during the end-Permian (Beerling *et al.*, 2007; Black *et al.*, 2014; see also Chapter 20).

Most large explosive eruptions occur in subduction-zone settings (also called volcanic arcs), where the magmas are typically water-rich and reach fluid over-saturation at depth, which in turn represents a major driving force for the explosivity. The high volatile contents in subduction-zone magmas originate from dehydration and partial melting of the subducted oceanic plate. Subduction-zone volcanism occurs at nearly all latitudes, and probably represents the volumetrically

*Volcanism and Global Environmental Change*, eds. Anja Schmidt, Kirsten E. Fristad and Linda T. Elkins-Tanton. Published by Cambridge University Press. © Cambridge University Press 2015.

most important subaerial explosive volcanism in the tropics (Schmincke, 2004). This kind of explosive eruption also comprises higher amounts of the volatiles sulfur and halogens compared to hot-spot and rift-related volcanism, which is especially true for chlorine (Scaillet *et al.*, 2003; Pyle and Mather, 2009; Mather *et al.*, 2012; Shinohara, 2013). Large subduction-zone eruptions in the tropics therefore have a high potential to influence climate and atmospheric composition, including ozone depletion (Metzner *et al.*, 2012; Kutterolf *et al.*, 2013a).

Following an introduction on halogen outputs from Plinian eruptions, this chapter gives an overview about novel measurements to detect bromine (Br) and chlorine (Cl) masses emitted from large explosive subduction-zone eruptions in the past. Literature data of volcanic emissions are combined with detailed field data and chemical measurements of the halogen release from the entire Central American Volcanic Arc (CAVA) covering the last 200 kyr. Our mass budgets are based on petrological data obtained by microchemical analyses of Br and Cl in representative melt inclusions in minerals and the corresponding matrix glasses (Kutterolf *et al.*, 2013a; unpublished), and scaled according to mass estimates of the eruption products (Kutterolf *et al.*, 2008, and references therein). Finally, this chapter addresses the potential role of halogen release from large, tropical eruptions on the stratospheric halogen burden and evaluates their potential impact on the ozone layer. The influence of volcanic emissions on the troposphere is addressed in Chapters 13 and 14.

## 16.2  Halogen output from Plinian eruptions

A number of studies have been performed on the Cl output of large explosive eruptions (Devine *et al.*, 1984; Symonds *et al.*, 1988; Gerlach *et al.*, 1996). Volatile masses released from single eruptions have typically been estimated using the petrologic method (Johnston, 1980), in which the difference between pre-eruptive volatile concentrations measured in glass inclusions in minerals and those in the degassed tephra particles are scaled up using the erupted mass of previously mapped volcanic products. Another approach used for modern eruptions is to estimate the erupted sulfur mass in the gas phase with satellite-based instruments (Krueger, 1983; Carn *et al.*, 2003; Theys *et al.*, 2013), and thereafter derive the Cl contents using an assumed Cl/S ratio (Giggenbach, 1996; Halmer *et al.*, 2002). Mass balance calculations using prescribed Cl/S ratios are problematic due to the large natural variations of Cl/S ratios in magmas, which are further enhanced by degassing processes. Thus, global estimates of Cl output from large eruptions based on a combination of both petrologic Cl data and indirect estimates are hampered with comparatively large errors, which may reach one to two orders of magnitude (Aiuppa *et al.*, 2009). Recent reviews of the data used for such global compilations are given by Scaillet *et al.* (2003), Fischer (2008), Pyle and Mather (2009) and Shinohara (2013).

Nearly all data on Br contents and Br/Cl ratios in volcanic emanations originate from condensate samples collected at volcanic arcs (Gerlach, 2004, and references therein). Very little data exist on the Br inventory of explosive eruptions, and most values are estimated from measured Cl concentrations using experimentally determined Cl/Br ratios for fluids in magmatic systems (Bureau *et al.*, 2000), or using halogen ratios measured in fumaroles (Symonds *et al.*, 1994; Gerlach, 2004). Each of the approaches may give ambiguous results, as both degassing processes and reactions with surrounding rocks (in the case of fumarole gases) influence the Cl/Br ratios. In order to achieve consistent petrologic data on released masses of Br from large eruptions, refined analytical tools were developed using focused synchrotron radiation (Hansteen *et al.*, 2000). Using an optimized synchrotron micro-X-ray fluorescence (XRF) setup, detection limits for Br of about 0.5 ppm in typical glass inclusions have been achieved, which enables the analyses of a glass inclusion within minutes (Kutterolf *et al.*, 2013a). A combination of electron microprobe and synchrotron micro-XRF (for Br only) thus allows for semi-routine measurements of S, Cl and Br in typical glass inclusions and glassy matrix originating from Plinian eruptions. This method has so far been used to determine the halogen inventory of 31 Quaternary eruptions in Central America, covering an along-arc distance of 1200 km (Kutterolf *et al.*, 2013a; unpublished).

## 16.3  Volcanic halogen entrainment into the present-day stratosphere

Emissions of Cl and Br play an important role for the catalytic destruction of ozone in the present-day stratosphere. Most studies concentrate on the anthropogenic release of these gases in the form of chlorofluorocarbons and their substitute halons, which caused the ozone hole over Antarctica since the 1980s (Chubachi, 1984; Farmann *et al.*, 1985). More recently, the role of biogenic oceanic halogen sources has been investigated for the stratospheric halogen and ozone loading, revealing a small but not negligible role under the present-day high Cl loading (Montzka *et al.*, 2011, and references therein). However, the relevance of other natural halogen sources such as from present-day large volcanic eruptions for the stratospheric ozone layer has hardly been discussed in the past. This may be partly due to the fact that the amount of halogens emitted from the respective volcanic vent into the stratosphere is highly uncertain. Observations after the El Chichón (1982) eruption revealed a 40% increase in total chloride column 3 to 6 months after the eruption in the region of the volcanic cloud (Mankin and Coffey, 1984), while after the Pinatubo (1991) eruption no significant increase in the HCl concentration was detected (Mankin *et al.*, 1992; Wallace and Livingston, 1992). On the other hand, investigation of both volcanogenic halogens Cl and Br together are restricted to observations from the Plinian El Chichón 1982 eruption and relatively small volcanic eruptions since 2000, and encompass the stratosphere only in parts (Kotra *et al.*, 1983; Mankin and Coffey, 1984; Millard *et al.*, 2006; Rose *et al.*, 2006; Hörmann *et al.*, 2013). While the Cl release from large eruptions has been better studied (Gerlach *et al.*, 1996; Gerlach, 2004; Wallace, 2005), the release of Br remains largely unknown. Although Br is typically more than two orders of magnitude less abundant than Cl in volcanic gases (Bureau *et al.*, 2000), its efficiency in halogen-catalyzed ozone destruction is ~ 60 times higher than for Cl (Montzka *et al.*, 2011, and references therein), if Br does reach the stratosphere. Only recently, ground-based, aircraft and satellite measurements revealed that small volcanic eruptions at low and high latitudes delivered reactive Br into tropospheric and up to stratospheric levels (Bobrowski *et al.*, 2003; Millard *et al.*, 2006; Rose *et al.*, 2006; Theys *et al.*, 2009; Heue *et al.*, 2011; Hörmann *et al.*, 2013).

Model calculations of the effects of additional Br in the stratosphere strongly indicate that Plinian eruptions represent a special case. The combination of sulfuric acid aerosols together with elevated Br and Cl concentrations is particularly efficient for ozone destruction in the mid-latitude lower stratosphere, where a large fraction of the Br-induced ozone loss is dominated by the $BrO_x/HO_x$ catalytic cycle, and BrO/ClO ratios are additionally relevant with regard to ozone loss in polar latitudes (Salawitch *et al.*, 2005; Sinnhuber *et al.*, 2009). Halogens from Plinian eruptions therefore seem to have a potential impact on the stratospheric ozone concentration as well (Stolarski and Butler, 1978; von Glasow, 2009; Kutterolf *et al.*, 2013a), aside from heterogeneous reactions on volcanic aerosols leading to enhanced stratospheric ozone depletion during high Cl background levels.

## 16.4 The CAVA background

The Central American subduction zone is characterized by the subduction of the Cocos beneath the Caribbean plate since the early Miocene at convergence rates of 70–90 mm/year (Barckhausen *et al.*, 2001; DeMets, 2001). In the resulting CAVA, extending ~ 1400 km from Guatemala in the northwest to Panama in the southeast, numerous felsic tephras from highly explosive Plinian eruptions have been emplaced during the Pliocene and Holocene, where the varying magmatic compositional diversity along the CAVA results from systematically changing subducting parameters (Syracuse and Abers, 2006; Carr *et al.*, 2007, and references therein).

Quaternary, andesitic to rhyolitic highly explosive eruptions, originating from large caldera volcanoes in the north (Guatemala and El Salvador) and stratovolcanoes in the south (Nicaragua and Costa Rica) (Figure 16.1), produced 92 widespread tephras, which occur > 20 km from the source (Kutterolf *et al.*, 2008 and references therein). Kutterolf *et al.* (2013a; unpublished) and Metzner

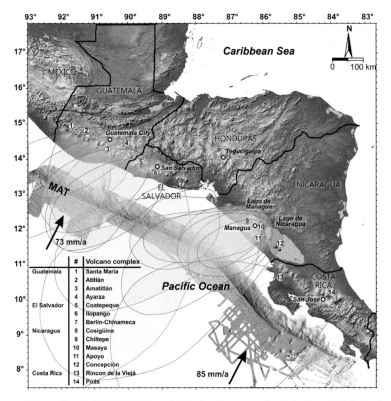

Figure 16.1    Shaded SRTM (Shuttle Radar Topography Mission, NASA) elevation model of Central America (NASA/JPL/NGA, 2000) and high-resolution
bathymetry along the Middle America Trench (MAT) modified from Kutterolf
*et al.* (2008). Numbers represent the names of source volcanoes (14) of the
92 eruptions in the last 200 kyr that produced the widespread tephras mentioned
in the text; 31 of those eruptions were suitable for a complete set of halogen
analyses.

*et al.* (2012) recently determined volatile emissions (Br, Cl, S, $SO_2$) for over
30 of these eruptions. The voluminous, widespread tephras account for more than
half of the total late Pleistocene–Holocene erupted magma mass of the volcanic
front (Kutterolf *et al.*, 2008). Tephra volumes generally range from 1 $km^3$ to
~ 100 $km^3$ (Metzner *et al.*, 2012). An exception is the largest known eruption in
Central America, the 84 ka old Los Chocoyos eruption from Atitlán Caldera in
Guatemala, which generated deposits with about 800 $km^3$ tephra volume.

## 16.5  Halogen release from CAVA eruptions

In contrast to the first 'exotic' intra-plate case study from the Laacher See tephra
(Textor *et al.*, 2003a), the data set of Kutterolf *et al.* (2013a, b; unpublished)

provides a nearly complete time series along an entire volcanic arc for Br and Cl release from palaeo-Plinian eruptions during the last 450 kyr, and can therefore be used to constrain global halogen budgets from large explosive eruptions. Such extrapolations, however, rely on the representativity of Cl emissions calculated for CAVA eruptions. Data on degassing during volcanic eruptions worldwide, as reviewed by Shinohara (2013), show large variations in Cl emissions determined by the petrologic method. The concentration differences between pre-eruptive melt inclusions and matrix glasses range mostly between 500 and 1500 ppm of degassed Cl. This overlaps well with the average degassed Cl of 1070 ppm for the CAVA eruptions (Kutterolf *et al.*, 2013a), which we use as a typical value for Plinian eruptions worldwide.

All matrix glasses of the CAVA tephras are degassed and have low Br concentrations between 0.4 and 1.3 ppm, independent of their chemical composition (basalt to rhyolite). These values are comparable to the 1 to 3 ppm Br found in bulk rock-powder samples from the 1902 Santa Maria eruption determined by pyrohydrolysis and ion chromatography (Balcone-Boissard *et al.*, 2010) as well as to the Laacher See eruption (Textor *et al.*, 2003a). Because degassing during eruption is highly effective, the concentrations of residual Br in the matrix glasses are similar for all eruptions investigated (Kutterolf *et al.*, 2013a). According to Kutterolf *et al.* (unpublished), the melt inclusions, which should preserve the halogen contents sampled from the respective magma chambers, if they did not leak after their formation, in contrast have maximum Br concentrations ranging between 0.9 and 17.9 ppm, and are strongly bromine-enriched compared to the corresponding matrix glasses. Since fluid-oversaturation of the magmas prior to the eruptions can be demonstrated in the form of fluid inclusions in phenocrysts from all eruptions (Kutterolf *et al.*, 2013a), we consider the highest Br concentration value determined from each sample as our best estimate for the initial Br abundance in the melts at magma chamber depths before the onset of eruptive degassing. Consequently, we estimated the Br fraction released during magmatic degassing as the difference between the maximum Br abundance in the most undegassed inclusions and the average concentration in matrix glass, and converted these fractions into erupted Br masses, using the respective erupted magma masses for each eruption (Kutterolf *et al.*, unpublished; this study). Erupted Br masses of the 31 eruptions investigated range from roughly 2 to 1100 kt (i.e. 2 to 1100 $\times$ $10^6$ kg) and Cl masses from 12 to 80 700 kt (Table 16.1).

## 16.6 Discussion

The values for erupted Br masses for a typical CAVA eruption (95 kt/eruption) lie within the range of present annual anthropogenic and natural Br emissions into

Table 16.1 *Volcanic output and stratospheric input of halogens from CAVA eruptions.*

| Tephra | Age (ka) | Degassed mass of Br (kg) | Degassed mass of Cl (kg) | Br concentration added to stratosphere 1% (ppt) | (% rel) | 10% (ppt) | (% rel) | 25% (ppt) | (% rel) | Cl concentration added to stratosphere 1% (ppt) | (% rel) | 10% (ppt) | (% rel) | 25% (ppt) | (% rel) | EESC added to stratosphere 1% (ppt) | (% rel) | 10% (ppt) | (% rel) | 25% (ppt) | (% rel) |
|---|---|---|---|---|---|---|---|---|---|---|---|---|---|---|---|---|---|---|---|---|---|
| Santa Maria (GUA) | 0.1 | 4.72E+06 | 2.24E+09 | 0.022 | 0.36 | 0.2 | 4 | 0.5 | 9 | 23 | 4.2 | 234 | 42 | 584 | 106 | 25 | 4.5 | 247 | 45 | 617 | 112 |
| Cosigüina 1835 (NIC) | 0.18 | 1.26E+08 | 5.06E+08 | 0.058 | 0.96 | 0.6 | 10 | 1.4 | 24 | 5 | 1.0 | 53 | 10 | 132 | 24 | 9 | 1.6 | 87 | 16 | 218 | 40 |
| Terra Blanca Joven (SAL) | 1.55 | 5.25E+08 | 1.39E+11 | 2.402 | 40.04 | 24.0 | 400 | 60.1 | 1001 | 1447 | 263.1 | 14 472 | 2631 | 36 179 | 6578 | 1591 | 289.3 | 15 913 | 2893 | 39 782 | 7233 |
| Masaya Tuff (NIC) | 1.7 | 2.27E+07 | 7.99E+08 | 0.104 | 1.73 | 1.0 | 17 | 2.6 | 43 | 83 | 15.2 | 833 | 152 | 2083 | 379 | 90 | 16.3 | 896 | 163 | 2239 | 407 |
| Chiltepe Tephra (NIC) | 1.8 | 1.25E+08 | 3.13E+10 | 0.572 | 9.53 | 5.7 | 95 | 14.3 | 238 | 326 | 59.3 | 3263 | 593 | 8158 | 1483 | 361 | 65.6 | 3606 | 656 | 9015 | 1639 |
| Masaya Triple Layer (NIC) | 2.1 | 6.15E+06 | 5.54E+08 | 0.028 | 0.47 | 0.3 | 5 | 0.7 | 12 | 58 | 10.5 | 578 | 105 | 1445 | 263 | 59 | 10.8 | 595 | 108 | 1487 | 270 |
| Rio Blanco Tephra (CRC) | 3.5 | 3.20E+07 | 1.57E+09 | 0.146 | 2.43 | 1.5 | 24 | 3.7 | 61 | 16 | 3.0 | 164 | 30 | 409 | 74 | 25 | 4.6 | 251 | 46 | 629 | 114 |
| Mateare Tephra (NIC) | c. 4 | 1.11E+07 | 1.27E+09 | 0.051 | 0.85 | 0.5 | 8 | 1.3 | 21 | 13 | 2.4 | 133 | 24 | 332 | 60 | 16 | 3.0 | 163 | 30 | 408 | 74 |
| San Antonio Tephra (NIC) | 6 | 5.59E+07 | 9.44E+09 | 0.256 | 4.26 | 2.6 | 43 | 6.4 | 107 | 98 | 17.9 | 985 | 179 | 2462 | 448 | 114 | 20.7 | 1138 | 207 | 2845 | 517 |
| Xiloa Tephra (NIC) | 6.1 | 8.03E+06 | 9.13E+08 | 0.037 | 0.61 | 0.4 | 6 | 0.9 | 15 | 10 | 1.7 | 95 | 17 | 238 | 43 | 12 | 2.1 | 117 | 21 | 293 | 53 |
| Laguna de Haule (CRC) | 6.2 | 2.34E+06 | 1.44E+09 | 0.011 | 0.18 | 0.1 | 2 | 0.3 | 4 | 15 | 2.7 | 150 | 27 | 376 | 68 | 16 | 2.8 | 157 | 28 | 392 | 71 |
| U.Apoyeque Tephra (NIC) | 12.4 | 1.67E+07 | 8.91E+09 | 0.076 | 1.27 | 0.8 | 13 | 1.9 | 32 | 93 | 16.9 | 930 | 169 | 2325 | 423 | 98 | 17.7 | 976 | 177 | 2439 | 444 |
| L.Apoyeque Tephra (NIC) | 14.5 | 2.63E+07 | 8.13E+09 | 0.120 | 2.00 | 1.2 | 20 | 3.0 | 50 | 85 | 15.4 | 848 | 154 | 2121 | 386 | 92 | 16.7 | 920 | 167 | 2301 | 418 |
| U.Ometepe Tephra (NIC) | 15.5 | 4.74E+07 | 2.12E+09 | 0.217 | 3.61 | 2.2 | 36 | 5.4 | 90 | 22 | 4.0 | 222 | 40 | 554 | 101 | 35 | 6.4 | 352 | 64 | 879 | 160 |
| Pinos Altos (GUA) | 23 | 2.83E+07 | 7.40E+09 | 0.129 | 2.15 | 1.3 | 22 | 3.2 | 54 | 77 | 14.0 | 772 | 140 | 1930 | 351 | 85 | 15.4 | 850 | 154 | 2124 | 386 |
| U.Apoyo Tephra (NIC) | 24.5 | 6.02E+08 | 1.07E+11 | 2.753 | 45.88 | 27.5 | 459 | 68.8 | 1147 | 1112 | 202.2 | 11119 | 2022 | 27 796 | 5054 | 1277 | 232.2 | 12 770 | 2322 | 31 925 | 5805 |
| L.Apoyo Tephra (NIC) | 25 | 1.51E+07 | 1.38E+10 | 0.069 | 1.15 | 0.7 | 12 | 1.7 | 29 | 144 | 26.2 | 1439 | 262 | 3596 | 654 | 148 | 26.9 | 1480 | 269 | 3700 | 673 |
| TB4 (ESA/SLV) | 36 | 8.89E+07 | 2.43E+10 | 0.407 | 6.78 | 4.1 | 68 | 10.2 | 169 | 254 | 46.2 | 2538 | 462 | 6346 | 1154 | 278 | 50.6 | 2782 | 506 | 6956 | 1265 |
| Mixta Tephra (GUA) | 39 | 1.18E+07 | 8.70E+09 | 0.054 | 0.90 | 0.5 | 9 | 1.3 | 22 | 91 | 16.5 | 908 | 165 | 2269 | 413 | 94 | 17.1 | 940 | 171 | 2350 | 427 |
| Conacaste (SAL) | 51 | 1.36E+07 | 2.31E+10 | 0.062 | 1.03 | 0.6 | 10 | 1.6 | 26 | 241 | 43.8 | 2411 | 438 | 6028 | 1096 | 245 | 44.5 | 2448 | 445 | 6121 | 1113 |
| E-Fall (GUA) | 51 | 5.18E+07 | 5.59E+09 | 0.237 | 3.95 | 2.4 | 39 | 5.9 | 99 | 58 | 10.6 | 583 | 106 | 1458 | 265 | 73 | 13.2 | 725 | 132 | 1813 | 330 |
| Congo (SAL) | 53 | 1.89E+07 | 2.45E+11 | 0.087 | 1.44 | 0.9 | 14 | 2.2 | 36 | 2560 | 465.4 | 25598 | 4654 | 63996 | 11636 | 2565 | 466.4 | 25650 | 4664 | 64126 | 11659 |
| Fontana Tephra (NIC) | 60 | 4.33E+06 | 1.11E+09 | 0.020 | 0.33 | 0.2 | 3 | 0.5 | 8 | 12 | 2.1 | 116 | 21 | 291 | 53 | 13 | 2.3 | 128 | 23 | 320 | 58 |
| Twins (ESA/SLV) | 61 | 1.45E+08 | 1.36E+10 | 0.661 | 11.02 | 6.6 | 110 | 16.5 | 275 | 142 | 25.9 | 1422 | 259 | 3555 | 646 | 182 | 33.1 | 1819 | 331 | 4547 | 827 |
| Unicit Tephra (NIC) | c. 70 | 3.92E+06 | 1.17E+08 | 0.018 | 0.30 | 0.2 | 3.0 | 0.4 | 7.5 | 1 | 0.2 | 12 | 2.2 | 30 | 5.5 | 2 | 0.4 | 23 | 4.2 | 57 | 10.4 |
| Arce (SAL) | 72 | 3.37E+08 | 3.30E+10 | 1.542 | 25.71 | 15.4 | 257 | 38.6 | 643 | 344 | 62.6 | 3441 | 626 | 8602 | 1564 | 437 | 79.4 | 4366 | 794 | 10 916 | 1985 |
| Blanca Rosa (SAL) | 75 | 9.44E+07 | 1.10E+10 | 0.432 | 7.19 | 4.3 | 72 | 10.8 | 180 | 115 | 20.9 | 1150 | 209 | 2874 | 523 | 141 | 25.6 | 1409 | 256 | 3522 | 640 |
| Los Chocoyos (GUA) | 84 | 1.10E+09 | 8.07E+11 | 5.021 | 83.68 | 50.2 | 837 | 125.5 | 2092 | 8420 | 1531 | 84203 | 15310 | 210509 | 38274 | 8722 | 1586 | 87 216 | 15 857 | 218 040 | 39 644 |
| L-Fall (GUA) | 191 | 3.51E+08 | 1.05E+11 | 1.607 | 26.78 | 16.1 | 268 | 40.2 | 670 | 1098 | 199.6 | 10979 | 1996 | 27447 | 4990 | 1194 | 217.1 | 11 943 | 2171 | 29 857 | 5429 |
| Averages ignoring Los Chocoyos: | | 9.51E+07 | 2.93E+10 | 0.435 | 7.25 | 4.35 | 72 | 10.87 | 181 | 305 | 55.5 | 3052 | 555 | 7629 | 1387 | 331 | 60.2 | 3313 | 602 | 8281 | 1506 |
| Averages including Los Chocoyos: | | 1.30E+08 | 5.61E+10 | 0.593 | 9.88 | 5.93 | 99 | 14.82 | 247 | 585 | 106.4 | 5850 | 1064 | 14625 | 2659 | 621 | 112.8 | 6206 | 1128 | 15 514 | 2821 |

Assumptions: 1, 10 and 25 wt% of emitted halogens reach the stratosphere; the global stratosphere has 2.7 ×10¹⁹ moles, 15% of 1.8 × 10²⁰ moles of total atmosphere. Pre-industrial stratosphere concentrations approximated as: Br = 6 ppt, Cl = 550 ppt, equivalent effective stratospheric chlorine (EESC) = 900 ppt (Montzka *et al.*, 2011). GUA = Guatemala, NIC = Nicaragua, SAL = El Salvador, and CRC = Costa Rica.

the troposphere of about 37 kt/yr and 107 kt/yr, respectively, in 2008 (derived from Figure 1–3, Montzka *et al.*, 2011). In comparison, Br emissions by quiescent volcanic degassing and from smaller eruptive events (considering short- and long-lived volcanogenic Br species) have been estimated as 3 to 40 kt/yr (Aiuppa *et al.*, 2005 and references therein; see also Chapter 8). Ground-based and satellite monitoring of recent eruptions at Soufrière Hills (Montserrat) 2002 and Kasatochi (Aleutians) 2008 revealed 350 and 50–120 t/yr of Br injection into the troposphere and lower stratosphere, respectively (Bobrowski *et al.*, 2003; Theys *et al.*, 2009). In contrast, the recent study by Hörmann *et al.* (2013) roughly quantified a total BrO mass up to 1 kt during the sub-Plinian eruption of Kasatochi (9–11 August 2008) using GOME-2 satellite data, which is in the range of Br masses from CAVA eruptions (Table 16.1). All CAVA eruptions considered in the present study developed eruption columns of > 18 km altitude that reached well into the stratosphere (Kutterolf *et al.*, 2008). In order to evaluate the potential effect on ozone of such eruptions, we need to estimate the proportion of the originally erupted halogens actually entering the stratosphere.

Different numerical models, considering microphysical and partly chemical processes in an ascending eruption column, simulate that interaction with hydro-meteors results in a partial scavenging of halogens, causing only a fraction of the halogen mass emitted at the vent to actually reach the stratosphere. Tabazadeh and Turco (1993) modelled that almost all halogen load is lost through scavenging by condensed water before an eruption column reaches the stratosphere. More complex plume model studies, in contrast, consider scavenging mainly by ice formed in the ascending eruption columns and suggest that 10% to more than 25% of emitted halogens may reach the stratosphere (Textor *et al.*, 2003a, b). The scavenging efficiency strongly depends on the composition of the gas phase, including its salinity, the formation of ice crystals and on the dry versus wet nature of the atmosphere (Textor *et al.*, 2003a). Additionally, the properties of simultan-eously emitted ash particles present in the atmosphere can influence the formation of halogen-rich precipitates and aerosols from the gas phase and therefore change the potential of eruption columns to deliver halogens into the stratosphere (Delmelle *et al.*, 2005; Gislason *et al.*, 2011). Two contrasting scenarios for the degree of preservation of halogens in the stratosphere have been observed: The 'wet' Pinatubo eruption 1991, which was comparatively halogen-rich (Shinohara, 2013), caused no measurable accumulation of stratospheric halogens since it occurred simultaneously with the typhoon Yunya, probably leading to efficient wash-out of volcanic halogens over a large region over the Philippine peninsula Luzon on 15 June 1991 (McCormick *et al.*, 1995). On the other hand, the Hekla 2000 eruption plume, which reached the upper troposphere/lower stratosphere at high latitudes (dry atmosphere, low tropopause height), generated local ozone depletion, thus testifying to the limited halogen scavenging (Millard *et al.*, 2006; Rose *et al.*, 2006). Due to the high ascent rates of large Plinian eruption columns, scavenging due to humid conditions in tropical regions can be partly compensated by the formation of halogen-bearing ice crystals and even solid salt particles (NaCl containing Br), preserving the halogens for stratospheric release (Woods *et al.*, 1985; Textor *et al.*, 2003b), which indeed has been observed for the tropical El Chichon 1982 eruption. We therefore assume that between 1% and 25% of the emitted halogens may have reached the stratosphere (Table 16.1), for which we consider the range between 1% and 10% as a conservative low estimate and the range above 25% as an upper limit in the light of recent observations and modelling results (Rose *et al.*, 2006; Textor *et al.*, 2003b). Applying the 10% estimate, the resulting stratospheric injections per CAVA eruption vary between 0.2–110 kt Br and 1–8070 kt Cl. In contrast to rapid scavenging from the wet troposphere, these inorganic halogen compounds can remain in the dry stratosphere for as long as the stratospheric turnover time of 3–6 years (Waugh and Hall, 2002). For the total effect of halogens on stratospheric ozone the equivalent effective stratospheric chlorine (EESC) quantity can be used, which is estimated from tropospheric emissions of chlorofluorocarbons and other halogens, assumptions on transport timescales into the upper atmosphere and by weighting the ozone-destroying potential of Br relative to Cl (EESC = Cl + (60 × Br); Danilin *et al.*, 1996; Montzka *et al.*, 2011).

Relating the hypothetical volcanic halogen burden from an average CAVA eruption to the observed pre-ozone hole background levels in 1980 for mid-latitudes, Figure 16.2 illustrates the effect of a single eruption in the past and future stratosphere. Pre-ozone hole 1980 EESC values are projected to occur in 2046 (Daniel *et al.*, 2011). This implies that an average CAVA eruption in the future may lead to a significant lofting of the EESC amounts probably for up to 3 to 6 years for all three scenarios. Such an effect becomes even more pronounced towards the end of the twenty-first century due to the amendment of anthropogenic chlorofluorocarbons and halons. Inspecting the total Cl and Br loadings it becomes obvious that the future EESC evolution is mainly controlled by the total Cl decline (1980 recovery in 2040), whereas total Br stays above pre-1980 values during the whole century. Figure 16.3a presents the percentage increase of each halogen component and the EESC compared to 1980 values, applying the 10% scenario from Figure 16.2. If an average CAVA volcano were to erupt, it would lead to a 50% increase of total Br in the present day (2015), which increases to almost 100% towards 2100 (this does not necessarily mean a correlated increase in ozone depletion). The relative Cl impact is even higher, given the projected more rapid decline of total Cl in the twenty-first century. Here the relative impact more than doubles between 2015 and 2100 changing from 370% to 770%, whereas

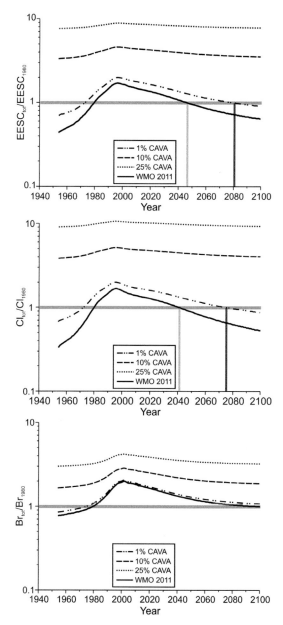

Figure 16.2   Ratio between average CAVA eruption (scenario 1%, 10% and 25% of emitted halogens reaching the stratosphere) and loading for: EESC, total Cl ($Cl_{tot}$), total Br ($Br_{tot}$) (Daniel *et al.*, 2011), applying 1980 values as a threshold for pre-ozone hole conditions.

the EESC rises from 300% to 550%. In absolute concentrations, an average CAVA eruption in 2015 would lead to 8 to 17 ppt total Br, assuming the 1% to 25% estimate range, and 1.5 up to 9 ppb total Cl loading in the mid-latitude stratosphere (Figure 16.3b), compared to only 7.5 ppt total Br and 1.1 ppb total Cl abundances

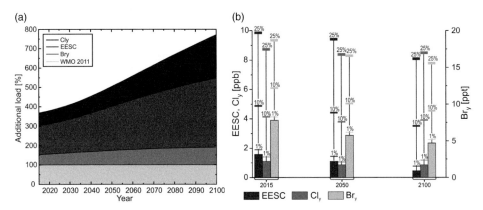

Figure 16.3    (a) Additional load (in %) of a future average CAVA eruption to the
projected anthropogenic halogen loading (Daniel *et al.*, 2011). (b) Scenarios (1%,
10% and 25%) for stratospheric loading of future EESC (in ppb), total Cl (Cl$_y$, in
ppb), total Br (Br$_y$, in ppt) applying anthropogenic halogen burden and average CAVA
eruptions during 2015, 2050 and 2100. A black and white version of this figure will
appear in some formats. For the colour version, please refer to the plate section.

from anthropogenic emissions in 2015. Towards the end of the twenty-first century
the potential role of volcanic Br increases compared to volcanic Cl resulting in a
total (anthropogenic and volcanic) EESC abundance of 1 to 8 ppb in 2100 compared to 2 to 10 ppb in 2015.

    If we want to evaluate the impact of large explosive subduction-zone eruptions on
the stratosphere, we need realistic estimates of their global eruption frequency.
A possible approach is to compare the two probably best studied subaerial arcs
worldwide, the CAVA and Japan, and scale the results to the global situation.
For the long-term eruption frequency, we selected the last 191 000 years of the CAVA
history, which covers 92 eruptions (Metzner *et al.*, 2012) probably larger than M = 6,
leading to an eruption frequency of one every 2100 years on average for this region.
The largest uncertainty for the global extrapolation is probably the poor coverage of
absolute dates for large eruptions. We thus use the comparatively well-studied
geological record of Japan, which covers about 10% of the subaerial arc volcanoes
worldwide (Mori *et al.*, 2013), and is distributed along about 2100 km of arc length. If
we assume that the time-averaged volcanic production rates (Völker *et al.*, 2011,
2014) and the amounts of large explosive eruptions per unit arc length are comparable
to CAVA, the CAVA arc length of 1200 km would account for about 6% of the global
arc large explosive eruptions. Taking these relations as representative, we get a global
eruption frequency of large explosive eruptions of 2100 years × (6/100) = 126 years.
This is consistent with the study of Deligne *et al.* (2010), which assumes a magnitude
6.5 to 7 eruption every 100 to 200 years. If we further follow Lallemand *et al.* (2005)
and note that about half of the global subduction zone length (about 28 000 km of
55 000 km) occurs within the tropical belt (30° N to 30° S), we can assume that 50% of

such future eruptions would occur in the tropics. The resulting global estimate would be one 'large stratosphere-endangering' subduction-zone eruption in the tropics every 200 to 400 years. Indeed, the last ~ 150 years were a rather quiet period concerning explosive volcanic activity compared to other periods, such as, for example, 1780–1850 (Lamb, 1970). In addition to a lack of available measurements, this may be a reason why there has been a gap in halogen-rich, ozone-depleting volcanic eruptions in the tropics during the past 150 years.

## 16.7 Summary

The effect of volcanic halogen release from present-day eruptions on the ozone layer is often neglected as it is believed not to play a significant role. This view may, however, change during the next large and halogen-rich volcanic eruption in the tropics, especially when considering the current decline of anthropogenic Cl loading in the twenty-first century back to pre-ozone hole conditions.

Since El Chichon 1982 and Pinatubo 1991, no large, sulfur-rich eruption took place in the tropics impacting climate globally, but statistically it could happen any day, given the above discussion. Measurement techniques have been significantly improved over the last decades, now enabling the detection of volcanic Br in extremely low concentration range (ppt) (see also Chapter 8). Future research should take into account the likelihood of volcanic halogen entrainment to the stratosphere and prepare *in situ* and satellite measurement techniques as well as chemistry climate modelling for these explosive events. Improving the understanding of natural, external climate factors is of importance for the present and future development of the ozone layer, especially during the current decline of the anthropogenic Cl background levels.

### Acknowledgements

We would like to thank Steven Montzka and Guus Velders for providing the data for the halogen and EESC projections from the 2010 WMO ozone assessment; and our colleagues from the SFB574 in particular Armin Freundt, Heidi Wehrmann, Wendy Planert-Perez as well as Karen Appel from DESY Hamburg for carrying out the intensive field and lab measurements. Part of this work was carried out within the SFB574 (DFG) project 'Volatiles and Fluids in Subduction Zones' at Kiel University.

### References

Aiuppa, A., Federico, C., Franco, A. *et al.*, 2005, Emission of bromine and iodine from Mount Etna volcano, *Geochem. Geophys. Geosys.*, **6**, Q08008.

Aiuppa, A., Baker, D.R. and Webster, J.D., 2009, Halogens in volcanic systems, *Chem. Geol.*, **263**, 1–18.

Balcone-Boissard, H., Villemant, B. and Boudon, G., 2010, Behavior of halogens during the degassing of felsic magmas, *Geochem. Geophys. Geosys.*, **11**, Q09005.

Barckhausen, U., Ranero, C.R., von Huene, R., Cande, S., Roeser, H., 2001, Revised tectonic boundaries in the Cocos plate off Costa Rica: implications for the segmentation of the convergent margin and for plate tectonic models, *J. Geophys. Res.*, **106**, 207–220.

Beerling, D.J., Harfoot, M., Lomax, B. and Pyle, J.A., 2007, The stability of the stratospheric ozone layer during the end-Permian eruption of the Siberian Traps, *Phil. Trans. R. Soc. London. Ser. A*, **365**, 1843–1866.

Black, A.B., Lamarque, J.F., Shields, C.A., Elkins-Tanton, L.T. and Kiehl, J.T., 2014, Acid rain and ozone depletion from pulsed Siberian Traps magmatism, *Geology*, **42**, 67–70.

Bobrowski, N., Honninger, G., Galle, B. and Platt, U., 2003, Detection of bromine monoxide in a volcanic plume, *Nature*, **423**, 273–276.

Bureau, H., Keppler, H. and Metrich, N., 2000, Volcanic degassing of bromine and iodine: experimental fluid/melt partioning data and applications to stratospheric chemistry, *Earth Planet. Sci. Lett.*, **183**, 51–60.

Carn, S.A., Krueger, A.I., Bluth, G.J.S. *et al.*, 2003, Volcanic eruption detection by the Total Ozone Mapping Spectrometer (TOMS) instruments: a 22-year record of sulphur dioxide and ash emissions, *Geol. Soc. London Spec. Publ.*, **213**, 1.

Carr, M.J., Patino, L.C. and Feigenson, M.D., 2007, Petrology and geochemistry of lavas, in Buntschuh, J., Alvarado, G.E. (eds.), *Central America – Geology, Resources and hazards*, Rotterdam, Balkema, pp. 565–591.

Chubachi, S., 1984, Preliminary result of ozone observations at Syowa station from February 1982 to January 1983, *Mem. Natl. Inst. Polar Res.*, **34**, 13–19.

Danilin, M., Sze, N., Ko, M., Rodriguez, J. and Prather, M., 1996, Bromine–chlorine coupling in the Antarctic ozone hole, *Geophys. Res. Lett.*, **23**, 153–156.

Daniel, J.S., Velders, G.J.M., Morgenstern, O. *et al.*, 2011, A focus on information and options for policymakers, scientific assessment of ozone depletion: 2010, *World Meteorological Organization Global Ozone Research and Monitoring Project Report*, **52**, 5.1–5.56.

Deligne, N.I., Coles, S.G. and Sparks, R.S.J., 2010, Recurrence rates of large explosive volcanic eruptions, *J. Geophys. Res.*, **115**, B06203.

Delmelle, P., Villieras, F., Pelletier, M., 2005, Surface area, porosity and water adsorption properties of fine volcanic ash particles, *Bull. Volcanol.*, **67**, 160–169.

DeMets, C., 2001, A new estimate for present-day Cocos–Caribbean plate motion: implications for slip along the Central American Volcanic Arc, *Geophys. Res. Lett.*, **28**, 4043–4046.

Devine, J.D., Sigurdsson, H., Davis, A.N. and Self, S., 1984, Estimates of sulfur and chlorine yield to the atmosphere from volcanic eruptions and potential climatic effects, *J. Geophys. Res.*, **89**, 6309–6325.

Farman, J.C., Gardiner, B.G. and Shanklin, J.D., 1985, Large losses of total ozone in Antarctica reveal seasonal $ClO_x/NO_x$ interaction, *Nature*, **315**, 207–210.

Fischer, T.P., 2008, Fluxes of volatiles ($H_2O$, $CO_2$, $N_2$, Cl, F) from arc volcanoes, *Geochem. J.*, **42**, 21–38.

Gerlach, T.M., Westrich, H.R., Symonds, R.B.,1996, Preeruption vapor in magma of the climactic Mount Pinatubo eruption: source of the giant stratospheric sulfur dioxide cloud, in Newhall, C.G. and Punongbayan, R.S. (eds.), *Fire and Mud: Eruptions and Lahars of Mount Pinatubo, Philippines*, Seattle, University of Washington Press, pp. 415–433.

Gerlach, T.M., 2004, Volcanic sources of tropospheric ozone-depleting trace gases, *Geochem. Geophys. Geosyst.*, **5**, Q09007.

Giggenbach, W.F., 1996, Chemical composition of volcanic gases, in Scarpa, R.I.S., Tilling, R.I. (eds.), *Monitoring and Mitigation of Volcano Hazards*, Heidelberg, Springer, pp. 221–256.

Gislason, S.R., Hassenkam, T., Nedel, S. *et al.*, 2011, Characterization of Eyjafjallajökull volcanic ash particles and a protocol for rapid risk assessment, *Proc. Natl. Acad. Sci. USA*, **108**, 7307–7312.

Halmer, M.M., Schmincke, H.-U. and Graf, H.-F., 2002, The annual volcanic gas input into the atmosphere in particular into the stratosphere: a global data set for the past 100 years, *J. Volcanol. Geotherm. Res.*, **115**, 511–528.

Hansteen, T.H., Sachs, P.M. and Lechtenberg, F., 2000, Synchrotron-XRF microprobe analysis of silicate reference standards using fundamental-parameter quantification. *Eur. J. Mineral.* **12**, 25–31.

Heue, K.P., Brenninkmeijer, C.A.M., Baker, A.K. *et al.*, 2011, $SO_2$ and BrO observation in the plume of the Eyjafjallajökull volcano: CARIBIC and GOME-2 retrievals, 2011, *Atmos. Chem. Phys.*, **11**, 2973–2989.

Hörmann, C., Sihler, H., Bobrowski, N. *et al.*, 2013, Systematic investigation of bromine monoxide in volcanic plumes from space by using the GOME-2 instrument, *Atmos. Chem. Phys.*, **13**, 4749–4781.

Johnston, D.A., 1980, Volcanic contribution of chlorine to the stratosphere: more significant to ozone than previously estimated?, *Science*, **209**, 491–493.

Kotra, J.P., Finnegan, D.L., and Zoller, W.H., 1983, El Chichón: composition of plume gases and particles, *Science*, **222**, 1018–1021.

Kutterolf, S., Freundt, A., and Pérez, W., 2008, Pacific offshore record of Plinian arc volcanism in Central America, part 2: tephra volumes and erupted masses, *Geochem. Geophys. Geosyst.*, **9**, Q02S02.

Kutterolf, S., Hansteen, T.H., Appel, K. *et al.*, 2013a, Combined bromine and chlorine release from large explosive volcanic eruptions: a threat to stratospheric ozone?, *Geology*, **41**, 707–710.

Kutterolf, S., Jegen, M., Mitrovica, J.X. *et al.*, 2013b, A detection of Milankovitch frequencies in global volcanic activity, *Geology*, **41**, 227–230.

Krueger, A.J., 1983, Sighting of El Chichon sulphur dioxide clouds with the Nimbus 7 Total Ozone Mapping Spectrometer, *Science*, **220**, 1377–1379.

Lallemand, S., Heuret, A., and Boutelier, D., 2005, On the relationships between slab dip, back-arc stress, upper plate ansolution motion, and crustal nature in subduction zones, *Geochem. Geophys. Geosys.*, **6**.

Lamb, H., 1970, Volcanic dust in the atmosphere; with a chronology and assessment of its meteorological significance, *Trans. Phil. Trans. R. Soc. Ser. A*, **266**, 426–533.

Mankin, W.G. and Coffey, M.T., 1984, Increased stratospheric hydrogen chloride in the El Chichón cloud, *Science*, **226**, 170–172.

Mankin, W.G., Coffey, M.T. and Goldman, A., 1992, Airborne observations of $SO_2$, HCl, and $O_3$ in the stratospheric plume of the Pinatubo Volcano in July 1991, *Geophys. Res. Lett.*, **19**, 179–182.

Mather, T.A, Witt, M.L.I., Pyle, D.M. *et al.*, 2012, Halogens and trace metal emissions from the ongoing 2008 summit eruption of Kilauea volcano, Hawai, *Geochim. Cosmochim. Acta*, **83**, 292–323.

McCormick, M.P., Thomason, L.W. and Trepte, C.R., 1995, Atmospheric effects of the Mt. Pinatubo eruption, *Nature*, **373**, 399–404.

Metzner, D., Kutterolf, S., Toohey, M. *et al.*, 2012, Radiative forcing and climate impact resulting from $SO_2$ injections based on a 200,000 year record of Plinian eruptions along the Central American Volcanic Arc, *Int. J. Earth Sci.*, doi10.1007/s00531-012-0814-z.

Millard, G.A., Mather, T.A., Pyle, D.M., Rose, W.I. and Thornton, B.F., 2006, Halogen emissions from a small volcanic eruption: modeling the peak concentrations, dispersion and volcanically induced ozone loss in the stratosphere, *Geophys. Res. Lett.*, **33**, L19815.

Montzka, S., Reimann, S., Engel, A. *et al.*, 2011, Ozone-depleting substances (ODSs) and related chemicals, scientific assessment of ozone depletion: 2010, *World Meteorological Organization Global Ozone Research and Monitoring Project Report No.* **52**, 1.1–1.86.

Mori, T., Shinohara, H., Kazahaya, K., Hirabayashi, J.-I. and Matsushima, T., 2013, Time-averaged $SO_2$ fluxes of subduction-zone volcanoes: example of a 32-year exhaustive survey for Japanese volcanoes, *J. Geophys. Res.*, **118**, 8662–8674.

Pyle, D.M. and Mather, T.A., 2009, Halogens in igneous processes and their fluxes to the atmosphere and oceans from volcanic activity: a review, *Chem. Geol.*, **263**, 110–121.

Robock, A., 2000, Volcanic eruptions and climate, *Rev. Geophys.*, **38**, 191–219.

Rose, W.I., Millard, G.A., Mather, T.A. *et al.*, 2006, Atmospheric chemistry of a 33–34 hour old volcanic cloud from Hekla volcano (Iceland): insights from direct sampling and the application of chemical box modelling, *J. Geophys. Res.*, **111**, D20206.

Salawitch, R.J., Weisenstein, D.K., Kovalenko, L.J. *et al.*, 2005, Sensitivity of ozone to bromine in the lower stratosphere, *Geophys. Res. Lett.*, **32**, L05811.

Scaillet, B., Luhr, J.F. and Carroll, M.R., 2003, Petrological and volcanological constraints on volcanic sulfur emissions to the atmosphere, *Geophys. Monogr.*, **139**, 11–40.

Schmincke, H.U., 2004, *Volcanism*, Heidelberg, Springer, 1–324.

Shinohara, 2013, Volatile flux from subduction zone volcanoes: insights from a detailed evaluation of the fluxes from volcanoes in Japan, *J. Volcanol. Geotherm. Res.*, **268**, 46–63.

Sinnhuber, B.-M., Chipperfield, M.P. and Feng, W., 2009, The contribution of anthropogenic bromine emissions to past stratospheric ozone trends: a modelling study, *Atmos. Chem. Phys.*, **9**, 2863–2871.

Stolarski, R. and Butler, D., 1978, Possible effects of volcanic eruptions on stratospheric minor constituent chemistry, *Pageoph.*, **117**, 486–497.

Symonds, R.B., Rose, W.I. and Reed, M.H., 1988, Contribution of Cl- and F-bearing gases to the atmosphere by volcanoes, *Nature*, **334**, 415–418.

Symonds, R.B., Rose, W.I., Bluth, G.J.S. and Gerlach, T.M., 1994, Volcanic-gas studies: methods, results and applications, *Rev. Mineral.*, **30**, 1–66.

Syracuse, E.M. and Abers, G.A., 2006, Global compilation of variations in slab depth beneath arc volcanoes and implications, *Geochem. Geophys. Geosys.*, **7**, Q05017.

Tabazadeh, A. and Turco, R.P., 1993, Stratospheric chlorine injection by volcanic eruptions: HCl scavenging and implications for ozone, *Science*, **260**, 1082–1086.

Textor, C., Sachs, P.M., Graf, H.F., Hansteen, T.H., 2003a, The Laacher See eruption: estimation of volatiles in magma and simulation of their fate in the plume, *Geol. Soc. London Spec. Publ.*, **213**, 307–328.

Textor, C., Graf, H.F., Herzog, M. and Oberhuber, J.M., 2003b, Injection of gases into the stratosphere by explosive volcanic eruptions, *J. Geophys., Res.*, **108**, 4606.

Theys, N., Van Roozendael, M., Dils, B., Hendrick, F. and De Mazière, M., 2009, First satellite detection of volcanic bromine monoxide emission after Kasatochi eruption, *Geophys. Res. Lett.*, **36**, L03809.

Theys, N., Campion, R., Clarisse, L. *et al.*, 2013, Volcanic $SO_2$ fluxes derived from satellite data: a survey using OMI, GOME-2, IASI and MODIS, *Atmos. Chem. Phys.*, **13**, 5945–5968.

Timmreck, C., 2012, Modeling the climatic effects of large explosive volcanic eruptions, *Wiley Interdisc. Rev. Clim. Change*, **3**, 545–564.

Toohey, M., Krüger, K., Niemeier, U. and Timmreck, C., 2011, The influence of eruption season on the global aerosol evolution and radiative impact of tropical volcanic eruptions, *Atmos. Chem. Phys.*, **11**, 12351–12367.

Völker, D., Kutterolf, S. and Wehrmann, H., 2011, Comparative mass balance of volcanic edifices at the Southern Volcanic Zone of the Andes between 33° S and 46° S, *J. Volcanol. Geotherm. Res.*, **205**, 114–129.

Völker, D., Wehrmann, H., Kutterolf, S. *et al.*, 2014, Constraining input and output fluxes of the southern central Chile subduction zone: water, chlorine, sulphur, *Int. J. Earth. Sci.*, **103**, 2129–2153.

von Glasow, R., Bobrowski, N. and Kern, C., 2009, The effects of volcanic eruptions on atmospheric chemistry, *Chem. Geol.*, **263**, 131–142.

Wallace, P.J., 2005, Volatiles in subduction zone magmas: concentrations and fluxes based on melt inclusion and volcanic gas data, *J. Volcanol. Geotherm. Res.*, **140**, 217–240.

Wallace, L. and Livingston, W., 1992, The effect of the Pinatubo cloud on hydrogen chloride and hydrogen fluoride, *Geophys. Res. Lett.*, **19**, 1209–1211.

Walker, G.P.L., 1973, Explosive volcanic eruptions – A new classification scheme, *Geol. Rdsch.*, **62**, 431–446.

Waugh, D.W. and Hall, T.M., 2002, Age of stratospheric air: theory, observations, and models, *Rev. Geophys.*, **40**, 1010.

Woods, D.C., Chuan, R.L. and Rose, W.I., 1985, Halite particles injected into the stratosphere by the 1982 El Chichon eruption, *Science*, **230**, 170–172.

WMO (World Meteorological Organization), 2011, Scientific Assessment of Ozone Depletion: 2010, Global Ozone Research and Monitoring Project – Report No. 52, Geneva.

# 17

# The environmental and climatic impacts of volcanic ash deposition

MORGAN T. JONES

## 17.1 Introduction

Explosive volcanic eruptions rapidly inject a mix of gases, aerosols and solids into the atmosphere, leading to the wide dispersal of ejecta. The solid particles erupted by a volcano consist of igneous minerals, fragments of country rock, and volcanic glass in varying abundances. The mineral species and volcanic glass compositions are determined by the chemistry of the parent magma. A considerable fraction of the airborne solids (tephra) is volcanic ash, defined as $< 2$ mm in size (Rose and Durant, 2009; Durant *et al.*, 2010). The fine-grained nature of volcanic ash allows for transportation thousands of kilometres from the source prior to deposition. The largest known eruptions can eject $> 10^{16}$ kg of material in a single event (Mason *et al.*, 2004), capable of blanketing whole continents and/or extensive swathes of the ocean floor with deposits several centimetres thick (Self, 2006; Figure 17.1).

Ejected volcanic volatiles cool as they mix with the atmosphere, forming salts, condensed gases, and aerosols (Figure 17.2). Co-erupted solids act as nucleation surfaces for such phases (Delmelle *et al.*, 2007; Oskarsson, 1980). These soluble surface coatings, termed ash-leachates, are comprised of sulfur species and halogens as the main anion donors and significant concentrations of many alkali and transition metals as cation donors (Frogner *et al.*, 2001). Up to 40% of the erupted $SO_2$ and 10–20% of HCl can be scavenged from volatile phases onto particle surfaces in this manner (Rose, 1977). Ash-leachate chemistry and volume varies substantially between different eruptions as a function of geological setting, eruption size, plume concentration, and magmatic chemistry (Jones and Gislason, 2008; Witham *et al.*, 2005).

Volcanic ash deposition can have substantial impacts on both terrestrial and oceanic ecosystems, the extents of which can be global and long-lived. Many of

*Volcanism and Global Environmental Change*, eds. Anja Schmidt, Kirsten E. Fristad and Linda T. Elkins-Tanton. Published by Cambridge University Press. © Cambridge University Press 2015.

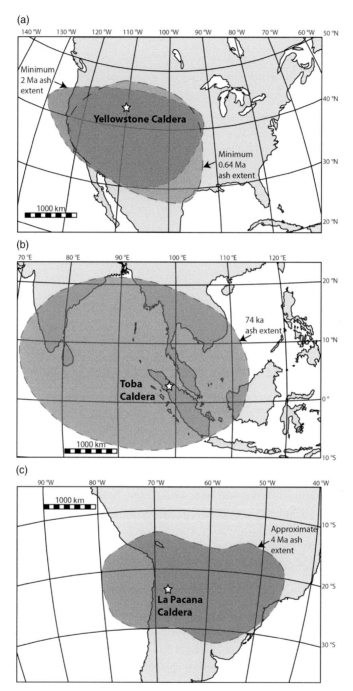

Figure 17.1    (a)–(c) Past examples of very large volcanic ash blankets (modified from Jones, 2008; and references therein).

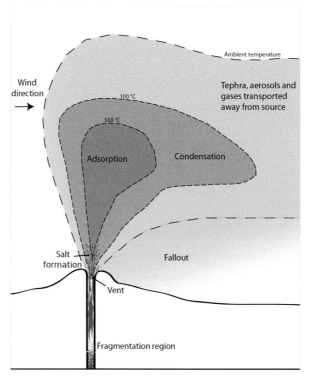

Figure 17.2  A schematic section of an eruption plume showing ash-leachate formation (modified from Oskarsson, 1980).

the environmental impacts are poorly understood due to a combination of the diversity of depositional environments and the complexity of multiple contemporaneous forcings. This chapter summarises the major environmental effects of volcanic ash deposition and how these can affect the climate over various timescales.

## 17.2  Terrestrial ash blankets

### 17.2.1  Environmental impact

Volcanic ash deposition can mechanically bury, overload or abrade vegetation, and/or induce chemical changes from nutrient and acid release (Figure 17.3A; Mack, 1981; Ayris and Delmelle, 2012). Ash blankets can be several metres thick proximal to the source, so the extent of destruction is dictated both by the size of the eruption and the proximity to the volcano. Moreover, flora varies considerably in their ability to survive ash deposition. Mosses and lichens can be inhibited by just 10 mm of ash (Antos and Zobel, 1985), while some shrubs and pine trees can survive tephra deposits > 750 mm (Eggler, 1948). The blanketing of soil

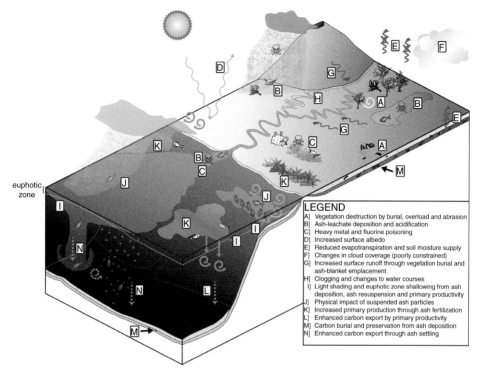

euphotic zone

LEGEND
A] Vegetation destruction by burial, overload and abrasion
B] Ash-leachate deposition and acidification
C] Heavy metal and fluorine poisoning
D] Increased surface albedo
E] Reduced evapotranspiration and soil moisture supply
F] Changes in cloud coverage (poorly constrained)
G] Increased surface runoff through vegetation burial and ash-blanket emplacement
H] Clogging and changes to water courses
I] Light shading and euphotic zone shallowing from ash deposition, ash resuspension and primary productivity
J] Physical impact of suspended ash particles
K] Increased primary production through ash fertilization
L] Enhanced carbon export by primary productivity
M] Carbon burial and preservation from ash deposition
N] Enhanced carbon export through ash settling

Figure 17.3    A schematic diagram illustrating the main terrestrial and oceanic responses to volcanic ash deposition. A black and white version of this figure will appear in some formats. For the colour version, please refer to the plate section.

modifies its ability to interact with the atmosphere, altering microbial activity and the breakdown of organic material.

The dissolution and release of ash-leachates can potentially cause acidification and toxic conditions in affected environments (Blackford *et al.*, 1992). Ash-leachates are very soluble, dissolving within minutes of first contact with water. As the upper atmosphere does not support liquid $H_2O$, much of these surface coatings are still adsorbed to the particle surfaces upon deposition, allowing rapid release into the surrounding environment (Duggen *et al.*, 2007). The subsequent dissolution of ash-leachates can have a strong impact on local biota. Ecological sensitivity to ash-leachate dispersal depends on the chemistry and volume of the ash-leachates and the ecosystem's ability to buffer itself from the effects. Systems with a low turnover rate, such as ponds, lakes and soils, are particularly vulnerable to the remobilization of acids and toxic heavy metals (Figure 17.3B; Frogner-Kockum *et al.*, 2006). Some soils have the capacity to buffer the effects of acid release (Delfosse *et al.*, 2005); others can experience microbial collapse and subsequent ecosystem stress (Kilian *et al.*, 2006). Acids, such as HF, and heavy metals tend to accumulate in top soils

(Shoji *et al.*, 1993), limiting the extent of groundwater contamination but increasing the incorporation into flora (e.g. Martin *et al.*, 2009). This exposes grazing animals to the effects of fluorosis (Figure 17.3C), such as the mortality of > 60% of the sheep livestock in Iceland following the 1783 eruption of Laki (Thordarson and Self, 2003).

## 17.2.2 Radiative changes

Terrestrial ash deposition is capable of initiating changes in atmosphere–ocean circulation through altering surface albedo and by removing vegetation. Volcanic ash is generally extremely reflective due to its silica-rich composition and high particle vesicularity (Jones *et al.*, 2007). As a result, most ash blankets will significantly increase the surface albedo (Figure 17.3D). The insulation and increased reflection of incoming radiation can lead to markedly reduced soil temperatures below an ash layer (Cook *et al.*, 1981). The burial of vegetation diminishes moisture supply to the atmosphere through evapotranspiration (Figure 17.3E), reducing the latent heat flux and therefore internal atmospheric convection. This both stabilizes the atmospheric boundary layer and leads to less recycling of precipitation (Trenberth, 1999). The combined effects of vegetation burial and surface-albedo increase become more prevalent with increasing eruption magnitude. In large-scale scenarios, local effects include a marked decrease in precipitation, an increase in reflected shortwave radiation, cooling from reduced heat retention, and changes to cloud coverage through increased convective heating (Figure 17.3F; Jones *et al.*, 2007). The climate response to a large ash blanket varies considerably with latitude, with strong seasonal amplifications to changes in surface conditions. These modifications manifest themselves as decadal variations in surface pressures, temperatures and precipitation patterns, which interplay with jet stream locations and strength (Jones *et al.*, 2007).

## 17.2.3 Residence times

The surface residence time of an ash blanket at the surface is governed by the rate of remobilization and erosion, which in turn is controlled by the local geomorphology and climate. Residence times will be shorter in areas of high rainfall and/or high wind speeds. Finer ash particles are more likely to be resuspended by aeolian or fluvial action. However, very fine ash is cohesive and capable of forming a crust through desiccation, inhibiting erosion, increasing surface runoff (Figure 17.3G) and hindering the re-emergence of vegetation in areas of high deposition (Nammah *et al.*, 1986). This can initially restrict removal of the ash blanket to incising

Figure 17.4   An rill erosion network established in tephra deposits around Ol Doinyo Lengai volcano, Tanzania. Image by Tom Pfeiffer / www.volcanodiscovery.com (©).

channel (rill) networks (Figure 17.4; Collins and Dunne, 1986), which stabilize to fewer and deeper master rills with time (Manville *et al.*, 2009). Slope erosion processes lead to the preferential denudation of tephra deposited on steep topography, leading to the concentration of ash in valleys and the choking of freshwater lakes and rivers. This can lead to changes in river courses for systems as big as the Blue Nile (Adamson *et al.*, 1982) and can lead to catastrophic break-out floods if catchments become dammed (Figure 17.3H; Manville, 2002).

### *17.2.4 Ecosystem recovery*

The recovery of ecosystems depends on the severity of the ash deposition. Recovery of flora will be slower at higher latitudes due to the shorter growing period. Studies of the 74 ka Younger Toba Tuff deposits in India (see Figure 17.1) indicated that tephra remained mobile for several years prior to burial and re-vegetation (Jones, 2010). Ecological recovery after the complete destruction of proximal vegetation from the 1883 eruption of Krakatau (Indonesia) took 15 years to re-establish trees and shrubs, and considerably longer to return to pre-eruption conditions (Bush, 2006; Figure 17.5). The re-establishment of pioneer species requires seeding by wind or animal transport, so in cases where large areas are completely deforested then the recovery of flora must begin at the ash-blanket edges. The speed of recovery can be further restricted or even halted by feeding insects (Fagan and Bishop, 2000; Knight and Chase, 2005). In extreme cases where ecosystems are particularly vulnerable the recovery time can take millennia (Kilian *et al.*, 2006).

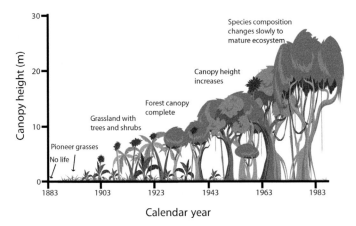

Figure 17.5   Ecological recovery of flora following destruction by ash deposition from the 1883 eruption of Krakatau in Indonesia (modified from Bush, 2006).

## 17.3  Oceanic ash deposition

### 17.3.1  Physical effects

Ash particles can affect marine ecosystems through light shading and thinning of the euphotic zone (Figure 17.3I). In the open ocean this effect is limited by the rate of settling through the water column, a process that appears to be rapid (Wiesner *et al.*, 1995). However, in coastal ecosystems the remobilization of ash by tides, waves and currents can continue to inhibit photosynthesis (Figure 17.6). This effect would be extremely long-lived in scenarios where the erosion of a terrestrial ash blanket affects benthic coastal ecosystems, such as coral reefs, previously unaccustomed to high sediment fluxes. The settling of particles through the water column may have a transient effect on planktic flora and fauna, by becoming tangled in the mucus nets of pteropods (Wall-Palmer *et al.*, 2011), by affecting micro-organism buoyancy, by clogging the gills of fish or by affecting larger filter-feeding animals such as whales (Figure 17.3J).

### 17.3.2  Fertilization

The growth of phytoplankton (the dominant marine primary producers) will continue unabated in sunlit areas until the water becomes deficient in one or several ingredients, commonly N, P or Fe. Ash surface coatings contain abundant macro- (N, P, Si) and micro-nutrients (e.g. Fe, Mn, Zn) integral to photosynthesis and other biological processes. The release of these elements can therefore potentially increase primary productivity (Figure 17.3K). It has been suggested that the

Figure 17.6   An ash-laden lahar/jökulhlaup arrives in the coastal waters of southern Iceland during the Eyjafjallajökull eruption, 14 April 2010. Image courtesy of Thórdís Högnadóttir (©). A black and white version of this figure will appear in some formats. For the colour version, please refer to the plate section.

deposition of volcanic ash could enhance $CO_2$ removal from the atmosphere by increasing the strength of the biological carbon pump (Spirakis, 1991; Sarmiento, 1993), i.e. the fixation of gaseous $CO_2$ as organic carbon and as solid $CaCO_3$, and subsequent sinking and removal from the surface ocean upon death. There are now a range of studies that support volcanic ash as a fertilizer in a range of ocean surface waters (e.g. Uematsu *et al.*, 2004; Hamme *et al.*, 2010; Achterberg *et al.*, 2013), with anecdotal evidence of the potential for a food-web response (Olgun *et al.*, 2013). The best candidates for a large increase in biological activity following ash deposition are surface waters with high nutrients and low chlorophyll (HNLC), essentially waters that are limited by the availability of one or several key nutrients (Duggen *et al.*, 2010). The utilization of nutrients depends on the biologically limiting factors within the ecosystem, and any subsequent impact will vary greatly with the depositional environment.

### 17.3.3 Poisoning

Some ingredients in the cocktail of elements released are inhibitory to biological growth, including many that are nutrients at lower concentrations (Brand *et al.*, 1986; Bruland *et al.*, 1991). Toxicity thresholds vary considerably between species, with diatoms appearing to fare better than coccolithophores and cyanobacteria in response to toxic-metal release (Hamme *et al.*, 2010; Hoffmann *et al.*,

2012). Explosive eruptions from the island of Montserrat in the Caribbean appear to be linked to considerable mortality events of pteropods and planktonic foraminifera (Wall-Palmer *et al.*, 2011). When ash-leachate fluxes are sufficiently elevated, the high buffering capacity of seawater can be overcome and induce transient acidic conditions (Frogner *et al.*, 2001; Jones and Gislason, 2008). This can increase the bioavailability of ions such as $Cu^{2+}$ and $Al^{3+}$, which are toxic to fish and other organisms. A drop in pH also affects the saturation states of $CaCO_3$ polymorphs, including calcite and aragonite. Consequently, organisms that utilize $CaCO_3$ for shell or skeleton formation experience reduced calcification rates, malformation and enhanced dissolution (Gattuso *et al.*, 1998; Riebesell *et al.*, 2000). Given that biogenic carbonate precipitation is dominated by phytoplankton, declines in these species would have a knock on effect on the food-web structure and health. There is, however, evidence that ecosystems can become acclimatized to such forcings, with some coccolithophore species able to produce exudates or store toxic metals intra-cellularly in response to high transition metal concentrations (e.g. Dupont and Ahner, 2005).

## 17.4 Long-term climate impacts

Large explosive eruptions have been hypothesized as possible catalysts for step changes in Earth's climate, although this remains contentious due to the paucity of data available (e.g. Oppenheimer, 2002). Terrestrial surface changes, while potentially devastating to local ecosystems, are predicted not to have a sustained impact on global climate (Jones *et al.*, 2007). However, there are numerous ways in which ash deposition on land and at sea can lead to long-term carbon removal from the atmosphere.

### 17.4.1 Primary productivity

One avenue for ash deposition to instigate long-term atmospheric $CO_2$ change is through increased photosynthesis (Figure 17.3L). The weathering and dissolution of volcanic products is an integral part of global element cycling, with the reflux of key nutrients to the ocean–atmosphere system essential to maintaining primary productivity in both terrestrial and marine environments. Periods of Earth's history with flare-ups of large-scale volcanism have been suggested as instigators and/or contributors to millennial climate change through increased fertilization of HNLC waters (e.g. Bay *et al.*, 2004; Cather *et al.*, 2009). While plausible, there are several factors that are poorly quantified in this hypothesis. Firstly, given the efficient recycling of organic material and associated respiration in the water column, it is unclear to what degree increased primary productivity results in elevated carbon export to depth. Secondly, eruption volume, duration and depositional environment will have a very large effect on any subsequent fertilization. A large-scale silicic eruption that is largely deposited in the open ocean will have a relatively limited immediate effect, given the speed of settling of ash through the water column (Wiesner *et al.*, 1995), volcanogenic atmospheric effects during the period of deposition (e.g. a $H_2SO_4$-induced volcanic winter; Self, 2006) and uncertainties of residence times of insoluble nutrients such as Fe in the euphotic zone in the absence of biological uptake. Marine ash deposition may, however, have a delayed effect because benthic waters, enriched by nutrient release, may be subsequently upwelled (Cather *et al.*, 2009). Sustained surface-water fertilization that decreases atmospheric $CO_2$ is more likely for episodes of many smaller-scale eruptions, or the remobilization of a large terrestrial ash blanket.

### 17.4.2 Silicate weathering

The abundance of easily weatherable, fine-grained material leads to the consumption of $CO_2$ through the breakdown of primary silicates, such as this reaction for plagioclase feldspar:

$$CaAl_2Si_2O_8 + 2\,CO_2 + 3\,H_2O \rightarrow Al_2Si_2O_5(OH)_4 + Ca^{2+} + 2\,HCO_3^- \quad (17.1)$$

The non-clay components are then transported to the oceans, where the vast majority of the alkali earth metals form carbonates by:

$$Ca^{2+} + 2\,HCO_3^- \rightarrow CaCO_3 + CO_2 + H_2O \quad\quad\quad (17.2)$$

The net effect over geological timescales is that nearly one mole of $CO_2$ is consumed for each mole of $Ca^{2+}$ and $Mg^{2+}$ weathered and released into seawater. The reworking, weathering and erosion of volcanic deposits will therefore result in a considerable drawdown of atmospheric $CO_2$ (Gislason *et al.*, 2009). Given that even moderate-sized eruptions, such as from Mount St Helens in 1980, can emit tephra with an estimated total specific surface area of $\sim 5 \times 10^8$ km$^2$ (equivalent to the Earth's surface area; Ayris and Delmelle, 2012), the subsequent weathering after a large eruption could markedly decrease atmospheric $CO_2$ concentrations over millennial timescales.

### 17.4.3 Carbon burial

A third factor affecting long-term climate impacts is the marine carbon export to sediments, which results from ash deposition (Figure 17.3M, N). Significant marine ash deposition instigates vertical density currents (Wiesner *et al.*, 1995; Manville and Wilson, 2004), allowing the transport of organic material in affected waters to depth without recycling. It is also possible, but currently unknown, that organic material may adhere to ash particles during settling. Given the strong efficiency of organic material recycling in the water column, these effects could be non-trivial. Ash undergoes relatively rapid diagenesis on the seafloor, leading to a fast depletion of oxygen in pore waters through the oxidation of silicate-bound $Fe^{2+}$ (Haeckel *et al.*, 2001; Hembury *et al.*, 2012). This reduces the breakdown of buried organic material, strongly increasing the preservation of $C_{org}$ below the ash-layer surface.

In terrestrial environments, a large ash blanket can permanently decouple a soil from the atmosphere, leading to the formation of a new soil over the coming years to millennia and the storage of C in the palaeosol. An extreme example of this comes from New Zealand, where an ignimbrite eruption from Taupo covered 20 000 $km^2$ and engulfed ~ 1 $km^3$ of mature podocarp forest (Hudspith *et al.*, 2010). An albedo-related drop in soil temperature would slow microbial activity and, therefore, the rate of organic-matter decomposition (Ayris and Delmelle, 2012). New soil formation is often sluggish, given the dearth of N and $C_{org}$ in freshly deposited ash, but over millennial timescales even fragile ecosystems can completely recover (Kilian *et al.*, 2006).

### *17.4.4 Long-term global cooling?*

The high correlation between volcanism and millennial-timescale cooling in ice-cores (Bay *et al.*, 2004), and between ignimbrite flare-ups and onsets of glaciations (Cather *et al.*, 2009), has led several authors to conclude that Fe fertilization of ocean surface waters could have accelerated global cooling. While each process is poorly constrained, primary productivity, silicate weathering and carbon burial all act as a sink for atmospheric $CO_2$. This suggests that a combination of these processes could lead to an extended global cooling by reducing atmospheric $CO_2$ after large-scale explosive volcanism, although further research is needed to constrain each process.

### 17.5 Summary

The deposition of volcanic ash can lead to numerous contemporaneous climate forcings, which in turn will vary with location, volume, chemistry, climate, season and depositional environment. Therefore, the environmental response will be unique for each eruption. Ash deposition on land leads to changes in vegetation coverage and surface albedo, which interplay with atmospheric and oceanic circulation. Acid and metal release can poison ecosystems, while nutrient addition can sustain or replenish affected ecosystems over longer periods. In marine settings, ash deposition can both fertilize and poison ecosystems, altering nutrient availability in both surface and benthic environments and affecting entire food-web structures. In the aftermath of a large explosive eruption, the combined effects of increased photosynthesis, weathering and carbon burial will lower atmospheric $CO_2$ levels, but to what extent is poorly quantified and will vary hugely between each volcanic eruption.

## Acknowledgements

Becca Neely, Ian Bailey, John Jones, Lev Kasicki, Tamsin Mather, Vernon Manville and an anonymous reviewer are warmly thanked for constructive comments. This work was partly supported by the Research Council of Norway through its Centres of Excellence funding scheme, project number 223272.

## References

Achterberg, E.P., Moore, C.M., Heson, S.A. *et al.* (2013): Natural iron fertilization by the Eyjafjallajökull volcanic eruption. *Geophys. Res. Lett.*, **40**, 1–6.

Adamson, D.A., Williams, M.A.J., Gillespie, R. (1982): Palaeogeography of the Gezira and of the lower blue and white Nile rivers, 165–219. In *A Land Between Two Niles: Quaternary Geology and Biology of the Central Sudan*. Balkema, Rotterdam.

Antos, J.A., Zobel, D.B. (1985): Recovery of forest under-stories buried by tephra from Mount St. Helens. *Vegetatio*, **642**, 103–111.

Ayris, P.M., Delmelle, P. (2012): The immediate environmental effects of tephra emission. *Bull. Volcanol.*, **74**, 1905–1936.

Bay, R.C., Bramall, N., Price, P.B. (2004): Bipolar correlation of volcanism with millennial climate change. *Proc. Natl. Acad. Sci. USA*, **101**, 6341–6345.

Blackford, J.J., Edwards, K.J., Dugmore, A.J., Cook, G.T., Buckland, P.C. (1992): Icelandic volcanic ash in the mid-Holocene Scots pine (*Pinus sylvestris*) pollen decline in northern Scotland. *Holocene*, **2**, 260–265.

Brand, L.E., Sunda, W.G., Guillard, R.R.L. (1986): Reduction of marine phytoplankton reproductive rates by copper and cadmium. *J. Exp. Mar. Biol. Ecol.*, **96**, 225–250.

Bruland, K.W., Donat, J.R., Hutchins, D.A. (1991): Interactive influences of bioactive trace metals on biological production in oceanic waters. *Limnol. Oceanogr.*, **36**, 1555–1577.

Bush, M.B. (2006): *Ecology of a Changing Planet*, 3rd edn. Prentice Hall, Upper Saddle River, NJ.

Cather, S.M., Dunbar, N.W., McDowell, F.W., McIntosh, W.C., Scholle, P.A. (2009): Climate forcing by iron fertilization from repeated ignimbrite eruptions: the icehouse–silicic large igneous province (SLIP) hypothesis. *Geosphere*, **5**, 315–324.

Collins, B.D., Dunne, T. (1986): Erosion of tephra from the 1980 eruption of Mount St. Helens. *Geol. Soc. Am. Bull.*, **97**, 896–905.

Cook, R.J., Barron, J.C., Papendick, R.I., Williams, G.J., III (1981): Impact on agriculture of the Mt. St. Helens eruptions. *Science*, **211**, 16–22.

Delfosse, T., Delmelle, P., Iserentant, A., Delvaux, B. (2005): Contribution of $SO_3$ to the acid neutralizing capacity of andosols exposed to strong volcanogenic acid and $SO_2$ deposition. *Eur. J. Soil Sci.*, **56**, 113–125.

Delmelle, P., Lambert, M., Dufrene, Y., Gerin, P., Oskarsson, N. (2007): Gas/aerosol–ash interaction in volcanic plumes: new insights from surface analyses of fine ash particles. *Earth Planet. Sci. Lett.*, **259**, 159–170.

Duggen, S., Croot, P., Schacht, U., Hoffmann, L. (2007): Subduction zone volcanic ash can fertilize the surface ocean and stimulate phytoplankton growth: evidence from biogeochemical experiments and satellite data. *Geophys. Res. Lett.*, **34**, L01612.

Duggen, S., Olgun, N., Croot, P. *et al.* (2010): The role of airborne volcanic ash for the surface ocean biogeochemical cycle: a review. *Biogeosci. Disc.*, **6**, 6441–6489.

Dupont, C.L., Ahner, B.A. (2005): Effects of copper, cadmium, and zinc on the production and exudation of thiols by *Emiliania huxleyi*. *Limnol. Oceanogr.*, **50**, 508–515.

Durant, A.J., Bonadonna, C., Horwell, C.J. (2010): Atmospheric and environmental impacts of volcanic particulates. *Elements*, **6**, 235–240.

Eggler, W.A. (1948): Plant communities in the vicinity of the volcano El Paricutin, Mexico, after two and a half years of eruption. *Ecology*, **29**, 415–436.

Fagan, W., Bishop, J. (2000): Trophic interactions during primary succession: herbivores slow a plant reinvasion at Mount St. Helens. *Am. Nat.*, **155**, 238–251.

Frogner-Kockum, P.C., Herbert, R.B., Gislason, S.R. (2006): A diverse ecosystem response to volcanic aerosols. *Chem. Geol.*, **231**, 57–66.

Frogner, P., Gislason, S.R., Oskarsson, N. (2001): Fertilizing potential of volcanic ash in ocean surface water. *Geology*, **29**, 487–490.

Gattuso, J.-P., Frankignoulle, M., Bourge, I., Romaine, S., Buddemeier, R. (1998): Effect of calcium carbonate saturation of seawater on coral calcification. *Global Planet. Change*, **18**, 37–46.

Gislason, S.R., Oelkers, E.H., Eiriksdottir, E.S. *et al.* (2009): Direct evidence of the feedback between climate and weathering. *Earth Planet. Sci. Lett.*, **277**, 213–222.

Haeckel, M., van Beusekom, J., Wiesner, M.G., König, I. (2001): The impact of the 1991 Mount Pinatubo tephra fallout on the geochemical environment of the deep-sea sediments in the South China Sea. *Earth Planet. Sci. Lett.*, **193**, 151–166.

Hamme, R.C., Webley, P.W., Crawford, W.R. *et al.* (2010): Volcanic ash fuels anomalous plankton bloom in subarctic northeast Pacific. *Geophys. Res. Lett.*, **37**, L19604.

Hembury, D.J., Palmer, M.R., Fones, G.R. *et al.* (2012): Uptake of dissolved oxygen during marine diagenesis of fresh volcanic material. *Geochim. Cosmochim. Acta*, **84**, 353–368.

Hoffmann, L.J., Breitbarth, E., Ardelan, M.V. *et al.* (2012): Influence of trace metal release from volcanic ash on growth of *Thalassiosira pseudonana* and *Emiliania huxleyi*. *Mar. Chem.*, **132–133**, 28–33.

Hudspith, V.A., Scott, A.C., Wilson, C.J.N., Collinson, M.E. (2010): Charring of woods by volcanic processes: an example from the Taupo ignimbrite, New Zealand. *Palaeogeogr. Palaeoclimatol. Palaeoecol.*, **291**, 40–51.

Jones, M.T. (2008): The climatic impact of supervolcanic eruptions. Ph.D. Thesis, University of Bristol, UK.

Jones, M.T., Gislason, S.R. (2008): Rapid releases of metal salts and nutrients following the deposition of volcanic ash into aqueous environments. *Geochim. Cosmochim. Acta*, **72**, 3661–3680.

Jones, M.T., Sparks, R.S.J., Valdes, P.J. (2007): The climatic impact of supervolcanic ash blankets. *Clim. Dyn.*, **29**, 553–564.

Jones, S. (2010): Palaeoenvironmental response to the 74 ka Toba ash-fall in the Jurreru and Middle Son valleys in southern and north-central India. *Quat. Res.*, **73**(2), 336–350.

Knight, T., Chase, J. (2005): Ecological succession: out of the ash. *Curr. Biol.*, **15**, R926–R927.

Kilian, R., Biester, H., Behrmann, J. *et al.* (2006): Millenium-scale volcanic impact on a superhumid and pristine ecosystem. *Geology*, **34**, 609–612.

Mack, R.N. (1981): Initial effects of ashfall from Mount St. Helens on vegetation in eastern Washington and adjacent Idaho. *Science*, **213**, 537–539.

Manville, V. (2002): Sedimentary and geomorphic responses to ignimbrite emplacement: readjustment of the Waikato River after the AD 181 Taupo eruption, New Zealand. *J. Geol.*, **110**, 519–541.

Manville, V., Wilson, C.J.N. (2004): Vertical density currents: a review of their potential role in the deposition and interpretation of deep-sea ash layers. *J. Geol. Soc.*, **161**, 947–958.

Manville, V., Németh, K., Kano, K. (2009): Source to sink: a review of three decades of progress in the understanding of volcaniclastic processes, deposits, and hazards. *Sediment. Geol.* **220**(3–4), 136–161.

Martin, R.S., Mather, T.A., Pyle, D.M. *et al.* (2009): Sweet chestnut (*Castanea sativa*) leaves as a bio-indicator of volcanic gas, aerosol and ash deposition onto the flanks of Mt Etna in 2005–2007: *J. Volcanol. Geotherm. Res.*, **179**, 107–119.

Mason, B.G., Pyle, D.M., Oppenheimer, C. (2004): The size and frequency of the largest explosive eruptions on Earth. *Bull. Volcanol.*, **66**, 735–748.

Nammah, H., Larsen, F.E., McCool, D.K., Fritts, R., Molnau, M. (1986): Mt. St. Helens volcanic ash: effect of incorporated and unincorporated ash of two particle sizes on runoff and erosion. *Agr. Ecosyst. Environ.*, **15**, 63–72.

Olgun, N., Duggen, S., Langmann, B. *et al.* (2013): Geochemical evidence of oceanic iron fertilization by the Kasatochi (2008) volcanic eruption and evaluation of the potential impacts on sockeye salmon population. *Mar. Ecol.-Prog. Ser.*, **488**, 81–88.

Oppenheimer, C. (2002): Limited global change due to the largest known Quaternary eruption, Toba ~ 74 kyr BP? *Quat. Sci. Rev.*, **21**, 1593–1609.

Oskarsson, N. (1980): The interaction between volcanic gases and tephra: fluorine adhering to tephra of the 1970 Hekla eruption. *J. Volcanol. Geotherm. Res.*, **8**, 251–266.

Riebesell, U., Zondervan, I., Rost, B. *et al.* (2000): Reduced calcification of marine phytoplankton in response to increased atmospheric $CO_2$. *Nature*, **407**, 364–367.

Rose, W.I. (1977): Scavenging of volcanic aerosol by ash: atmospheric and volcanologic implications. *Geology*, **5**, 621–624.

Rose, W.I., Durant, A.J. (2009): Fine ash content of explosive eruptions. *J. Volcanol. Geotherm. Res.*, **186**, 32–39.

Sarmiento, J.L. (1993): Atmospheric $CO_2$ stalled. *Nature*, **365**, 697–698.

Self, S. (2006): The effects and consequences of very large explosive volcanic eruptions. *Phil. Trans. R. Soc. A.*, **364**, 2073–2097.

Shoji, S., Nanzyo, M., Dahlgren, R.A. (1993): *Volcanic Ash Soils. Genesis, Properties, and Utilization*. Elsevier, Amsterdam.

Spirakis, C.S. (1991): Iron fertilization with volcanic ash? *EOS*, **2**, 525.

Thordarson, T., Self, S. (2003): Atmospheric and environmental effects of the 1783–1784 Laki eruption: a review and reassessment. *J. Geophys. Res.*, **108**, D1, 4011.

Trenberth, K. (1999): Atmospheric moisture recycling: role of advection and local evaporation. *J. Climate*, **12**, 1368–1381.

Uematsu, M., Toratani, M., Kajino, M. *et al.* (2004): Enhancement of primary productivity in the western North Pacific caused by the eruption of the Miyake-jima volcano. *Geophys. Res. Lett.*, **31**, L06106.

Wall-Palmer, D., Jones, M.T., Hart, M.B. *et al.* (2011): Explosive volcanism as a cause for mass mortality of pteropods. *Mar. Geol.*, **282**, 231–239.

Wiesner, M., Wang, Y., Lianfu, Z. (1995): Fallout of volcanic ash to the deep South China Sea induced by the 1991 eruption of Mount Pinatubo (Philippines). *Geology*, **23**, 885–888.

Witham, C.S., Oppenheimer, C., Horwell, C.J. (2005): Volcanic ash-leachates. A review and recommendations for sampling methods. *J. Volcanol. Geotherm. Res.*, **141**, 299–326.

# 18

# Oceanic anoxia during the Permian–Triassic transition and links to volcanism

ELLEN K. SCHAAL, KATJA M. MEYER, KIMBERLY V. LAU, JUAN CARLOS
SILVA-TAMAYO AND JONATHAN L. PAYNE

## 18.1 Introduction

One of the best-documented environmental perturbations at the Permian–Triassic transition is the development of extensive oceanic anoxia, a prime suspect in the mass extinction of marine organisms at the end-Permian (Wignall and Hallam, 1992). Ever since the eruption of the Siberian Traps flood basalt province was recognized as contemporaneous with the Permian–Triassic boundary (PTB; Renne *et al.*, 1995), there has been ongoing research on how the Siberian Traps may have caused changes in ocean chemistry and, ultimately, mass extinction. Massive volcanism could affect ocean oxygen content through both greenhouse warming and reduced solubility of oxygen in seawater (Wignall and Twitchett, 1996) and continental weathering feedbacks on nutrient delivery, primary productivity and oxygen demand (Meyer *et al.*, 2008).

The development of new proxies for ocean redox chemistry over the last decade has enabled rapid increases in empirical constraints on Permian–Triassic ocean anoxia. Consequently, numerous proxy records have been produced since the last systematic review of the topic by Wignall and Twitchett (2002). Here, we review the constraints provided by each proxy (summarized in Table 18.1) and discuss insights from Earth system models and geological observations in evaluating the potential role for Siberian Traps volcanism in driving ocean anoxia.

## 18.2 Lithological evidence for anoxia

### 18.2.1 Bioturbation intensity and black shales

Poorly bioturbated, fine-grained sediments are typical of dysaerobic to anoxic depositional settings (reviewed in Wignall, 1994). Such sediments have long been

*Volcanism and Global Environmental Change*, eds. Anja Schmidt, Kirsten E. Fristad and Linda T. Elkins-Tanton.
Published by Cambridge University Press. © Cambridge University Press 2015.

Table 18.1 Sensitivity of various proxies to paleoredox conditions. The categories for degree of deoxygenation are: suboxic (0.3–1.5 mL/L $O_2$), dysoxic (0.1–0.3 mL/L $O_2$), anoxic (0–0.1 mL/L $O_2$) and euxinic (anoxic and sulfidic) (Kaiho, 1994).

| Proxy | Degree of deoxygenation | Water depth of deoxygenation | Spatial extent of deoxygenation |
|---|---|---|---|
| Poorly bioturbated to laminated shales | Dysoxic to anoxic | Bottom water | Local |
| Framboidal pyrite | Euxinic | Water column | Local |
| Isorenieratane and aryl isoprenoids | Euxinic | Photic zone | Local |
| Sulfur isotopes | Euxinic | Unclear | Unclear |
| Cerium anomaly (Ce/Ce*) | Suboxic to anoxic | Bottom water | Local |
| Uranium concentration | Anoxic | Bottom water | Local/global* |
| Uranium isotopes | Anoxic | Bottom water | Global |
| Molybdenum isotopes | Anoxic to euxinic | Bottom water | Global |

* Interpreting the spatial extent of deoxygenation from uranium concentrations depends on context. Anoxic facies can record local changes in anoxia and uranium preservation, but shallow-marine carbonates may approximate global ocean changes in uranium concentration and the extent of oceanic anoxia.

known from PTB sections, especially in the Tethys region (e.g. Logan and Hills, 1973), and they have been used to implicate ocean anoxia as a cause of the end-Permian mass extinction (Hallam, 1989). Numerous PTB sections from the Tethys and Panthalassic Oceans exhibit lithological characteristics typically of dysoxic to anoxic settings, such as reduced intensity of bioturbation and sizes of burrows, increased abundance of framboidal pyrite and increased prevalence of black shales (reviewed in Wignall and Twitchett, 2002). In shallow-water sections, the onset of these indicators is typically coincident with the main extinction interval in the fossil record (e.g. Wignall and Hallam, 1992; Wignall and Twitchett, 1996). However, several lines of evidence suggest that the variation in ocean redox chemistry was temporally and spatially complex. Changes in the lithology and chemistry of deep-marine shales from Japan and British Columbia suggest that the onset of anoxia was more gradual, beginning near the middle–late Permian transition, peaking in intensity near the PTB, and waning gradually through Early Triassic time (Isozaki, 1997). Variation in ichnofauna and ichnofabric across a palaeo-depth gradient in northwest Canada indicates shallow-water environments were typically better oxygenated than deeper-water environments during the earliest Triassic (Griesbachian) time (Beatty *et al.*, 2008). Precipitation of carbonate crystal fans in outer shelf deposits of the uppermost Lower Triassic (Spathian) Virgin Limestone in the western USA points toward local persistence or recurrence of anoxic conditions even several million years after the main extinction event (Woods *et al.*, 1999).

### 18.2.2 Framboidal pyrite

The presence and size distribution of framboidal pyrite grains can be particularly informative regarding ancient ocean chemistry because exclusively small framboidal pyrite is expected when the overlying water column is euxinic (anoxic and sulfidic) and pyrite framboids form in the water column (Wilkin *et al.*, 1996). Pyrite framboids have been widely reported from PTB sections, particularly in strata overlying the main extinction horizon (Figure 18.1; e.g. Wignall and Hallam, 1992; Bond and Wignall, 2010). Small framboids occur in abundance above the main extinction horizon on shallow-marine, tropical carbonate platforms (e.g. Bond and Wignall, 2010; Liao *et al.*, 2010); in shallow-marine, tropical, mixed carbonate and clastic systems (Wignall *et al.*, 2005); in deep-water, tropical carbonates (e.g. Shen *et al.*, 2007); and in high-latitude clastic deposits (e.g. Nielsen and Shen, 2004; Bond and Wignall, 2010; Nielsen *et al.*, 2010). Abundant framboids occur below the mass extinction horizon at some localities, suggesting euxinic conditions developed prior to the main extinction event in some regions (e.g. Nielsen and Shen, 2004). In deep-water sections, the co-occurrence of small framboidal pyrite grains with redox-sensitive metal abundance patterns, indicative of oxic to dysoxic bottom water conditions, suggests that a euxinic oxygen minimum zone existed within a water column oxygenated at the surface by wind mixing and in deep water via large-scale circulation (Algeo *et al.*, 2010). The frequency of framboidal pyrite in stratigraphic sections and trends toward smaller size within some sections point toward a greater prevalence of euxinic conditions above the end-Permian extinction horizon.

## 18.3 Geochemical evidence for anoxia

### 18.3.1 Organic biomarkers

Biomarker studies have primarily focused on the immediate PTB, with very limited data for the Changhsingian and Griesbachian. Isorenieratane and aryl isoprenoid derivatives occur in late-Permian and PTB rocks from the Perth Basin (Western Australia; Grice *et al.*, 2005) and Meishan (China; Cao *et al.*, 2009) in the Tethys region, reflecting the presence of green sulfur bacteria and indicating euxinic conditions within the photic zone (Figure 18.1). In addition, Panthalassic sediments from the Peace River Basin (Canada; Hays *et al.*, 2007) and Boreal Ocean sediments from Kap Stosch (Greenland; Hays *et al.*, 2012) also contain biomarker evidence for photic-zone euxinia (PZE) during late-Permian time.

Following the extinction, evidence for PZE persists into Griesbachian strata in all three of the major ocean basins. Again, isorenieratane occurs in the Perth Basin (Grice *et al.*, 2005), at the GSSP (Global Boundary Stratotype Section and Point)

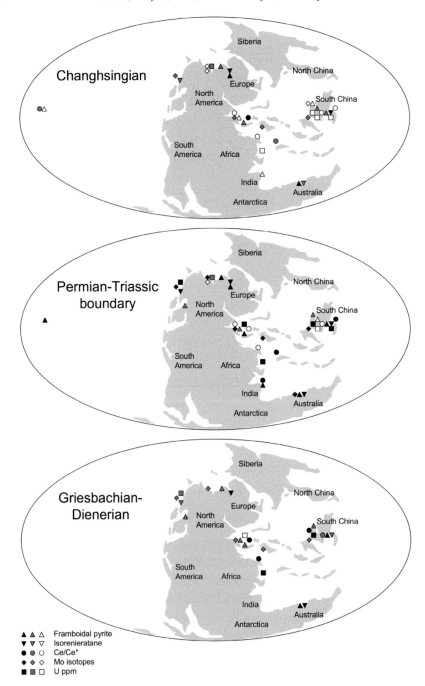

Figure 18.1　Global palaeogeography of the late Permian–Early Triassic, showing the locations and ocean redox conditions indicated by proxy records discussed in the text. Black symbols = anoxia/euxinia, grey symbols = dysoxia or intermittent anoxia, white symbols = oxic conditions. Note the global expansion of oceanic anoxia/euxinia seen in most proxy records at the PTB. Base map modified from R. Blakey (http://cpgeosystems.com/260moll.jpg).

at Meishan (Cao *et al.*, 2009), in Greenland (Hays *et al.*, 2012) and in British Columbia (Hays *et al.*, 2007). In the Peace River Basin and at Meishan, aromatic biomarkers persist into Dienerian strata, where PZE indicators wane and then disappear. The most common interpretation of isorenieratane and its aryl isoprenoid derivatives is that of water-column euxinia, but anoxygenic phototrophs can also produce these biomarkers within microbial mats underlying an oxic water column (Casford, 2011; Meyer *et al.*, 2011), leaving open the possibility that some of these biomarker records reflect benthic conditions.

### 18.3.2 Sulfur isotopes

Sulfur isotopes are sensitive to sulfate reduction and to the proportion of pyrite burial, thus linking them to euxinic ocean conditions. During sulfate reduction, the light isotope ($^{32}$S) is preferentially incorporated into sulfide unless the system is sulfate-limited (Habicht *et al.*, 2002). Therefore, as pyrite forms in euxinic water columns, and sulfur is preferentially buried as pyrite rather than sulfate, isotopically light sulfur is removed from the oceans (Wignall and Hallam, 1992). The first sulfur isotope records across the Permian–Triassic transition were from marine evaporites, which show a positive shift in $\delta^{34}$S values from +12‰ to +28‰ between the end-Permian and the late Early Triassic (e.g. Cortecci *et al.*, 1981), suggesting a prolonged interval of extensive anoxia and pyrite formation.

More recent and higher-resolution $\delta^{34}$S records from pyrite and carbonate-associated sulfate (CAS) tend to show a rapid short-term variation, with adjacent samples differing by many per mil (e.g. Riccardi *et al.*, 2006; Algeo *et al.*, 2008; Nielsen *et al.*, 2010). Such large changes over such a short timescale are difficult to explain by changing the isotopic composition of the whole ocean. Instead, they may reflect a vertically stratified ocean with isotopically distinct sulfur reservoirs in shallow and deep water; for such an ocean, variation in $\delta^{34}$S records would reflect the movement of water masses (e.g. upwelling; Kajiwara *et al.*, 1994; Newton *et al.*, 2004; Algeo *et al.*, 2008). A potential problem with this mechanism is that any $\delta^{34}$S gradient generated by sulfate reduction in the deep ocean would not be preserved during upwelling and mixing of isotopically light sulfide and isotopically heavy sulfate from deep waters (Riccardi *et al.*, 2006). Alternatively, the excursions could represent rapid changes in the proportion of pyrite to sulfate burial in an ocean with low sulfate concentrations (Luo *et al.*, 2010). Because these two distinct interpretations of $\delta^{34}$S variation have inherent differences in the extent of anoxia they imply (local upwelling versus global redox shifts), it is at present unclear whether sulfur isotope records tell us about local or global ocean oxygen conditions.

### 18.3.3  Cerium and REEs

The concentration of cerium relative to other rare earth elements (REEs) is redox sensitive (reviewed in German and Elderfield, 1990). A positive cerium anomaly (Ce/Ce*) recorded in carbonate sediments is interpreted to reflect suboxic–anoxic conditions in the local water column (Alibo and Nozaki, 1999).

There is spatial variation in Ce/Ce* throughout the Changhsingian and Griesbachian (Figure 18.1). Records from Panthalassa have been interpreted as a shift from oxic conditions in the middle Permian to suboxic–anoxic conditions during the Changhsingian (Kato et al., 2002). In the Tethys region, some sections indicate anoxic conditions at the extinction horizon (Kakuwa and Matsumoto, 2006; Algeo et al., 2007), whereas others indicate oxygenated conditions (e.g. Dolenec et al., 2001; Fio et al., 2010; Brand et al., 2012). These PTB records could either reflect spatially heterogeneous bottom-water redox conditions or suggest that Ce/Ce* records have been misinterpreted; Loope et al. (2013) argue that, in some sections, the discrepancy between geochemical evidence for anoxia and very fossiliferous strata indicates that the Ce/Ce* records have been compromised by silicilastic influence or dolomitization. Following the extinction, Ce/Ce* measured in conodont apatite from south China suggests multiple pulses of anoxia occurring from the PTB to the earliest Dienerian, the basal Smithian to the earliest Spathian, and in the mid Spathian (Song et al., 2012).

### 18.3.4  Uranium

The marine uranium cycle offers the potential to constrain the global extent of seafloor anoxia, as the uranium concentration and the $^{238}U/^{235}U$ isotope ratio are each sensitive to redox conditions at the sediment–water interface (Stirling et al., 2007; Weyer et al., 2008). Changes in uranium concentrations and isotope ratios may represent global conditions because the residence time of uranium in the modern ocean ($3.2$–$5.6 \times 10^5$ yr) is much longer than the ocean mixing time (Dunk et al., 2002).

The majority of Permian–Triassic uranium data are confined to the extinction horizon. Uranium concentrations decrease abruptly at the extinction horizon in shallow-marine carbonates in Saudi Arabia (Ehrenberg et al., 2008), Iran (Tavakoli and Rahimpour-Bonab, 2012) and south China (Brennecka et al., 2011; Song et al., 2012), indicating a global drawdown of uranium concentrations in seawater caused by ocean anoxia (Figure 18.1). The only sections that show an increase in uranium concentration are localities where deposition switches to black shales and other dysaerobic facies (e.g. British Columbia and Italy), where uranium would be concentrated due to local anoxia (Wignall and Twitchett, 2002). Records from clastic sections are more complicated due to the local effects of bottom-water redox changes superimposed on the global shift in seawater uranium concentrations (Algeo *et al.*, 2012; Shen *et al.*, 2012). The sole $\delta^{238}U$ record for the PTB demonstrates a rapid negative isotopic shift coincident with the extinction horizon, also consistent with a global increase in seafloor anoxia (Brennecka *et al.*, 2011).

Studies that extend beyond the extinction horizon indicate that anoxia continued, at least episodically, through much of Early Triassic time. Uranium concentrations in carbonates suggest that anoxic conditions were prevalent throughout the Griesbachian into the Dienerian (Ehrenberg *et al.*, 2008). Conodont Th/U data from south China point toward multiple peaks in anoxia, in the Griesbachian, Smithian–earliest Spathian, and mid Spathian, after which the ocean became oxygenated through the Middle Triassic (Song *et al.*, 2012).

### 18.3.5 Molybdenum

The Mo isotope fractionations between seawater and modern marine sediments deposited under different redox conditions allows the use of ancient sediments to track the evolution of the Mo isotope composition of ancient oceans; to quantify the partitioning of Mo isotopes by oxic, suboxic–anoxic and euxinic sinks in ancient oceans; and, consequently, to quantify oceanic oxygen levels (Siebert *et al.*, 2003; Neubert *et al.*, 2008; Dahl *et al.*, 2010).

For most of the late Palaeozoic, seawater $\delta^{98/95}Mo$ values remained at around +2.0‰, which is slightly lower than those of modern seawater and implies a similar oxygen content (Dahl *et al.*, 2010). At the end-Permian mass extinction horizon, seawater $\delta^{98/95}Mo$ values decreased to ~ 0‰ (Voegelin *et al.*, 2009; Silva-Tamayo *et al.*, 2013; Zhou *et al.*, 2012; Proemse *et al.*, 2013). This decrease, which is recorded by successions located along the Tethys and the Panthalassic oceans, has been interpreted as the global expansion of oceanic euxinia during the main extinction event (Figure 18.1). Low $\delta^{98/95}Mo$ values (+1.5‰) characterize the Early Triassic oceans and suggest the return to anoxic conditions (Silva-Tamayo *et al.*, 2013).

## 18.4 Pattern of anoxia – a summary

High-resolution records and novel palaeoredox proxies have sharpened and altered our picture of oceanic anoxia across the Permian–Triassic transition. Rather than being prolonged and persistent, anoxia appears to have waxed and waned in several discrete episodes during Early Triassic time (Wignall and Twitchett, 2002; Song *et al.*, 2012). Biomarkers for green sulfur bacteria in PTB sediments have

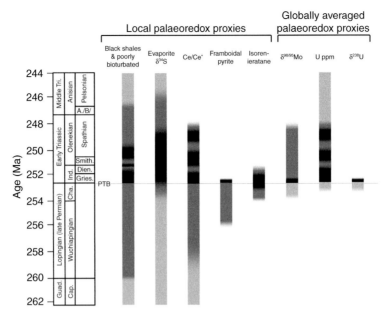

Figure 18.2    Summary of the prevalence of anoxia through time (by proxy). Blue = oxic oceans, grey = some evidence for anoxia–suboxia, black = strong evidence for widespread anoxia, white = no data. The $\delta^{34}$S evaporite record is interpreted to reflect changes in the proportion of sulfur buried as pyrite under euxinic conditions. Note the abrupt change from more oxic to more anoxic oceans at the PTB, especially in globally averaged palaeoredox proxies. See also colour plates section.

confirmed the existence of euxinic conditions in the photic zone (Grice *et al.*, 2005; Hays *et al.*, 2007). Uranium and molybdenum proxies are opening opportunities to quantify the extent of anoxia globally from data in local sections (Brennecka *et al.*, 2011; Proemse *et al.*, 2013).

Taken together, redox proxy data provide evidence of a clear trend toward rapid deoxygenation across the PTB (Figure 18.2). New proxies that represent globally averaged redox conditions tend to contradict the idea that widespread end-Permian anoxia began early in the late Permian (Isozaki, 1997). Uranium and molybdenum records indicate generally oxic ocean conditions right up to the main extinction horizon and a rapid expansion of anoxia at that time. Pyrite framboids, biomarkers and cerium-anomaly records require some localized deoxygenation in the Changhsingian, especially in deeper-water settings in Panthalassa, the Boreal Ocean and the Perth Basin (Figure 18.1). In the Tethys region, typically fossiliferous Changhsingian strata suggest that any dysoxic conditions were not extreme enough to restrict animal life. By the PTB, however, the only remaining palaeoredox records indicating oxic conditions come from shallow-marine sections, where the surface waters may have been oxygenated by wind mixing despite the expansion of oxygen minimum zones onto the continental shelves.

After the PTB, palaeoredox constraints indicate intermittent oxic and anoxic conditions throughout the Early Triassic moving toward more oxic conditions in the Middle Triassic. Lithologic observations, cerium anomalies and uranium concentrations together point toward recurring intervals of widespread anoxia in the Early Triassic: the most widely recognized at the PTB through the Griesbachian, another at the Smithian/Spathian boundary event, and another in the Spathian.

Apparent disagreements among proxies about the redox state of Permian–Triassic oceans likely result from differences in sensitivity across proxies as well as spatial heterogeneity in marine redox conditions (Figure 18.2). For example, even though globally averaged proxies support a well-ventilated late-Permian ocean, the Peace River Basin (British Columbia) could have experienced upwelling, photic-zone euxinia and preservation of isorenieratane due to local controls on circulation patterns. In addition, variation in redox conditions on geologically short timescales could create disagreements among proxies. For example, seasonal or other short-term anoxic pulses might leave biomarker evidence of euxinia within beds that also contain animal fossils. Most local proxies agree with widespread anoxia at the PTB, if not the specific timing of anoxic pulses in the Early Triassic. Although the $\delta^{34}$S evaporite record seems to indicate persistent anoxia throughout the Early Triassic (Figure 18.2), this is likely a consequence of low temporal resolution of this proxy, rather than a contradiction of other, more punctuated records. Conflicts among the proxies recording a globally averaged palaeoredox signal can be resolved due to differing sensitivities to the degree of deoxygenation; Mo is more sensitive to euxinia while U is sensitive to anoxia, explaining the widespread episodes of Early Triassic anoxia evident in U records but not in Mo records.

Palaeoredox proxies that record a globally averaged signal point toward a major expansion of anoxic waters at the end-Permian extinction horizon, but these records have not yet been developed much beyond the PTB interval (Figure 18.2). With primarily records of local geochemical conditions below the boundary, the baseline redox state of Permian oceans remains poorly constrained. Likewise, while there is evidence for reoccurring widespread anoxia later in the Early Triassic, we do not yet know if these episodes were as extensive or persistent as that at the PTB.

## 18.5 Volcanism and anoxia

The potential for massive volcanism to drive ocean anoxia is well established (reviewed in Meyer and Kump, 2008). Because oxygen solubility is temperature-dependent, the release of volcanic $CO_2$ and resulting greenhouse warming decrease oxygen supply to ocean water. In addition, global warming enhances the hydrologic cycle, continental weathering and nutrient delivery to the oceans (e.g. phosphate), which, in turn, stimulates biological production and increases oxygen demand. The development of anoxic conditions creates a positive feedback, liberating more phosphate from sediments and increasing nutrient fluxes (Van Cappellen and Ingall, 1994).

Numerical Earth system modelling has been used to investigate the factors that led to widespread ocean anoxia during the Permian and Triassic. While ocean stagnation has long been linked to anoxia in the palaeoceanographic literature, most modelling studies have shown that end-Permian stagnation was physically unlikely. In an early review of modelling studies, Winguth and Maier-Reimer (2005) concluded that the end-Permian oceans were likely well-ventilated and that anoxia was more likely a result of changes in carbon cycling than the physical mixing of the oceans. A fully coupled, high-resolution global climate simulation of end-Permian conditions also showed reduced, but not absent, overturning circulation (a ~ 25% increase in the ideal age of water at 3000 m), and that marine biotic changes were likely influenced by rapid increases in atmospheric $p$CO$_2$ (Kiehl and Shields, 2005).

Many numerical models require elevated nutrient contents of the ocean to generate anoxic conditions. In multiple studies, anoxia and euxinia are more widespread in the Palaeo-Tethys Ocean than the Panthalassic Ocean due to the nutrient trapping circulation of the Tethys Ocean (Meyer *et al.*, 2008; Winguth and Winguth, 2012). Simulations using Earth system models of intermediate complexity that also included a marine sulfur cycle required at least a tripling of phosphate (the limiting nutrient over long timescales; Tyrrell, 1999) to generate PZE in the Palaeo-Tethys Ocean and in areas of upwelling (Figure 18.3; Meyer *et al.*, 2008). However, the extent of oxygen depletion and hydrogen sulfide build-up varies widely between model parameterizations. At least one model generates oxygen depletion but no extreme anoxia–euxinia even under 10 × modern phosphate concentrations (Winguth and Winguth, 2012). However, primary production in this model was limited by Fe supply, and euxinia would have surely resulted from a combination of high Fe and high phosphate conditions.

Siberian Traps volcanism makes an attractive trigger for PTB anoxia, because of the potential for rapid, massive $CO_2$ release (Kamo *et al.*, 2003) and weathering–nutrient feedbacks. There have been multiple efforts to estimate total $CO_2$ release from the volume of erupted basalt and degassing country rocks (see Chapters 10–12). However, we have yet to quantify the corresponding amount and duration of greenhouse warming, or the expected increase in nutrient delivery to the oceans. Using Earth system models that incorporate

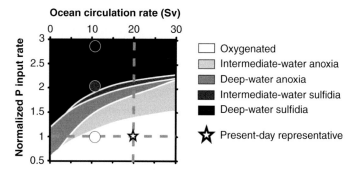

Figure 18.3    Ocean redox conditions based on changes in ocean circulation rate (in sverdrups, Sv) and marine phosphate input in a one-dimensional ocean biogeochemical model (modified after Ozaki *et al.*, 2011). Circles represent Earth system modelling results of end-Permian euxinia from Meyer *et al.* (2008). Note that the decrease in ocean circulation alone is insufficient to cause ocean anoxia.

biogeochemical representations of the marine C, N, P, S and other cycles, future work should test the hypothesis that Siberian Traps eruption was large enough to cause the observed deoxygenation.

Geochronology supports the link between volcanism and the pattern of anoxia at the Permian–Triassic transition. The intensification of anoxia–euxinia at the PTB suggested by both local and global palaeoredox proxies is coeval with the initiation of Siberian Traps eruption, especially the early explosive phase with phreatomagmatic pipes (Mundil *et al.*, 2004; Svensen *et al.*, 2009; Burgess *et al.*, 2014). It is not clear whether records of local-scale anoxia in the late Permian represent the baseline redox state of late Palaeozoic oceans or enhanced deoxygenation. Another open question is how well the later Siberian Traps eruptive history matches the pulses of anoxia in the Early Triassic. Radiometric ages for late Siberian Traps magmatism extend into the Spathian (Kamo *et al.*, 2003), so anoxic pulses could be driven by later eruptive phases. However, the timing of potential Early Triassic peaks in volcanic activity remains unknown, preventing correlation with perturbations in marine geochemistry. The emerging view of Early Triassic anoxia as punctuated rather than persistent fits better with the proposed volcanic driver because the residence time of $CO_2$ in the ocean–atmosphere system is on the scale of 100–200 kyr (Archer *et al.*, 1997), making it difficult to maintain elevated rates of nutrient input for millions of years. If flood basalt volcanism was the main trigger of Permian–Triassic anoxia, one would expect a contraction of anoxic conditions soon after cessation of volcanism, as weathering draws down $CO_2$ and the climate cools. Siberian Traps volcanism can explain the widespread anoxia and euxinia at the PTB, but its role as a potential driver for Early Triassic anoxia deserves more attention.

## 18.6 Summary

Our review of palaeoredox records from the Permian–Triassic transition suggests a rapid shift from relatively well-ventilated late-Permian oceans to widespread anoxic and euxinic conditions coincident with the extinction horizon. Redox constraints from above the PTB provide evidence that anoxia recurred episodically during Early Triassic time, particularly at the Smithian–Spathian boundary and during the mid Spathian. Numerical models support the hypothesis that a rapid increase in atmospheric $pCO_2$ and nutrient fluxes could generate end-Permian oceanic anoxia–euxinia, and the timing and volume of Siberian Traps eruption provide a mechanism for the expansion of anoxia at the PTB. High-resolution records using geochemical proxies that capture globally averaged marine redox conditions hold particular promise for constraining the baseline redox state of Permian oceans, the spatial and temporal distribution of Early Triassic anoxia, and for testing the temporal correspondence between Siberian Traps volcanism and Early Triassic variation in marine redox chemistry.

## Acknowledgements

We thank L. Kump and K. Fristad for thoughtful reviews. This work was supported by the National Science Foundation (EAR-0807377 to J.L.P.).

## References

Algeo, T. J., Henderson, C. M., Ellwood, B. B. *et al.* (2012). Evidence for a diachronous late Permian marine crisis from the Canadian Arctic region. *Geological Society of America Bulletin*, **124**, 1424–1448.

Algeo, T. J., Hannigan, R., Rowe, H. *et al.* (2007). Sequencing events across the Permian–Triassic boundary, Guryul Ravine (Kashmir, India). *Palaeogeography, Palaeoclimatology, Palaeoecology*, **252**, 328–346.

Algeo, T. J., Shen, Y., Zhang, T. *et al.* (2008). Association of $^{34}$S-depleted pyrite layers with negative carbonate $\delta^{13}$C excursions at the Permian–Triassic boundary: evidence for upwelling of sulfidic deep-ocean water masses. *Geochemistry, Geophysics, Geosystems*, **9**, 1–10.

Algeo, T. J., Hinnov, L., Moser, J. *et al.* (2010). Changes in productivity and redox conditions in the Panthalassic Ocean during the latest Permian. *Geology*, **38**, 187–190.

Alibo, D. S. and Nozaki, Y. (1999). Rare earth elements in seawater: particle association, shale-normalization, and Ce oxidation. *Geochimica et Cosmochimica Acta*, **63**, 363–372.

Archer, D., Kheshgi, H. and Maier-Reimer, E. (1997). Multiple timescales for neutralization of fossil fuel $CO_2$. *Geophysical Research Letters*, **24**, 405–408.

Beatty, T. W., Zonneveld, J. P. and Henderson, C. M. (2008). Anomalously diverse Early Triassic ichnofossil assemblages in northwest Pangea: a case for a shallow-marine habitable zone. *Geology*, **36**, 771–774.

Bond, D. P. G. and Wignall, P. B. (2010). Pyrite framboid study of marine Permian–Triassic boundary sections: a complex anoxic event and its relationship to contemporaneous mass extinction. *Geological Society of America Bulletin*, **122**, 1265–1279.

Brand, U., Posenato, R., Came, R. *et al.* (2012). The end-Permian mass extinction: a rapid volcanic $CO_2$ and $CH_4$ climatic catastrophe. *Chemical Geology*, **322–323**, 121–144.

Brennecka, G. A., Herrmann, A. D., Algeo, T. J. and Anbar, A. D. (2011). Rapid expansion of oceanic anoxia immediately before the end-Permian mass extinction. *Proceedings of the National Academy of Sciences of the United States of America*, **108**, 17631–17634.

Burgess, S. D., Bowring, S. and Shen, S.-Z. (2014). High-precision timeline for Earth's most severe extinction. *Proceedings of the National Academy of Sciences of the United States of America*, doi: 10.1073/pnas.1317692111.

Cao, C., Love, G. D., Hays, L. E. *et al.* (2009). Biogeochemical evidence for euxinic oceans and ecological disturbance presaging the end-Permian mass extinction event. *Earth and Planetary Science Letters*, **281**, 188–201.

Casford, J. (2011). Isorenieratene; biomarker for photic layer anoxia? 2011 Fall Meeting, AGU, San Francisco, CA, 5–9 December, abstract B21E-0293.

Cortecci, G., Reyes, E., Berti, G. and Casati, P. (1981). Sulfur and oxygen isotopes in Italian marine sulfates of Permian and Triassic ages. *Chemical Geology*, **34**, 65–79.

Dahl, T. W., Hammarlund, E. U., Anbar, A. D. *et al.* (2010). Devonian rise in atmospheric oxygen correlated to the radiations of terrestrial plants and large predatory fish. *Proceedings of the National Academy of Sciences of the United States of America*, **107**, 17911–17915.

Dolenec, T., Lojen, S. and Ramovš, A. (2001). The Permian–Triassic boundary in western Slovenia (Idrijca Valley section): magnetostratigraphy, stable isotopes, and elemental variations. *Chemical Geology*, **175**, 175–190.

Dunk, R. M., Mills, R. A. and Jenkins, W. J. (2002). A reevaluation of the oceanic uranium budget for the Holocene. *Chemical Geology*, **190**, 45–67.

Ehrenberg, S. N., Svånå, T. A. and Swart, P. K. (2008). Uranium depletion across the Permian–Triassic boundary in Middle East carbonates: signature of oceanic anoxia. *AAPG Bulletin*, **92**, 691–707.

Fio, K., Spangenberg, J. E., Vlahović, I. *et al.* (2010). Stable isotope and trace element stratigraphy across the Permian–Triassic transition: a redefinition of the boundary in the Velebit Mountain, Croatia. *Chemical Geology*, **278**, 38–57.

German, C. R. and Elderfield, H. (1990). Application of the Ce anomaly as a paleoredox indicator: the ground rules. *Paleoceanography*, **5**, 823–833.

Grice, K., Cao, C., Love, G. D. *et al.* (2005). Photic zone euxinia during the Permian–Triassic superanoxic event. *Science*, **307**, 706–709.

Habicht, K. S., Gade, M., Thamdrup, B., Berg, P. and Canfield, D. E. (2002). Calibration of sulfate levels in the Archean ocean. *Science*, **298**, 2372–2374.

Hallam, A. (1989). The case for sea-level change as a dominant causal factor in mass extinction of marine invertebrates. *Philosophical Transactions of the Royal Society of London, Series B: Biological Sciences*, **325**, 437–455.

Hays, L. E., Beatty, T. W. and Henderson, C. M. (2007). Evidence for photic zone euxinia through the end-Permian mass extinction in the Panthalassic Ocean (Peace River Basin, Western Canada). *Palaeoworld*, **16**, 39–50.

Hays, L. E., Grice, K., Foster, C. B. and Summons, R. E. (2012). Biomarker and isotopic trends in a Permian–Triassic sedimentary section at Kap Stosch, Greenland. *Organic Geochemistry*, **43**, 67–82.

Isozaki, Y. (1997). Permo-Triassic boundary superanoxia and stratified superocean: records from lost deep sea. *Science*, **276**, 235–238.

Kaiho, K. (1994). Benthic foraminiferal dissolved-oxygen index and dissolved-oxygen levels in the modern ocean. *Geology*, **22**, 719–722.

Kajiwara, Y., Yamakita, S., Ishida, K., Ishiga, H. and Imai, A. (1994). Development of a largely anoxic stratified ocean and its temporary massive mixing at the Permian/ Triassic boundary supported by the sulfur isotopic record. *Palaeogeography, Palaeoclimatology, Palaeoecology*, **111**, 367–379.

Kakuwa, Y. and Matsumoto, R. (2006). Cerium negative anomaly just before the Permian and Triassic boundary event: the upward expansion of anoxia in the water column. *Palaeogeography, Palaeoclimatology, Palaeoecology*, **229**, 335–344.

Kamo, S. L., Czamanske, G. K., Amelin, Y. *et al.* (2003). Rapid eruption of Siberian flood-volcanic rocks and evidence for coincidence with the Permian–Triassic boundary and mass extinction at 251 Ma. *Earth and Planetary Science Letters*, **214**, 75–91.

Kato, Y., Nakao, K. and Isozaki, Y. (2002). Geochemistry of late Permian to Early Triassic pelagic cherts from southwest Japan: implications for an oceanic redox change. *Chemical Geology*, **182**, 15–34.

Kiehl, J. T. and Shields, C. A. (2005). Climate simulation of the latest Permian: implications for mass extinction. *Geology*, **33**, 757–760.

Liao, W., Wang, Y., Kershaw, S., Weng, Z. and Yang, H. (2010). Shallow-marine dysoxia across the Permian–Triassic boundary: evidence from pyrite framboids in the microbialite in South China. *Sedimentary Geology*, **232**, 77–83.

Logan, A. and Hills, L. V., ed. (1973). *The Permian and Triassic Systems and their Mutual Boundary*. Calgary: Canadian Society of Petroleum Geologists.

Loope, G. R., Kump, L. R. and Arthur, M. A. (2013). Shallow water redox conditions from the Permian–Triassic boundary microbialite: the rare earth element and iodine geochemistry of carbonates from Turkey and South China. *Chemical Geology*, **351**, 195–208.

Luo, G., Kump, L. R., Wang, Y. *et al.* (2010). Isotopic evidence for an anomalously low oceanic sulfate concentration following end-Permian mass extinction. *Earth and Planetary Science Letters*, **300**, 101–111.

Meyer, K. M. and Kump, L. R. (2008). Oceanic euxinia in Earth history: causes and consequences. *Annual Review of Earth and Planetary Sciences*, **36**, 251–288.

Meyer, K. M., Kump, L. R. and Ridgwell, A. (2008). Biogeochemical controls on photic-zone euxinia during the end-Permian mass extinction. *Geology*, **36**, 747–750.

Meyer, K. M., Macalady, J. L., Fulton, J. M. *et al.* (2011). Carotenoid biomarkers as an imperfect reflection of the anoxygenic phototrophic community in meromictic Fayetteville Green Lake. *Geobiology*, **9**, 321–329.

Mundil, R., Ludwig, K. R., Metcalfe, I. and Renne, P. R. (2004). Age and timing of the Permian mass extinctions: U/Pb dating of closed-system zircons. *Science*, **305**, 1760–1763.

Neubert, N., Nägler, T. F. and Böttcher, M. E. (2008). Sulfidity controls molybdenum isotope fractionation into euxinic sediments: evidence from the modern Black Sea. *Geology*, **36**, 775–778.

Newton, R. J., Pevitt, E. L., Wignall, P. B. and Bottrell, S. H. (2004). Large shifts in the isotopic composition of seawater sulphate across the Permo-Triassic boundary in northern Italy. *Earth and Planetary Science Letters*, **218**, 331–345.

Nielsen, J. K. and Shen, Y. (2004). Evidence for sulfidic deep water during the late Permian in the East Greenland Basin. *Geology*, **32**, 1037–1040.

Nielsen, J. K., Shen, Y., Piasecki, S. and Stemmerik, L. (2010). No abrupt change in redox condition caused the end-Permian marine ecosystem collapse in the East Greenland Basin. *Earth and Planetary Science Letters*, **291**, 32–38.

Ozaki, K., Tajima, S. and Tajika, E. (2011). Conditions required for oceanic anoxia/ euxinia: constraints from a one-dimensional ocean biogeochemical cycle model. *Earth and Planetary Science Letters*, **304**, 270–279.

Proemse, B. C., Grasby, S. E., Wieser, M. E., Mayer, B. and Beauchamp, B. (2013). Molybdenum isotopic evidence for oxic marine conditions during the latest Permian extinction. *Geology*, **41**, 967–970.

Renne, P. R., Black, M. T., Zichao, Z., Richards, M. A. and Basu, A. R. (1995). Synchrony and causal relations between Permian–Triassic boundary crises and Siberian flood volcanism. *Science*, **269**, 1413–1416.

Riccardi, A. L., Arthur, M. A. and Kump, L. R. (2006). Sulfur isotopic evidence for chemocline upward excursions during the end-Permian mass extinction. *Geochimica et Cosmochimica Acta*, **70**, 5740–5752.

Shen, J., Algeo, T. J., Zhou, L. *et al.* (2012). Volcanic perturbations of the marine environment in South China preceding the latest Permian mass extinction and their biotic effects. *Geobiology*, **10**, 82–103.

Shen, W., Lin, Y., Xu, L. *et al.* (2007). Pyrite framboids in the Permian–Triassic boundary section at Meishan, China: evidence for dysoxic deposition. *Palaeogeography, Palaeoclimatology, Palaeoecology*, **253**, 323–331.

Siebert, C., Nägler, T. F., von Blanckenburg, F. and Kramers, J. D. (2003). Molybdenum isotope records as a potential new proxy for paleoceanography. *Earth and Planetary Science Letters*, **211**, 159–171.

Silva-Tamayo, J. C., Payne, J. L., Wignall, P. B. *et al.* (2013). Ca, Mo and U isotopes suggest Neoproterozoic-like ocean conditions during the late Permian mass extinction. *Mineralogical Magazine*, **77**, 2213.

Song, H., Wignall, P. B., Tong, J. *et al.* (2012). Geochemical evidence from bio-apatite for multiple oceanic anoxic events during Permian–Triassic transition and the link with end-Permian extinction and recovery. *Earth and Planetary Science Letters*, **353–354**, 12–21.

Stirling, C. H., Andersen, M. B., Potter, E.-K. and Halliday, A. N. (2007). Low-temperature isotopic fractionation of uranium. *Earth and Planetary Science Letters*, **264**, 208–225.

Svensen, H., Planke, S., Polozov, A. G. *et al.* (2009). Siberian gas venting and the end-Permian environmental crisis. *Earth and Planetary Science Letters*, **277**, 490–500.

Tavakoli, V. and Rahimpour-Bonab, H. (2012). Uranium depletion across Permian–Triassic boundary in Persian Gulf and its implications for paleooceanic conditions. *Palaeogeography, Palaeoclimatology, Palaeoecology*, **350–352**, 101–113.

Tyrrell, T. (1999). The relative influences of nitrogen and phosphorus on oceanic primary production. *Nature*, **400**, 525–531.

Van Cappellen, P. and Ingall, E. D. (1994). Benthic phosphorus regeneration, net primary production, and ocean anoxia: a model of the coupled marine biogeochemical cycles of carbon and phosphorus. *Paleoceanography*, **9**, 677–692.

Voegelin, A. R., Nägler, T. F., Samankassou, E. and Villa, I. M. (2009). Molybdenum isotopic composition of modern and Carboniferous carbonates. *Chemical Geology*, **265**, 488–498.

Weyer, S., Anbar, A. D., Gerdes, A. *et al.* (2008). Natural fractionation of $^{238}U/^{235}U$. *Geochimica et Cosmochimica Acta*, **72**, 345–359.

Wignall, P. B. (1994). *Black Shales*. Oxford: Oxford University Press.

Wignall, P. B. and Hallam, A. (1992). Anoxia as a cause of the Permian/Triassic mass extinction: facies evidence from northern Italy and the western United States. *Palaeogeography, Palaeoclimatology, Palaeoecology*, **93**, 21–46.

Wignall, P. B. and Twitchett, R. J. (1996). Oceanic anoxia and the end-Permian mass extinction. *Science*, **272**, 1155–1158.

Wignall, P. B. and Twitchett, R. J. (2002). Extent, duration, and nature of the Permian–Triassic superanoxic event. *Geological Society of America Special Paper*, **356**, 395–413.

Wignall, P. B., Newton, R. J. and Brookfield, M. E. (2005). Pyrite framboid evidence for oxygen-poor deposition during the Permian–Triassic crisis in Kashmir. *Palaeogeography, Palaeoclimatology, Palaeoecology*, **216**, 183–188.

Wilkin, R. T., Barnes, H. L. and Brantley, S. L. (1996). The size distribution of framboidal pyrite in modern sediments: an indicator of redox conditions. *Geochimica et Cosmochimica Acta*, **60**, 3897–3912.

Winguth, A. M. E. and Maier-Reimer, E. (2005). Causes of the marine productivity and oxygen changes associated with the Permian–Triassic boundary; a reevaluation with ocean general circulation models. *Marine Geology*, **217**, 283–304.

Winguth, C. and Winguth, A. M. E. (2012). Simulating Permian–Triassic oceanic anoxia distribution: implications for species extinction and recovery. *Geology*, **40**, 127–130.

Woods, A. D., Bottjer, D. J., Mutti, M. and Morrison, J. (1999). Lower Triassic large seafloor carbonate cements: their origin and a mechanism for the prolonged biotic recovery from the end-Permian mass extinction. *Geology*, **27**, 645–648.

Zhou, L., Wignall, P. B., Su, J. *et al.* (2012). U/Mo ratios and $\delta^{98/95}$Mo as local and global redox proxies during mass extinction events. *Chemical Geology*, **324–325**, 99–107.

# 19

# Spatial and temporal patterns of ocean acidification during the end-Permian mass extinction – an Earth system model evaluation

YING CUI, LEE R. KUMP AND ANDY RIDGWELL

## 19.1 Introduction

The end-Permian extinction was a geologically abrupt (~ 100 000-year duration) event that occurred ~ 252 million years ago (Ma) (Burgess *et al.*, 2014; Joachimski *et al.*, 2012; Shen *et al.*, 2011; Sun *et al.*, 2012). The main phase of the extinction event was characterized by 8 to 10 °C of global warming (Joachimski *et al.*, 2012), driven by the massive release of greenhouse gases, as reflected in a contemporaneous negative C isotope excursion (CIE) of ~ 6‰ (Shen *et al.*, 2011) in the ocean. This suggests that either the source or its oxidation product was $CO_2$. Recent analyses of calcium isotopes of marine sediments and the pattern of extinction selectivity (Clapham and Payne, 2011; Hinojosa *et al.*, 2012; Payne *et al.*, 2010), and volatile studies on melt inclusions from the Siberian Traps (Black *et al.*, 2012; Black *et al.*, 2014), suggest that the Siberian Traps volcanism might be the trigger for the end-Permian extinction.

Besides global warming, one other consequence of $CO_2$ emission is ocean acidification, known as "the other $CO_2$ problem" (Doncy *et al.*, 2009). Rising atmospheric $CO_2$ causes a decrease in ocean pH and adjustments in carbonate chemistry, leading to a reduction in carbonate ion concentration and the saturation state of calcite and aragonite (Zeebe, 2012). In addition to impacts on their physiology, calcifying organisms are also susceptible to dissolution of their carbonate skeletons (Kleypas *et al.*, 2006; Turley *et al.*, 2010).

Although direct proxy data of ocean carbonate chemistry for the end-Permian are not available, geological evidence suggests that ocean acidification, possibly caused by the large quantity of $CO_2$ from Siberian Traps volcanism, triggered the extinction event in the ocean (Payne and Clapham, 2012). Studies of the end-Permian extinction selectivity indicate that heavily calcified organisms and

*Volcanism and Global Environmental Change*, eds. Anja Schmidt, Kirsten E. Fristad and Linda T. Elkins-Tanton. Published by Cambridge University Press. © Cambridge University Press 2015.

other groups with limited ability to buffer calcifying fluids have higher extinction rates, suggesting hypercapnia and/or ocean acidification played a prominent role in the observed extinction patterns (Knoll and Fischer, 2011). Sedimentological features, such as erosional truncations, have been found in uppermost Permian shallow-marine carbonate deposits in South China, Turkey and Japan, suggesting widespread submarine carbonate dissolution (Payne *et al.*, 2007). However, whether the observed truncation surface occurred during subaerial exposure or submarine dissolution is controversial (Payne *et al.*, 2009; Wignall *et al.*, 2009).

Here we present model simulations of ocean acidification driven by $CO_2$ emission from seven distinct sources, ranging from biogenic methane ($-60‰$ $\delta^{13}C$) to volcanic $CO_2$ ($-9‰$). We invert the model by forcing the surface ocean dissolved inorganic carbon (DIC) to conform to the prescribed $\delta^{13}C$ values by adding depleted C at each time-step. Due to the uncertainty concerning the end-Permian ocean chemistry, we explore the ocean buffering capacity for $CO_2$ addition with three different initial ocean saturation states of calcite ($\Omega_{cal}$), and quantitatively evaluate the response of the extent and pattern of ocean acidification.

## 19.2 Methods

### 19.2.1 *"Neritan" ocean with reef deposition*

Because deep-sea calcifiers did not evolve until the mid Mesozoic, carbonates are presumably only deposited in shallow "neritic" regions of less than 200 m depth (Zeebe and Westbroek, 2003). Modeling studies using reconstructed sea level, atmospheric $pCO_2$, [$Ca^{2+}$] and weathering flux suggest that surface ocean saturation state with respect to calcite ($\Omega_{cal}$) can be as high as 9–11 in a "Neritan" Permian ocean (Ridgwell, 2005; Riding and Liang, 2005). The Earth system model we use, cGENIE, has been used in a previous study to simulate changes in the global carbon cycle and climate during the end-Permian extinction (Cui *et al.*, 2013). However, the mean ocean-surface saturation is not *a priori* known so sensitivity analysis must be carried out with a range of reasonable estimates. Prior to the perturbation, we assume an atmospheric $pCO_2$ of 2800 ppm (10 × PAL (preindustrial atmospheric level, 280 ppm)) as we did in a previous study (Cui *et al.*, 2013). This value is within the rather large range of proxy estimates and previous assumptions concerning the generally warm, ice-free late-Permian climate states (Cui *et al.*, 2013). We consider three different surface saturation states: (1) $\Omega_{calcite} = 10$ represents the late Paleozoic estimate based on "Neritan" mode modeling (Ridgwell, 2005), (2) $\Omega_{calcite} = 5$ is similar to today's value, and (3) $\Omega_{calcite} = 2.5$ is used in a previous carbon-cycle modeling study on the

Table 19.1 *Initial conditions specified for cGENIE in this study*

| Parameters | Description | Value | Units |
|---|---|---|---|
| $\Omega_{\text{calcite, init}}$ | Calcite surface saturation state | 2.5, 5 and 10 | Dimensionless |
| $pH_{\text{init}}$ | Ocean surface pH | 7.7 ($\Omega_{\text{calcite, init}} = 10$) | Dimensionless |
| | | 7.55 ($\Omega_{\text{calcite, init}} = 5$) | |
| | | 7.4 ($\Omega_{\text{calcite, init}} = 2.5$) | |
| $T_0$ | Global average ocean surface temperature | 21 | °C |
| $F_{\text{w, calcite}}$ | Weathering flux of calcite | 17 | Tmol yr$^{-1}$ |
| $F_{\text{w, silicate}}$ | Weathering flux of silicate | 10.5 | Tmol yr$^{-1}$ |
| $\delta^{13}C_{\text{w}}$ | $\delta^{13}C$ of weathering calcite | 3.6 | ‰ |
| $\delta^{13}C_{\text{outgassing}}$ | $\delta^{13}C$ of $CO_2$ from volcanic outgassing | 3.6 | ‰ |
| $\delta^{13}C_{\text{atm}}$ | $\delta^{13}C$ of atmosphere | −5 | ‰ |
| $\delta^{13}C_{\text{DIC}}$ | $\delta^{13}C$ of DIC in surface ocean | 2.9 | ‰ |
| $[Ca^{2+}]$ | $Ca^{2+}$ concentrations in the ocean | 13 | mmol kg$^{-1}$ |
| $pCO_{2\text{init}}$ | Initial atmospheric $pCO_2$ | 2800 | ppm |
| | | $0.5 \times 10^{18}$ | mol |
| $[DIC]_{\text{surf}}$ | Surface ocean DIC concentration | 6035 ($\Omega_{\text{calcite, init}} = 10$) | µmol kg$^{-1}$ |
| | | 4230 ($\Omega_{\text{calcite, init}} = 5$) | |
| | | 2970 ($\Omega_{\text{calcite, init}} = 2.5$) | |
| $[DIC]_{\text{deep}}$ | Deep ocean DIC concentration | 6304 ($\Omega_{\text{calcite, init}} = 10$) | µmol kg$^{-1}$ |
| | | 4450 ($\Omega_{\text{calcite, init}} = 5$) | |
| | | 3160 ($\Omega_{\text{calcite, init}} = 2.5$) | |

end-Permian extinction (Cui *et al.*, 2013). With a desired global mean surface saturation value (e.g. $\Omega_{\text{calcite}} = 10$, 5 and 2.5), cGENIE can be configured to determine the appropriate initial ocean alkalinity (Alk) and DIC value (the initial conditions of the model are summarized in Table 19.1).

### 19.2.3 Forcing the DIC with $CO_2$ addition

We used the carbonate C isotope record of the GSSP (Global Boundary Stratotype Section and Point) in Meishan, South China (Shen *et al.*, 2011) to perform a loess fit (Matlab® smoothing function "loess"), where a span parameter of 8% was specified, lower than that used in Cui *et al.* (2013), to capture a larger magnitude of CIE (Figure 19.1). We then forced the ocean-surface DIC $\delta^{13}C$ to conform to the prescribed $\delta^{13}C_{\text{DIC}}$ derived from the statistically treated $\delta^{13}C_{\text{carb}}$, using an offset of 0.6‰ meant to account for the difference between the DIC and sedimentary $CaCO_3$ $\delta^{13}C$. We varied the $\delta^{13}C$ values of added $CO_2$ such that they represent several proposed sources, including plume-released $CO_2$ (−9‰ and −12‰), $CO_2$ from mixed sources (−15‰ and −20‰), contact metamorphism of coal and kerogen (−25‰), $CO_2$ from oxidation of thermogenic methane (−40‰) and

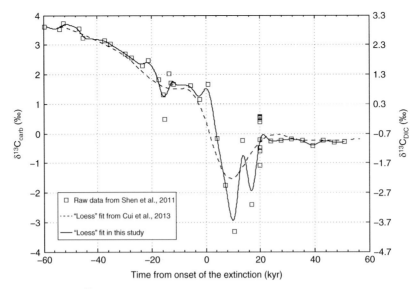

Figure 19.1   $\delta^{13}C$ value of carbonate from Shen *et al.* (2011) (squares), the statistical "loess" fit to the $\delta^{13}C$ in this study (solid line) and the "loess" fit in Cui *et al.*, (2013) (dotted line); the right *y*-axis gives the inferred $\delta^{13}C$ value of DIC used to force the model.

biogenic methane (–60‰). We conducted 21 runs in total with three initial saturation states and seven different sources (Table 19.2).

## 19.3  Results

### *19.3.1  Flux and total amount of C addition, pCO₂, ocean-surface temperature, Ω_calcite and pH*

The model results for peak flux and total amount of C addition, peak $pCO_2$, ocean-surface temperature, and minimum $\Omega_{calcite}$ and pH are summarized in Table 19.2. We show how two important environmental variables, the flux of the C addition and $\Omega_{calcite}$, change with time for various scenarios in Figure 19.2. The C release appears to have come in two multimillennial-duration pulses, with the highest peak rate of about 66 Gt C yr$^{-1}$ (cf. 9–10 Gt C yr$^{-1}$ today, Le Quéré *et al.*, 2009, 2013 ) for scenario 1 ($\delta^{13}C_{source} = -9$‰ and $\Omega_{calcite, init} = 10$) and the lowest peak rate of about 0.7 Gt C yr$^{-1}$ for scenario 21 (the scenario with $\delta^{13}C_{source} - -60$‰ and $\Omega_{calcite, init} = 2.5$). More $^{13}C$-depleted sources and lower initial saturation states cause dramatic declines in modeled peak rates of C addition (Figure 19.2a–c and Table 19.2). It is also interesting to note that the two pulsed C-release events are followed by two C-burial events, expressed as negative C flux (Figure 19.2a–c), required by the model to reproduce the rapid rate of C isotope recovery from the minimum values. The total

Table 19.2 *Model results for peak rate and total amount of C addition, peak pCO$_2$ and ocean surface temperature, and minimum $\Omega_{calcite}$ and pH for all the 21 scenarios. Also included are calculations of total amount of C needed in a simple mixing model.*

| Initial C (mol) | $\delta^{13}C_{source}$ (‰) | Scenario | Peak rate (Gt C yr$^{-1}$) in cGENIE | Peak amount released (Gt C) in cGENIE | Peak amount released (Gt C) in simple mixing model |
|---|---|---|---|---|---|
| $\Omega_{calcite,\ init}$ =10  9.3×10$^{18}$ | −9 | 1 | 66 | 338 003 | 93 000 |
| | −12 | 2 | 25 | 152 337 | 69 750 |
| | −15 | 3 | 13 | 92 820 | 55 800 |
| | −20 | 4 | 7 | 55 587 | 41 850 |
| | −25 | 5 | 5 | 39 746 | 33 480 |
| | −40 | 6 | 2 | 21 592 | 20 925 |
| | −60 | 7 | 1 | 13 496 | 13 950 |
| $\Omega_{calcite,\ init}$ = 5  6.7×10$^{18}$ | −9 | 8 | 50 | 256 561 | 67 000 |
| | −12 | 9 | 19 | 115 620 | 50 250 |
| | −15 | 10 | 10 | 70 271 | 40 200 |
| | −20 | 11 | 5 | 42 008 | 30 150 |
| | −25 | 12 | 3 | 29 958 | 24 120 |
| | −40 | 13 | 2 | 16 175 | 15 075 |
| | −60 | 14 | 1 | 10 058 | 10 050 |
| $\Omega_{calcite,\ init}$ = 2.5  4.9×10$^{18}$ | −9 | 15 | 38 | 195 203 | 49 000 |
| | −12 | 16 | 14 | 87 708 | 36 750 |
| | −15 | 17 | 7 | 53 366 | 29 400 |
| | −20 | 18 | 4 | 31 837 | 22 050 |
| | −25 | 19 | 3 | 22 671 | 17 640 |
| | −40 | 20 | 1 | 12 183 | 11 025 |
| | −60 | 21 | 1 | 7554 | 7350 |

| | $\delta^{13}C_{source}$ (‰) | Maximum pCO$_2$ (ppm) | Maximum sea surface $T$ (°C) | Minimum surface saturation state of calcite ($\Omega_{calcite}$) | Minimum pH |
|---|---|---|---|---|---|
| $\Omega_{calcite,\ init}$ = 10 | −9 | 126 837 | 37.1 | 0.95 | 6.26 |
| | −12 | 54 307 | 33.0 | 1.72 | 6.59 |
| | −15 | 31 978 | 30.6 | 2.45 | 6.79 |
| | −20 | 18 817 | 28.4 | 3.43 | 7.00 |
| | −25 | 13 589 | 27.1 | 4.19 | 7.12 |
| | −40 | 8067 | 25.0 | 5.68 | 7.32 |
| | −60 | 5877 | 23.9 | 6.78 | 7.43 |
| $\Omega_{calcite,\ init}$ = 5 | −9 | 95 964 | 35.7 | 0.66 | 6.25 |
| | −12 | 41 334 | 31.8 | 1.13 | 6.56 |
| | −15 | 24 607 | 29.5 | 1.55 | 6.76 |
| | −20 | 14 833 | 27.4 | 2.09 | 6.95 |
| | −25 | 10 948 | 26.2 | 2.47 | 7.06 |
| | −40 | 6829 | 24.4 | 3.20 | 7.23 |

Table 19.2 (*cont.*)

|  | $\delta^{13}C_{source}$ (‰) | Maximum $pCO_2$ (ppm) | Maximum sea surface $T$ (°C) | Minimum surface saturation state of calcite ($\Omega_{calcite}$) | Minimum pH |
|---|---|---|---|---|---|
|  | −60 | 5186 | 23.4 | 3.69 | 7.33 |
| $\Omega_{calcite,\ init} = 2.5$ | −9 | 72 945 | 34.4 | 0.46 | 6.23 |
|  | −12 | 31 598 | 30.6 | 0.74 | 6.54 |
|  | −15 | 19 163 | 28.5 | 0.98 | 6.72 |
|  | −20 | 111 876 | 26.5 | 1.25 | 6.89 |
|  | −25 | 8966 | 25.5 | 1.45 | 6.99 |
|  | −40 | 5893 | 23.9 | 1.77 | 7.14 |
|  | −60 | 4645 | 23.0 | 1.97 | 7.22 |

amount of C added during peak C addition is ∼ 330 000 Gt C for scenario 1, and becomes smaller as both $\delta^{13}C_{source}$ and $\Omega_{calcite,init}$ decreases (Table 19.2). The minimum $\Omega_{calcite}$ value during the peak C addition is a function of $\delta^{13}C_{source}$ and the initial $\Omega_{calcite}$ (Figure 19.2d–f). The $\Omega_{calcite}$ is elevated compared to the initial value at 20 kyr after the onset of the extinction, mostly clearly seen for a smaller $\Omega_{calcite,\ init}$. This is an expected response to an increased delivery of alkalinity from the higher weathering rate due to warming.

We select scenario 12 ($\Omega_{calcite} = 5$ and $\delta^{13}C_{source} = -25$‰) for a more comprehensive analysis of the overall environmental response to C addition (Figure 19.3). With reference to Figure 19.2, we note that the response of the other scenarios follows the trends shown in Figure 19.3 with increased or diminished amplitude. We present the time-series results relative to the onset of the extinction, which is defined as the time where the C isotopes initiate the sharp decline at bed 24e (age 252.3 Ma) according to Shen *et al.* (2011). The peak flux of C addition is ∼ 3 Gt C yr$^{-1}$, appearing at ∼ 9 kyr after the onset of the extinction, and is followed by a substantial C burial event (Figure 19.3a). The total amount of C added peaks at ∼ 30 000 Gt C, and stabilizes at about 17 000 Gt C by the end of the simulation (Figure 19.3b). The peak $pCO_2$ level reaches 11 000 ppm, and stabilizes at about 5000 ppm at 60 kyr after the onset of the extinction event (into the earliest Triassic) (Figure 19.3c). The globally averaged sea-surface temperature follows the shape of $pCO_2$, peaks at a little more than 26 °C ($\Delta T = 5$ °C), and stabilizes at about 23 °C into the Early Triassic (Figure 19.3d). Both globally averaged $\Omega_{calcite}$ and pH drop to their minimum value ($\Omega_{calcite,\ min} = 2.5$ and $pH_{min} = 7.06$) during the peak C addition. However, $\Omega_{calcite}$ returns to its pre-extinction level by the end of the simulation, whereas the pH remains low (Figures 19.3e, f).

We also group the specified three initial $\Omega_{cal}$ values and plot the maximum response of the six environmental parameters relative to the $\delta^{13}C_{source}$ (Figure 19.4).

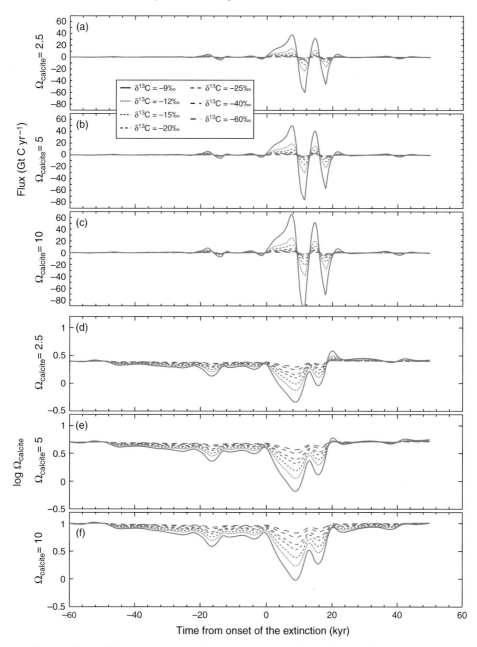

Figure 19.2   Time-series plot of carbon emission fluxes for the 21 experiments (a, $\Omega_{\text{calcite, init}} = 2.5$; b, $\Omega_{\text{calcite, init}} = 5$; c, $\Omega_{\text{calcite, init}} = 10$), and the log function of the surface ocean saturation state for the 21 experiments (d, $\Omega_{\text{calcite, init}} = 2.5$; e, $\Omega_{\text{calcite, init}} = 5$; f, $\Omega_{\text{calcite, init}} = 10$).

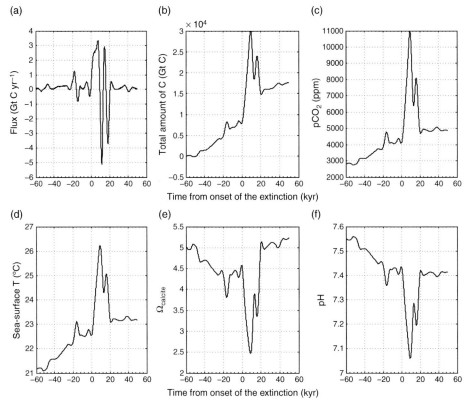

Figure 19.3    Time-series of environmental parameters for our favorite scenario ($\Omega_{calcite} = 5$, $\delta^{13}C = -25‰$), including (a) flux of C; (b) total amount of C released; (c) atmospheric $p$CO$_2$; (d) Global average sea surface temperature (°C); (e) Sea-surface saturation state of calcite; (f) Global average ocean surface pH.

Higher initial $\Omega_{calcite}$ values yield higher peak flux and amount of C added, as well as higher peak $p$CO$_2$ and sea-surface temperature (Figures 19.4a–d). $\Omega_{calcite}$ of the ocean surface is most sensitive to initial saturation state for the smaller perturbation associated with methane release; the initial buffering capacity is more rapidly overwhelmed for larger perturbations (Figures 19.4e, f).

### 19.3.2 Surface $\Omega_{calcite}$ and ocean-saturation-horizon response

To evaluate the spatial pattern of ocean-surface acidification, we show the map view of $\Omega_{calcite}$ during peak C addition (~ 10 kyr after the onset of extinction) for four selected scenarios (Figure 19.5a–o). Before we applied the perturbation, high-latitude regions had lower saturation states (Figure 19.5a–c), but no undersaturation is observed for the three prescribed initial saturation states. Under elevated $p$CO$_2$ (Figure 19.5d–o), the high latitudes tend to become undersaturated

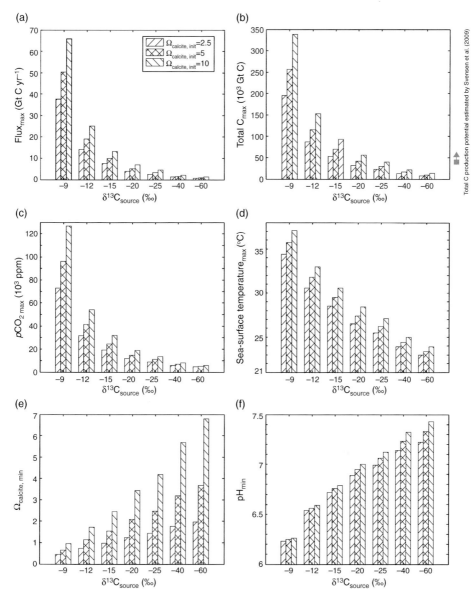

Figure 19.4   Bar plot of all 21 run results for (a) maximum C flux; (b) peak amount of C released; (c) maximum $pCO_2$ level in the atmosphere; (d) maximum temperature in the surface ocean; (e) minimum global average saturation state of calcite in the surface ocean; and (f) minimum ocean surface pH as a function of the initial saturation state and the $\delta^{13}C$ value of the C source.

first, similar to modern-day observation (Orr *et al.*, 2005). We find that for a plume-released C source, nearly all ocean-surface cells are undersaturated during peak C addition for initial $\Omega_{calcite}$ = 2.5 and 5 (Figures 19.5d–f), and much of the high-latitude area goes undersaturated for initial $\Omega_{calcite}$ = 10. High-latitude ocean

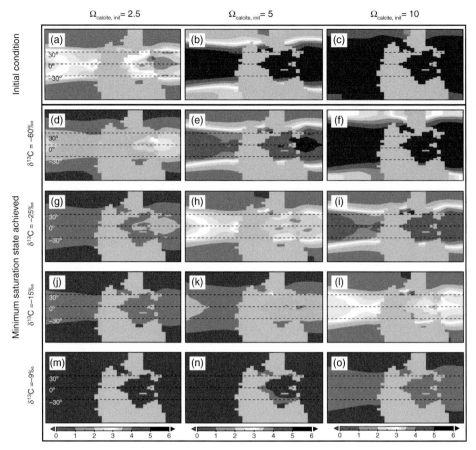

Figure 19.5   Map view of the initial ocean-surface saturation state of calcite (a, $\Omega_{calcite,init} = 2.5$; b, $\Omega_{calcite,init} = 5$; c, $\Omega_{calcite,init} = 10$) and map view of the ocean-surface saturation state for the twelve selected scenarios (d–o) when minimum ocean saturation is achieved. A black and white version of this figure will appear in some formats. For the colour version, please refer to the plate section.

undersaturation is also observed for the more [13]C-depleted sources with initial $\Omega_{calcite} = 2.5$, and the case where $\delta^{13}C_{source} = -15‰$ and initial $\Omega_{calcite} = 5$. No large regions of undersaturation are observed for the rest of the scenarios.

The depth of the saturation horizon before and during the perturbation for the selected 12 scenarios is shown in Figures 19.6a–o. The depth of the saturation horizon is shallower than 1000 m for an initial $\Omega_{calcite}$ of 2.5 and deeper than 4000 m for an initial $\Omega_{calcite}$ of 10. The depth of the saturation horizon during peak C addition is dependent upon the initial $\Omega_{calcite}$ and the $\delta^{13}C_{source}$. In the most severe C addition case (scenario 1), the depth of the saturation horizon rises up from a fully saturated ocean (Figure 19.6c) to < 500 m in mid- to low-latitude regions during peak C addition (Figure 19.6o).

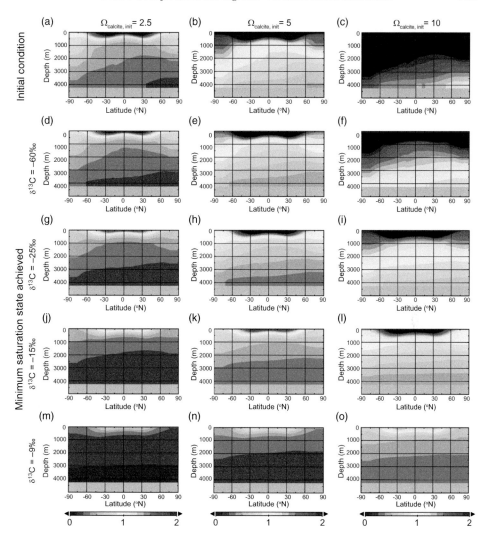

Figure 19.6   Depth profiles of the initial ocean saturation state of calcite (a, $\Omega_{calcite,init}$ = 2.5; b, $\Omega_{calcite,init}$ = 5; c, $\Omega_{calcite,init}$ = 10) and the selected four scenarios as in Figure 19.5, when minimum ocean saturation is acheived. A black and white version of this figure will appear in some formats. For the colour version, please refer to the plate section. The red coloration indicates supersaturation, and the blue, undersaturation.

To put the spatial pattern of ocean acidification in context, we compare our $\Omega_{calcite}$ results to an ocean-only simulation that spans preindustrial time to the future. Using a version of GENIE described in Cao *et al.* (2009), following the IPCC (2007) RCP8.5 trajectories for atmospheric $CO_2$, this simulation yields the following responses (Figure 19.7). We see a clear top-down acidification from preindustrial to modern time, and further deterioration in supersaturation to 2100, as atmospheric $p$CO$_2$ approaches ~ 900 ppm (Figures 19.7a–f). The decrease in

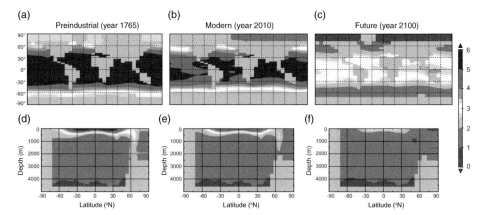

Figure 19.7    (a) and (d) map view and depth profile of $\Omega_{calcite}$ reconstruction for preindustrial time (year 1765); (b) and (e) map view and depth profile of $\Omega_{calcite}$ reconstruction for modern times (year 2010); and (c) and (f) map view and depth profile of $\Omega_{calcite}$ reconstruction for the projected future (year 2100) under IPCC RCP8.5 trajectories of $CO_2$. A black and white version of this figure will appear in some formats. For the colour version, please refer to the plate section.

$\Omega_{calcite}$ occurs first in high-latitude regions, consistent with previous modeling studies on high-latitude saturation-state change in response to climate change (e.g. Steinacher *et al.*, 2009), and then progresses to lower latitudes. On these short timescales there is little response of the ocean's saturation horizon to the perturbation. If we compare this to our results, the prominent top-down ocean acidification associated with fossil-fuel burning compares most favorably with our larger emission scenarios simulated for the end-Permian mass extinction.

## 19.4  Discussion

### *19.4.1 Estimates of C addition based on C isotope balance*

Estimates of C addition during the end-Permian mass extinction have been performed elsewhere using simple box models (Grard *et al.*, 2005; Payne *et al.*, 2010; Rampino and Caldeira, 2005; Retallack and Jahren, 2008) and an Earth system model of intermediate complexity (Cui *et al.*, 2013). These estimates are functions of the magnitude of assumed global CIE based on estimates from particular geologic sections, generally yielding a higher amount of C release with a higher magnitude of CIE, ranging from 3‰ (Rampino and Caldeira, 2005) to 8‰ (Berner, 2002). Payne *et al.* (2010) used a simple mixing model to predict that the upper estimate of the total amount of C added during the extinction event is 43 200 Gt C, using 3.6‰ CIE and assuming the source $\delta^{13}C = \sim -8‰$.

It is important to note that the calculated maximum flux and total amount of C added that we report herein are higher than simple mixing model and box model calculations (Table 19.2), especially for the more enriched sources. Time-continuous simulations using box models of the open-ocean C cycle, depending on the duration of the simulation, give slightly more C compared to simple mixing models because they take into account the damping effects of the weathering input. The Earth system model, which also calculates the effect of saturation state on carbonate deposition and seafloor dissolution, provides additional buffering capacity, and thus demands even more light C input to generate a given excursion.

### *19.4.2 Ocean acidification*

Geologic evidence, including the extinction selectivity of heavily calcified marine organisms with limited buffering capacity (Clapham and Payne, 2011), the deposition of carbonate-rich oolite and microbialite immediately after the extinction event (Kershaw *et al.*, 2002; Payne *et al.*, 2007), the submarine carbonate dissolution observed underlying the microbialite (Payne *et al.*, 2007) and calcium isotope constraints (Payne *et al.*, 2010), suggests ocean acidification might have been associated with the extinction event.

Montenegro *et al.* (2011) presented model simulations of the end-Permian event; surface ocean aragonite saturation values dropped below 1.15 and were thus unsuitable for present-day reef-forming coral species (Kleypas *et al.*, 1999). The difference between their study and ours arises because Montenegro *et al.* (2011) did not treat the ocean's carbonate system as an open system; i.e. there is no weathering delivery of alkalinity to the ocean to balance the burial of carbonate in their model; they were therefore unable to capture this important component of the oceanic response to acidification.

Overall, $CO_2$ build-up from the larger emission scenario due to the Siberian Traps volcanism is consistent with the ocean acidification from the top down shown in our model and can explain both the observed sedimentological features and calcium isotopes. This is different from the ocean acidification from bottom up proposed for the Paleocene–Eocene thermal maximum. This occurred in an ocean that had a better buffering capacity to changes in $pCO_2$ because of its deep-sea $CaCO_3$ sediment accumulation and for which rates of emission were apparently slower.

### 19.5 Conclusion

We find it most likely that the Siberian Traps volcanism released $CO_2$ in two major multimillennial pulses. The modeled rates of C release are dependent on the $\delta^{13}C$ of the source and the initial saturation state. Ocean acidification is most severe for a plume-related $CO_2$ release because it requires the most total C to explain the CIE, and polar regions develop undersaturated conditions at the lowest level of perturbation. We find that the initial buffering capacity of the ocean is quickly overwhelmed for many of the plausible scenarios for C release. However, for smaller releases (i.e. < 30 000 Gt C over 60 000 years), the buffering capacity of the ocean is capable of mitigating against severe ocean acidification in response to $CO_2$ release. For a geologically feasible C source, i.e. metamorphism of organic-rich sediment ($\delta^{13}C = -25‰$), the total amount of C released could be higher than 32 000 Gt C, but the upper limit is uncertain. The global warming and the likely ocean acidification due to the Siberian Traps volcanism might have pitched the end-Permian Earth system over a critical threshold and caused the mass extinction and subsequent long recovery.

## Acknowledgements

Y.C. and L.R.K. acknowledge support from NSF grant EAR-0807744, NASA Astrobiology funds. Y.C. acknowledges travel support from Krynine funds at Penn State and a GSA research grant.

## References

Berner, R.A., 2002. Examination of hypotheses for the Permo-Triassic boundary extinction by carbon cycle modeling. *Proceedings of the National Academy of Sciences*, **99**: 4172.

Black, B.A., Elkins-Tanton, L.T., Rowe, M.C. and Peate, I.U., 2012. Magnitude and consequences of volatile release from the Siberian Traps. *Earth and Planetary Science Letters*, **317**: 363–373.

Black, B.A., Lamarque, J.-F.O., Shields, C.A., Elkins-Tanton, L.T. and Kiehl, J.T., 2014. Acid rain and ozone depletion from pulsed Siberian Traps magmatism. *Geology*: **42**, 67–70.

Burgess, S.D., Bowring, S. and Shen, S.-Z., 2014. High-precision timeline for Earth's most severe extinction. *Proceedings of the National Academy of Sciences*, **111**: 3316–3321.

Cao, L., Eby, M., Ridgwell, A. *et al.*, 2009. The role of ocean transport in the uptake of anthropogenic $CO_2$. *Biogeosciences*, **6**: 375–390.

Clapham, M.E. and Payne, J.L., 2011. Acidification, anoxia, and extinction: a multiple logistic regression analysis of extinction selectivity during the middle and late Permian. *Geology*, **39**: 1059–1062.

Cui, Y., Kump, L.R. and Ridgwell, A., 2013. Initial assessment of the carbon emission rate and climatic consequences during the end-Permian mass extinction. *Palaeogeography, Palaeoclimatology, Palaeoecology*, **389**: 128–136.

Doney, S.C., Fabry, V.J., Feely, R.A. and Kleypas, J.A., 2009. Ocean acidification: the other $CO_2$ problem. *Annual Review of Marine Science*, **1**: 169–192.

Grard, A., Francois, L., Dessert, C., Dupré, B. and Godderis, Y., 2005. Basaltic volcanism and mass extinction at the Permo-Triassic boundary: environmental impact and modeling of the global carbon cycle. *Earth and Planetary Science Letters*, **234**: 207–221.

Hinojosa, J.L., Brown, S.T., Chen, J. *et al.*, 2012. Evidence for end-Permian ocean acidification from calcium isotopes in biogenic apatite. *Geology*, **40**: 743–746.

IPCC., 2007. *Climate Change The Physical Science Basis* (eds. Solomon, S., Qin, D., Manning, M. *et al.*). Cambridge University Press.

Joachimski, M.M., Lai, X., Shen, S. *et al.*, 2012. Climate warming in the latest Permian and the Permian–Triassic mass extinction. *Geology*, **40**: 195–198.

Kershaw, S., Guo, L., Swift, A. and Fan, J., 2002. Microbialites in the Permian–Triassic boundary interval in central China: structure, age and distribution. *Facies*, **47**: 83–90.

Kleypas, J., Buddemeier, R., Archer, D. *et al.*, 1999. Geochemical consequences of increased atmospheric carbon dioxide on coral reefs. *Science*, **284**: 118.

Kleypas, J.A., Feely, R.A., Fabry, V.J. *et al.*, 2006. Impacts of ocean acidification on coral reefs and other marine calcifiers: a guide for future research. Report of a workshop held 18–20 April 2005, St Petersburg, FL, sponsored by NSF, NOAA and the US Geological Survey.

Knoll, A.H. and Fischer, W.W., 2011. Skeletons and ocean chemistry: the long view. *Ocean Acidification* (ed. Gattuso, J.P.). Oxford University Press, pp. 67–82.

Le Quéré, C., Raupach, M.R., Canadell, J.G. *et al.*, 2009. Trends in the sources and sinks of carbon dioxide. *Nature Geoscience*, **2**: 831–836.

Le Quéré, C., Peters, G.P., Andres, R.J. *et al.*, 2013. The global carbon budget 1959–2011. *Earth Systems Science Data*, **5**, 1107–1157.

Montenegro, A., Spence, P., Meissner, K. *et al.*, 2011. Climate simulations of the Permian–Triassic boundary: ocean acidification and the extinction event. *Paleoceanography*, **26**: 3207.

Orr, J.C., Fabry, V.J., Aumont, O. *et al.*, 2005. Anthropogenic ocean acidification over the twenty-first century and its impact on calcifying organisms. *Nature*, **437**: 681–686.

Payne, J.L. and Clapham, M.E., 2012. End-Permian mass extinction in the oceans: an ancient analog for the twenty-first century? *Annual Review of Earth and Planetary Sciences*, **40**: 89–111.

Payne, J.L., Lehrmann, D.J., Follett, D. *et al.*, 2007. Erosional truncation of uppermost Permian shallow-marine carbonates and implications for Permian–Triassic boundary events. *Geological Society of America Bulletin*, **119**: 771–784.

Payne, J.L., Lehrmann, D.J., Follett, D. *et al.*, 2009. Erosional truncation of uppermost Permian shallow-marine carbonates and implications for Permian–Triassic boundary events: Reply. *Geological Society of America Bulletin*, **121**: 957–959.

Payne, J.L., Turchyn, A.V., Paytan, A. *et al.*, 2010. Calcium isotope constraints on the end-Permian mass extinction. *Proceedings of the National Academy of Sciences*, **107**(19): 8543–8548.

Rampino, M.R. and Caldeira, K., 2005. Major perturbation of ocean chemistry and a 'Strangelove Ocean' after the end-Permian mass extinction. *Terra Nova*, **17**: 554–559.

Retallack, G.J. and Jahren, A.H., 2008. Methane release from igneous intrusion of coal during late Permian extinction events. *Journal of Geology*, **116**: 1–20.

Ridgwell, A., 2005. A mid Mesozoic revolution in the regulation of ocean chemistry. *Marine Geology*, **217**: 339–357.

Riding, R. and Liang, L., 2005. Geobiology of microbial carbonates: metazoan and sea-water saturation state influences on secular trends during the Phanerozoic. *Palaeogeography, Palaeoclimatology, Palaeoecology*, **219**: 101–115.

Shen, S.-Z., Crowley, J.L., Wang, Y. *et al.*, 2011. Calibrating the end-Permian mass extinction. *Science*, **334**: 1367–1372.

Steinacher, M., Joos, F., Frölicher, T.L., Plattner, G.-K. and Doney, S.C., 2009. Imminent ocean acidification in the Arctic projected with the NCAR global coupled carbon cycle-climate model. *Biogeosciences*, **6**, 515–533.

Sun, Y., Joachimski, M.M., Wignall, P.B. *et al.*, 2012. Lethally hot temperatures during the Early Triassic greenhouse. *Science*, **338**: 366–370.

Turley, C., Findlay, H.S., Mangi, S., Ridgwell, A. and Schmidt, D.N., 2010. Ocean Acidification in MCCIP Annual Report Card 2010–2011, MCCIP Science Review. See: http://www.mccip.org.uk/arc.

Wignall, P.B., Kershaw, S., Collin, P.-Y. and Crasquin-Soleau, S., 2009. Erosional truncation of uppermost Permian shallow-marine carbonates and implications for Permian–Triassic boundary events: comment. *Geological Society of America Bulletin*, **121**: 954–956.

Zeebe, R.E., 2012. History of seawater carbonate chemistry, atmospheric $CO_2$, and ocean acidification. *Annual Review of Earth and Planetary Sciences*, **40**: 141–165.

Zeebe, R.E. and Westbroek, P., 2003. A simple model for the $CaCO_3$ saturation state of the ocean: the "Strangelove," the "Neritan," and the "Cretan" Ocean. *Geochemistry Geophysics Geosystems*, **4**: 1104.

# 20

# Environmental effects of large igneous province magmatism: a Siberian perspective

BENJAMIN A. BLACK, JEAN-FRANÇOIS LAMARQUE, CHRISTINE SHIELDS,
LINDA T. ELKINS-TANTON AND JEFFREY T. KIEHL

## 20.1 Introduction

Even relatively small volcanic eruptions can have significant impacts on global climate. The eruption of El Chichón in 1982 involved only 0.38 $km^3$ of magma (Varekamp *et al.*, 1984); the eruption of Mount Pinatubo in 1993 involved 3–5 $km^3$ of magma (Westrich and Gerlach, 1992). Both these eruptions produced statistically significant climate signals lasting months to years. Over Earth's history, magmatism has occurred on vastly larger scales than those of the Pinatubo and El Chichón eruptions. Super-eruptions often expel thousands of cubic kilometres of material; large igneous provinces (LIPs) can encompass millions of cubic kilometres of magma. The environmental impact of such extraordinarily large volcanic events is controversial. In this work, we explore the unique aspects of LIP eruptions (with particular attention to the Siberian Traps), and the significance of these traits for climate and atmospheric chemistry during eruptive episodes.

As defined by Bryan and Ernst (2008), LIPs host voluminous ($> 100,000$ $km^3$) intraplate magmatism where the majority of the magmas are emplaced during short igneous pulses. The close temporal correlation between some LIP eruptions and mass extinction events has been taken as evidence supporting a causal relationship (Courtillot, 1994; Rampino and Stothers, 1988; Wignall, 2001); as geochronological data become increasingly precise, they have continued to indicate that this temporal association may rise above the level of coincidence (Blackburn *et al.*, 2013).

Several obstacles obscure the mechanisms that might link LIP magmatism with the degree of global environmental change sufficient to trigger mass extinction. First, for Mount Pinatubo-style sulfur cooling, explosive eruption is crucial. The radiative effects of sulfur aerosols are much greater when the aerosols reach the stratosphere via explosive plumes (Robock, 2000). Second, as the magnitude of

*Volcanism and Global Environmental Change*, eds. Anja Schmidt, Kirsten E. Fristad and Linda T. Elkins-Tanton.
Published by Cambridge University Press. © Cambridge University Press 2015.

sulfur release increases, aerosols form larger particles (with lower optical depth per unit mass and more rapid settling rates), effectively limiting the magnitude of sulfur-driven cooling (Pinto *et al.*, 1989). Third, while they are geologically rapid, LIP eruptions can still span a million years or more, potentially allowing the climate system to recover from intermittent fluxes of relatively short-lived volcanic gases such as $SO_2$. Finally, each mass extinction is associated with a distinctive pattern of environmental stress as recorded in the geological, geochemical and palaeontological records (Knoll *et al.*, 2007; Wignall, 2001); these distinctive patterns pose informative but challenging hurdles for any overarching model.

In recent years, volcanological understanding of LIP eruptions has grown more nuanced. Abundant volcaniclastic deposits associated with the Ferrar, Emeishan and Siberian Traps LIPs support episodes of phreatomagmatic and/or magmatic explosivity (Chapter 1; Black *et al.*, 2011; White *et al.*, 2009; Wignall *et al.*, 2009). As detailed by Self *et al.* in Chapter 11, under certain conditions plumes from LIP eruptions may intermittently breach the tropopause. Palaeomagnetic secular variation studies (Chapter 5; Chenet *et al.*, 2009; Chenet *et al.*, 2008; Pavlov *et al.*, 2011) and ultra-high-precision geochronology (Blackburn *et al.*, 2013) indicate that LIP magmatism often occurs in pulses, compressing gas emissions into several concentrated convulsions.

Current research has focused on $CO_2$ release (Sobolev *et al.*, 2011), sediment degassing (Beerling *et al.*, 2007; Ganino and Arndt, 2009; Svensen *et al.*, 2009), acid rain (Campbell *et al.*, 1992), ozone depletion (Beerling *et al.*, 2007; Visscher *et al.*, 2004) and shifts in continental weathering (Dessert *et al.*, 2003; Schaller *et al.*, 2012) as potential links between LIP magmatism and global environmental change. Possible changes in ocean circulation from prolonged sulfur cooling (Black *et al.*, 2012; Miller *et al.*, 2012) also warrant further investigation. In this contribution we will employ the ~ 252 Ma Siberian Traps as a test case. Where possible, we will evaluate these proposed climate forcing mechanisms using a global model of atmospheric chemistry and climate (Black *et al.*, 2014; Kiehl and Shields, 2005), and we will compare the conditions of these simulations to those expected for other Phanerozoic large igneous provinces.

## 20.2 The Siberian Traps: a case study

The Siberian Traps LIP constitutes a strong test case for the environmental effects of LIP magmatism because of its unusual size, its continental setting and its association with the catastrophic end-Permian mass extinction (Kamo *et al.*, 2003; Shen *et al.*, 2011; Burgess *et al.*, 2014). The total original volume of the Siberian Traps is difficult to estimate due to erosion and uncertain intrusive volumes, but has been estimated as > 3,000,000 km$^3$ (Reichow *et al.*, 2009). Volcaniclastic rocks characterize the lowermost parts of the volcanic section in many regions of the Siberian Traps (Black *et al.*, 2011; Ross *et al.*, 2005); some authors have estimated that they comprise as much as 25% of the total volume, though this figure may be inflated (Ross *et al.*, 2005). Chapter 1 in this book discusses in more detail the evidence for and importance of explosive fragmentation during LIP eruptions. Explosive delivery of volatiles to the stratosphere would strongly influence the potential climate effects of Siberian Traps magmatism.

Geologic and palaeontologic proxies for environmental conditions across the Permian–Triassic boundary provide a key constraint on the role of volcanic forcing during the end-Permian mass extinction. The pattern of extinction in the oceans – where Erwin (1994) notes that extinction rates exceeded 90% at the species level – suggests that organisms with less capacity to adapt to changes in partial pressure of $CO_2$ ($pCO_2$) may have suffered preferentially (Knoll *et al.*, 2007). Terrestrial sections record changes in the diversity and health of floral populations. Plant fossils directly reflect prevailing atmospheric conditions. While the statistical significance of extinction rates among some plant populations during the end-Permian has been disputed (Rees, 2002), aberrant pollen remains provide convincing evidence for widespread atmospheric stress (Foster and Afonin, 2005; Visscher *et al.*, 2004). Ward *et al.* (2000) interpret a shift in fluvial style as evidence for catastrophic devegetation.

## 20.3 Gas flux from large igneous provinces

The flux of gases such as $CO_2$, $CH_4$, $CH_3Cl$, HCl and $SO_2$ governs the range and magnitude of the climate response to volcanism. The atmospheric lifetime of $CO_2$ is between $10^3$ and $10^5$ years (Archer, 2005); the lifetimes of other important volcanic and thermogenic gases are much shorter, ranging from $10^{-2}$ to $10^1$ years (Seinfeld and Pandis, 1997). Because the hiatus between eruptive events may exceed the lifetime of many of these gases, climate forcing may occur in jolts determined by the magnitude and duration of gas emission.

Sobolev *et al.* summarize the lines of evidence available to substantiate estimates of volatile release in Chapter 10. Based on melt inclusion data, Black *et al.* (2012) found that magmatic degassing from the Siberian Traps may have accounted for the atmospheric release of 6,300–7,800 Gt S, 3,400–8,700 Gt Cl and 7,100–13,600 Gt F, depending on the efficiency of degassing. Sobolev *et al.* (2011) estimate that the Siberian Traps LIP could have released up to 170,000 Gt of mantle-derived $CO_2$. During intrusion and thermal metamorphism, the reaction of rock salt layers with hydrocarbons could also have generated $\sim 10^3$ Gt $CH_3Cl$.

Estimates of sulfur degassing from the Deccan Traps (Self *et al.*, 2008) and the Columbia River flood basalts (Blake *et al.*, 2010; Thordarson and Self, 1996) prescribe volumetric fluxes comparable to that of the Siberian Traps (Black *et al.*,

2012). Carbon dioxide release is poorly constrained in all cases, but depends on country-rock lithology and the composition of the mantle melt source. Siberian magmas transited a series of particularly volatile-rich sedimentary rocks, including evaporites, carbonates and hydrocarbon-bearing layers (Kontorovich *et al.*, 1997; Meyerhoff, 1980). This geologic setting may have contributed to the overall $CO_2$ and methane flux, and is a necessary precondition for the production of $CH_3Cl$ and other organohalogens (Svensen *et al.*, 2009).

To simulate the effects of both $CO_2$ updraw and pulsed release of sulfur and other relatively short-lived gases as estimated by Black *et al.* (2012), we employ a comprehensive global model of climate and chemistry (Black *et al.*, 2014; Kiehl and Shields, 2005; Lamarque *et al.*, 2012). In the next section, we describe the set-up, physical processes, chemistry and assumptions implicit in this model.

## 20.4 Model and simulation descriptions

The Community Earth System Model is a comprehensive global climate model that includes ocean, land, sea-ice and atmosphere components. It is capable of simulating fully interactive atmospheric chemistry with the CAM-Chem module (Lamarque *et al.*, 2012). The chemistry scheme includes reactions between 134 species, with rate constants as in Sander *et al.* (2006).

Here we present results from two types of simulations: one configured to capture climate change, and one configured to capture atmospheric chemistry. The first configuration is that of Kiehl and Shields (2005) who ran equilibrium simulations at 3.75° by 3.75° resolution, with online ocean circulation coupled to a land model, a sea-ice model, and an atmosphere with 26 vertical levels. Atmospheric chemistry is not simulated in this set-up, but the radiative effects of $CO_2$ and other green-house gases are allowed to percolate completely through the Earth system until a new equilibrium state is reached.

We use ocean sea-surface temperatures from an updated Community Climate System Model 4 version of the Kiehl and Shields (2005) equilibrium simulations as a boundary condition for the second configuration, in which we simulate atmospheric chemistry at 1.9° by 2.5° horizontal resolution in the atmosphere. The primary targets of these atmospheric chemistry simulations are rainfall pH and ozone chemistry.

For both configurations, we employ the model of palaeotopography from Kiehl and Shields (2005) as shown in Figure 20.1. Defining a realistic palaeobathymetry for the pre-Mesozoic is particularly challenging, and likely constitutes one of the major uncertainties for the simulation of ocean circulation. Another set of major uncertainties involves the composition of the Permian–Triassic atmosphere and the magnitude of magmatic and thermogenic fluxes. In all simulations, we hold $CO_2$

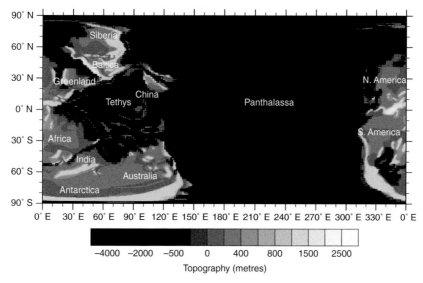

Figure 20.1   End-Permian palaeotopographic reconstruction, after Kiehl and Shields (2005), shown here at 0.5° by 0.5° horizontal resolution.

constant at either 355 ppm or 3550 ppm, following Kiehl and Shields (2005). Volcanic emissions of sulfur derive from Black *et al.* (2012), and thermogenic emissions of $CH_3Cl$ and $CH_4$ derive from Aarnes *et al.* (2011). Complete emissions and boundary conditions are tabulated in Black *et al.* (2014).

## 20.5 Global warming

Studies of oxygen isotopes in conodont microfossils suggest that sea-surface temperatures began to warm rapidly in the latest Permian. Joachimski *et al.* (2012) report a low-latitude temperature increase of +8 °C spanning a short interval across the Permian–Triassic boundary. Sun *et al.* (2012) consider oxygen isotopic variations from the middle Permian to the Middle Triassic, and estimate a +15 °C increase in low-latitude sea-surface temperatures during this time. This strong warming has been attributed to volcanic or thermogenic release of greenhouse gases associated with Siberian Traps magmatism, though the correspondence of carbon and oxygen isotopic variations with pulses of magmatism has not been conclusively demonstrated geochronologically (Joachimski *et al.*, 2012; Sun *et al.*, 2012).

Comparing the warming predicted from oxygen isotopic records with results from global climate models presents an interesting opportunity (Figure 20.2). Global climate model simulations have shown that the palaeotectonic configuration in the Permian was conducive to localized warm sea-surface temperatures in the Tethys Ocean (Kiehl and Shields, 2005). As shown in Figure 20.2, a ten-fold

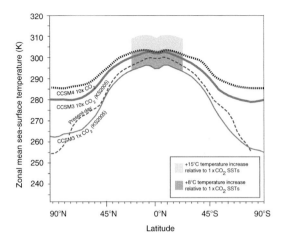

Figure 20.2   Zonal mean sea-surface temperatures (SSTs) as a function of $CO_2$ levels and latitude. KS2005 denotes simulations from Kiehl and Shields (2005). CCSM3 runs were completed with Community Climate System Model 3; CCSM4 runs were completed with Community Climate System Model 4. Based on oxygen isotope records from conodonts, Joachimski *et al.* (2012) report a +8 °C warming of low-latitude sea-surface temperatures across the Permian–Triassic boundary; Sun *et al.* (2012) report a +15 °C warming of low-latitude sea-surface temperatures from the middle Permian up to the Middle Triassic. For reference, both potential temperature increases are shown here, relative to a simulation of Permian climate with 1 × present-day $CO_2$ (355 ppm). Temperatures below −1.8 °C (271.3 K) correspond to sea-ice.

increase in $CO_2$ leads to a zonally averaged low-latitude sea-surface temperature increase that is highly consistent with the +8°C warming reported by Joachimski *et al.* (2012).

A +15 °C warming of the equatorial sea surface (Sun *et al.*, 2012) is more difficult to achieve, and exceeds the temperature increase associated with a ten-fold increase in $CO_2$ to 3550 ppm (Kiehl and Shields, 2005). Based on a survey of climate models, an approximately 2–4 °C global temperature increase is expected for each doubling of $CO_2$ (Rohling *et al.*, 2012). The magnitude of the long-term warming reported by Sun *et al.* (2012) thus requires higher climate sensitivity, extreme $CO_2$ levels or large quantities of other greenhouse gases (such as $CH_4$) in addition to $CO_2$.

## 20.6 Global cooling

Stratospheric volcanic emissions of short-lived sulfur gas could produce transient aerosol-induced cold spells within a long-term greenhouse warming trend from longer-lived $CO_2$. The eruption of the Siberian Traps occurred at relatively high

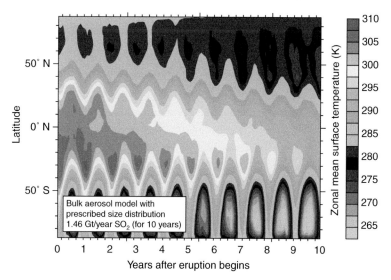

Figure 20.3   Zonal mean surface temperature during a prolonged Siberian Traps explosive eruption, with 1.46 Gt/year $SO_2$ release (corresponding to 240 km$^3$/year erupted volume).This simulation uses a prescribed aerosol size distribution that is inaccurate for sulfur release of this magnitude (Pinto *et al.*, 1989; Timmreck *et al.*, 2010); the magnitude of cooling in the simulation is therefore unrealistically severe. However, the asymmetric pattern of cooling, with larger temperature decreases in the northern hemisphere, should still be valid. A black and white version of this figure will appear in some formats. For the colour version, please refer to the plate section.

latitude – around 60° N (Cocks and Torsvik, 2007) – with several implications for sulfur aerosols. The high-latitude tropopause dips to an annual mean altitude of 10 km, versus 18 km at the equator (Grise *et al.*, 2010); as a result, high-latitude eruption plumes are more likely to breach the tropopause (Stothers *et al.*, 1986). The climate effects of high-latitude eruptions are asymmetric (Oman *et al.*, 2005), because the patterns of stratospheric circulation largely confine aerosols from high-latitude eruptions to the hemisphere of origin. Because of downward transport in the polar vortex, the residence time of stratospheric sulfur injected near the poles is significantly shorter than that of sulfur injected in the tropics (Hamill *et al.*, 1997).

Our bulk aerosol simulations allow us to track volcanogenic sulfate in the atmosphere, though with many of the relevant aerosol physics greatly simplified or neglected. Because of this simplification, the radiative effects of large sulfur emissions are vulnerable to significant overestimation. While our simulations may overestimate the magnitude of sulfur-induced cooling, they do reveal the geographic pattern of climate change. As expected, our simulations of a Siberian Traps pyroclastic eruption (illustrated in Figure 20.3) show a much stronger decrease in temperature in the northern hemisphere than in the southern hemisphere. Timmreck *et al.* (2010) find that during the 74 ka Toba super-eruption, equatorial release of 850 Mt S (comparable to the sulfur release during a 240 km$^3$ Siberian Traps pyroclastic episode) could have generated a maximum –3.5 K global temperature decrease, with more intense cooling locally. During extended Siberian Traps eruptive episodes, cooling in the northern hemisphere could have exceeded this value.

Miller *et al.* (2012) suggest that volcanic sulfur emissions from several closely spaced eruptions produced sea-ice/ocean feedbacks that ultimately triggered the Little Ice Age. During a LIP eruption, sustained eruptive episodes that include significant pyroclastic volcanism could likewise power prolonged periods of climatic cooling (Black *et al.*, 2012). The potential effect of such cooling on ocean circulation is a significant outstanding question, and represents a strong motivation for the future adaption of fully coupled Earth system models with realistic aerosol physics to configurations appropriate for deep time.

## 20.7 Acid rain

Campbell *et al.* (1992) noted that widespread acid rain likely accompanied the eruption of the Siberian Traps. Black *et al.* (2014) used a comprehensive model of global climate and chemistry to map the distribution of that acid rain throughout the end-Permian world (Figure 20.4).

Among volcanogenic products, $CO_2$ and $SO_4$ contribute most forcefully to the large-scale acidity of rain. Large proportions of HCl and HF are likely removed from the eruption plume proximally (Pinto *et al.*, 1989; Tabazadeh and Turco, 1993; Textor *et al.*, 2003). These strong acids may produce a local spike in acidity without altering the global pH of rain. As discussed previously, temperature proxies (Joachimski *et al.*, 2012; Sun *et al.*, 2012) and palaeontological analyses (Knoll *et al.*, 2007) are consistent with strongly increasing $CO_2$ levels across the Permian–Triassic boundary. Our model results show that this uptick in $CO_2$ translates to a long-term decrease in the pH of Permian–Triassic rain. The pH of the ocean is buffered by $CaCO_3$, but rapidly released $CO_2$ will equilibrate across the ocean–atmosphere system on timescales shorter than the dissolution of sea-floor $CaCO_3$, prompting ocean acidification (Archer, 2005; Kump *et al.*, 2009).

Even very large eruptions release insufficient sulfur to significantly affect the pH of the ocean. However, sulfuric rain can reach deeply acidic pH values during large eruptive episodes. Black *et al.* (2014) report that at atmospheric $CO_2$ levels of 3550 ppm, Permian–Triassic rain has pH $\approx 4$. As shown in Figure 20.4a, for a 1-year 240 km$^3$ eruption, annually averaged rainfall pH can drop one to two additional log units in the northern hemisphere (Black *et al.*, 2014). For an eruption of 2,400 km$^3$, a further decrease to annualized levels reaching pH $\approx 2$ over a wide

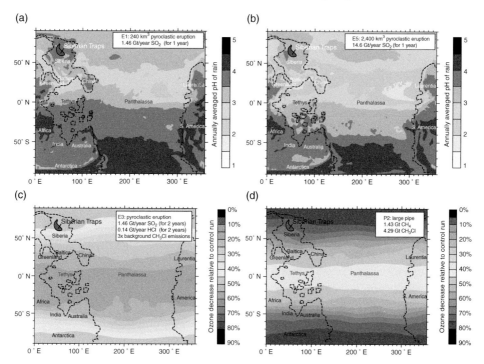

Figure 20.4    (a) Rainfall pH during a 1-year, 240 km$^3$ eruption. Simulation E1 from Black *et al.* (2014). (b) Rainfall pH during a 1-year, 2400 km$^3$ eruption. Simulation E5 from Black *et al.* (2014). (c) Ozone depletion during a 3-year, 720 km$^3$ eruption, assuming 75% of degassed HCl reaches the stratosphere, and with 10 × pre-industrial $CH_3Cl$ emissions. Simulation E3 from Black *et al.* (2014). (d) Ozone depletion after a large pipe eruption that releases accumulated $CH_3Cl$ and $CH_4$ from contact metamorphism (Aarnes *et al.*, 2011). Simulation P2 from Black *et al.* (2014). A black and white version of this figure will appear in some formats. For the colour version, please refer to the plate section.

area would be expected (Figure 20.4b). Such drastically acidic rain could produce highly stressful environmental conditions for northern hemisphere flora and fauna (Howells, 1990). $SO_4$ has an e-folding time in the stratosphere of around 1 year (Barnes and Hofmann, 1997); therefore, sulfuric acid rain, which is located in the troposphere, abates quickly after an eruption ceases.

The expected intensity of acid rain during other flood basalt eruptions depends largely on atmospheric $CO_2$ levels and the quantity of sulfur released to the stratosphere. Sulfur emissions from the Deccan Traps and the Columbia River flood basalts were volumetrically comparable to those during the Siberian Traps (Blake *et al.*, 2010; Self *et al.*, 2008). During the eruption of the Central Atlantic Magmatic Province (CAMP) at the Triassic–Jurassic boundary, $CO_2$ levels may have been even higher than during the Permian–Triassic (Schaller *et al.*, 2011).

## 20.8  Ozone

The stratospheric ozone layer provides a crucial shield against incoming ultraviolet radiation. Halogen-bearing compounds facilitate the catalytic destruction of stratospheric ozone through cycles that resemble:

$$Cl + O_3 \rightarrow ClO + O_2$$

$$ClO + O \rightarrow Cl + O_2.$$

Visscher *et al.* (2004) invoked ozone depletion as a potential causal mechanism for an observed spike in abnormal pollen during the end-Permian. Contact metamorphism of Siberian sedimentary rocks – including hydrocarbon reservoir strata and evaporates – may have produced abundant $CH_3Cl$ (Aarnes *et al.*, 2011; Beerling *et al.*, 2007; Svensen *et al.*, 2009). Beerling *et al.* (2007) demonstrated that atmospheric release of sufficient quantities of this thermogenic $CH_3Cl$ could drive near-total ozone layer collapse. Volcanic HCl can also trigger ozone depletion, especially when background $CH_3Cl$ emissions are elevated (Black *et al.*, 2014), but water and ice in eruptive plumes help to scavenge HCl and may limit its introduction to the stratosphere (Tabazadeh and Turco, 1993; Textor *et al.*, 2003).

Figure 20.4c and 20.4d show the distribution of ozone after several years of volcanic HCl or thermogenic gas release. While significant uncertainty accompanies estimates of metamorphic gas release during the eruption of the Siberian Traps, the presence of major evaporite and petroleum deposits in the Tunguska Basin differentiates the Siberian Traps from other LIP eruptions. The potential for $CH_3Cl$-driven ozone collapse is thus a distinctive attribute of the Siberian Traps.

## 20.9  Continental weathering

The eruption of the Siberian Traps and the end-Permian mass extinction both occurred near the onset of a major increase in seawater $^{87}Sr/^{86}Sr$ (Korte *et al.*, 2003; Korte *et al.*, 2004). This increase in radiogenic Sr, which spans the earliest Triassic, has been interpreted as evidence for a global increase in continental weathering. Warmer temperatures, elevated atmospheric $CO_2$, decimation of terrestrial vegetation, and acid rain are all potential drivers of accelerated weathering.

Flood basalt deposits may also provide an important long-term sink for $CO_2$. During weathering, the Ca contents in the basalts become available to bond with carbonate (Dessert *et al.*, 2001; Dessert *et al.*, 2003). The eruption of LIPs can create large new expanses of basaltic surface area, and chemical weathering of this basalt may be much more efficient than chemical weathering of granite or gneiss (Berner, 2006; Dessert *et al.*, 2001; Dessert *et al.*, 2003). On the basis of pedogenic carbonate measurements from the Newark Basin, Schaller *et al.* (2012) suggest that

LIP eruptions could initiate a cycle of increasing $CO_2$ and global warming followed by increased weathering, declining $CO_2$ and global cooling.

## 20.10 Summary

LIP eruptions are capable of perturbing Earth systems via many mechanisms that are not relevant to smaller eruptions. The geologic record provides increasing evidence that $CO_2$ and other greenhouse gases released by ancient LIP eruptions may have triggered episodes of global warming (Joachimski *et al.*, 2012; Schaller *et al.*, 2011; Sun *et al.*, 2012). Models and geochemical records also hint at the possibility of sustained sulfur-induced cooling, acid rain, ozone depletion, and/or shifts in silicate weathering.

The rapidly improving resolution of proxy records for palaeo-temperature, atmospheric conditions, biodiversity and geochemical cycling will support an improved understanding of the environmental consequences of LIPs through deep time. Comprehensive Earth system models must continue to work to fully reproduce and explain this complex geologic record.

### Acknowledgements

This work was funded by grant EAR-0807585 from NSF Continental Dynamics. The CESM project is supported by the National Science Foundation and the Office of Science (BER) of the US Department of Energy. NCAR is operated by the University Corporation of Atmospheric Research under sponsorship of the National Science Foundation. A review by Paul Wignall improved this manuscript. The authors gratefully thank Stephanie Shearer and Brenda Carbone.

### References

Aarnes, I., Fristad, K., Planke, S., Svensen, H., 2011. The impact of host-rock composition on devolatilization of sedimentary rocks during contact metamorphism around mafic sheet intrusions. *Geochemistry Geophysics Geosystems* **12**, Q10019.

Archer, D., 2005. Fate of fossil fuel $CO_2$ in geologic time. *Journal of Geophysical Research: Oceans (1978–2012)* **110**.

Barnes, J.E., Hofmann, D.J., 1997. Lidar measurements of stratospheric aerosol over Mauna Loa observatory. *Geophysical Research Letters* **24**, 1923–1926.

Beerling, D.J., Harfoot, M., Lomax, B., Pyle, J.A., 2007. The stability of the stratospheric ozone layer during the end-Permian eruption of the Siberian Traps. *Philosophical Transactions of the Royal Society a-Mathematical Physical and Engineering Sciences* **365**, 1843–1866.

Berner, R.A., 2006. Inclusion of the weathering of volcanic rocks in the Geocarbsulf model. *American Journal of Science* **306**, 295–302.

Black, B., Elkins-Tanton, L., Weiss, B., Veselovskiy, R., Latyshev, A., Pavlov, V., 2011. Emplacement temperatures and alteration histories of Siberian Traps volcaniclastic deposits. *AGU Fall Meeting Abstracts* **1**, 1042.

Black, B.A., Elkins-Tanton, L.T., Rowe, M.C., Peate, I.U., 2012. Magnitude and consequences of volatile release from the Siberian Traps. *Earth and Planetary Science Letters* **317–318**, 363–373.

Black, B.A., Lamarque, J.-F., Shields, C., Elkins-Tanton, L.T., Kiehl, J.T., 2014. Acid rain and ozone depletion from pulsed Siberian Traps Magmatism. *Geology,* **42**, 67–70.

Blackburn, T.J., Olsen, P.E., Bowring, S.A. *et al.*, 2013. Zircon U–Pb geochronology links the end-Triassic extinction with the Central Atlantic Magmatic Province. *Science* **340**, 941–945.

Blake, S., Self, S., Sharma, K., Sephton, S., 2010. Sulfur release from the Columbia River basalts and other flood lava eruptions constrained by a model of sulfide saturation. *Earth and Planetary Science Letters* **299**, 328–338.

Bryan, S.E., Ernst, R.E., 2008. Revised definition of large igneous provinces (LIPS). *Earth-Science Reviews* **86**, 175–202.

Burgess, S.D., Bowring, S., Shen, S., 2014. High-precision timeline for Earth's most severe extinction. *Proceedings of the National Academy of Sciences.*, doi: 10.1073/pnas.1317692111.

Campbell, I.H., Czamanske, G.K., Fedorenko, V.A., Hill, R.I., Stepanov, V., 1992. Synchronism of the Siberian Traps and the Permian–Triassic boundary. *Science* **258**, 1760–1763.

Chenet, A.L., Courtillot, V., Fluteau, F. *et al.*, 2009. Determination of rapid Deccan eruptions across the Cretaceous–Tertiary boundary using paleomagnetic secular variation: 2. Constraints from analysis of eight new sections and synthesis for a 3500-m-thick composite section. *Journal of Geophysical Research-Solid Earth* **114**.

Chenet, A.L., Fluteau, F., Courtillot, V., Gerard, M., Subbarao, K.V., 2008. Determination of rapid Deccan eruptions across the Cretaceous–Tertiary boundary using paleomagnetic secular variation: results from a 1200-m-thick section in the Mahabaleshwar Escarpment. *Journal of Geophysical Research-Solid Earth* **113**.

Cocks, L.R.M., Torsvik, T.H., 2007. Siberia, the wandering northern terrane, and its changing geography through the Palaeozoic. *Earth-Science Reviews* **82**, 29–74.

Courtillot, V., 1994. Mass extinctions in the last 300 million years: one impact and seven flood basalts. *Israel Journal of Earth Sciences* **43**, 255–266.

Dessert, C., Dupré, B., François, L.M. *et al.*, 2001. Erosion of Deccan Traps determined by river geochemistry: impact on the global climate and the $^{87}Sr/^{86}Sr$ ratio of seawater. *Earth and Planetary Science Letters* **188**, 459–474.

Dessert, C., Dupré, B., Gaillardet, J., François, L.M., Allegre, C.J., 2003. Basalt weathering laws and the impact of basalt weathering on the global carbon cycle. *Chemical Geology* **202**, 257–273.

Erwin, D.H., 1994. The Permo–Triassic extinction. *Nature* **367**, 231–236.

Foster, C.B., Afonin, S.A., 2005. Abnormal pollen grains: an outcome of deteriorating atmospheric conditions around the Permian–Triassic boundary. *Journal of the Geological Society* **162**, 653–659.

Ganino, C., Arndt, N.T., 2009. Climate changes caused by degassing of sediments during the emplacement of large igneous provinces. *Geology* **37**, 323–326.

Grise, K.M., Thompson, D.W.J., Birner, T., 2010. A global survey of static stability in the stratosphere and upper troposphere. *Journal of Climate* **23**, 2275–2292.

Hamill, P., Jensen, E.J., Russell, P., Bauman, J.J., 1997. The life cycle of stratospheric aerosol particles. *Bulletin of the American Meteorological Society* **78**, 1395–1410.

Howells, G.P., 1990. *Acid Rain and Acid Waters.* Ellis Horwood.

Joachimski, M.M., Lai, X., Shen, S. *et al.*, 2012. Climate warming in the latest Permian and the Permian–Triassic mass extinction. *Geology* **40**, 195–198.

Kamo, S.L., Czamanske, G.K., Amelin, Y. *et al.*, 2003. Rapid eruption of Siberian flood-volcanic rocks and evidence for coincidence with the Permian–Triassic boundary and mass extinction at 251 Ma. *Earth and Planetary Science Letters* **214**, 75–91.

Kiehl, J.T., Shields, C.A., 2005. Climate simulation of the latest Permian: implications for mass extinction. *Geology* **33**, 757–760.

Knoll, A.H., Barnbach, R.K., Payne, J.L., Pruss, S., Fischer, W.W., 2007. Paleophysiology and end-Permian mass extinction. *Earth and Planetary Science Letters* **256**, 295–313.

Kontorovich, A.E., Khomenko, A.V., Burshtein, L.M. *et al.*, 1997. Intense basic magmatism in the Tunguska petroleum basin, eastern Siberia, Russia. *Petroleum Geoscience* **3**, 359–369.

Korte, C., Kozur, H.W., Bruckschen, P., Veizer, J., 2003. Strontium isotope evolution of late Permian and Triassic seawater. *Geochimica et Cosmochimica Acta* **67**, 47–62.

Korte, C., Kozur, H.W., Joachimski, M.M. *et al.*, 2004. Carbon, sulfur, oxygen and strontium isotope records, organic geochemistry and biostratigraphy across the Permian/Triassic boundary in Abadeh, Iran. *International Journal of Earth Sciences* **93**, 565–581.

Kump, L.R., Bralower, T.J., Ridgwell, A., 2009. Ocean acidification in deep time. *Oceanography* **22**.

Lamarque, J.-F., Emmons, L., Hess, P. *et al.*, 2012. CAM-Chem: description and evaluation of interactive atmospheric chemistry in the Community Earth System Model. *Geoscientific Model Development* **5**, 369–411.

Meyerhoff, A.A., 1980. Geology and petroleum fields in Proterozoic and Lower Cambrian strata, Lena-Tunguska petroleum province, eastern Siberia, USSR, in Halbouty, M.T. (ed.), *Giant Oil and Gas Fields of the Decade 1968–1978*. American Association of Petroleum Geologists.

Miller, G.H., Geirsdóttir, Á., Zhong, Y. *et al.*, 2012. Abrupt onset of the Little Ice Age triggered by volcanism and sustained by sea-ice/ocean feedbacks. *Geophysical Research Letters* **39**.

Oman, L., Robock, A., Stenchikov, G., Schmidt, G.A., Ruedy, R., 2005. Climatic response to high-latitude volcanic eruptions. *Journal of Geophysical Research - Atmospheres* **110**.

Pavlov, V.E., Fluteau, F., Veselovskiy, R.V., Fetisova, A.M., Latyshev, A.V., 2011. Secular geomagnetic variations and volcanic pulses in the Permian–Triassic traps of the Norilsk and Maimecha–Kotui provinces. *Izvestiya - Physics of the Solid Earth* **47**, 402–417.

Pinto, J.P., Turco, R.P., Toon, O.B., 1989. Self-limiting physical and chemical effects in volcanic-eruption clouds. *Journal of Geophysical Research - Atmospheres* **94**, 11165–11174.

Rampino, M.R., Stothers, R.B., 1988. Flood basalt volcanism during the past 250 million years. *Science* **241**, 663–668.

Rees, P.M.A., 2002. Land-plant diversity and the end-Permian mass extinction. *Geology* **30**, 827–830.

Reichow, M.K., Pringle, M.S., Al'Mukhamedov, A.I. *et al.*, 2009. The timing and extent of the eruption of the Siberian Traps large igneous province: implications for the end-Permian environmental crisis. *Earth and Planetary Science Letters* **277**, 9–20.

Robock, A., 2000. Volcanic eruptions and climate. *Reviews of Geophysics* **38**, 191–219.

Rohling, E., Sluijs, A., Dijkstra, H. *et al.*, 2012. Making sense of palaeoclimate sensitivity. *Nature* **491**, 683–691.

Ross, P.S., Peate, I.U., McClintock, M.K. *et al.*, 2005. Mafic volcaniclastic deposits in flood basalt provinces: a review. *Journal of Volcanology and Geothermal Research* **145**, 281–314.

Sander, S.P., Golden, D., Kurylo, M. *et al.*, 2006. Chemical Kinetics and Photochemical Data for Use in Atmospheric Studies Evaluation Number 15. See http://jpldataeval. jpl.nasa.gov and http://jpldataeval.jpl.nasa.gov/pdf/JPL_15_AllInOne.pdf.

Schaller, M.F., Wright, J.D., Kent, D.V., 2011. Atmospheric $pCO_2$ perturbations associated with the Central Atlantic Magmatic Province. *Science* **331**, 1404–1409.

Schaller, M.F., Wright, J.D., Kent, D.V., Olsen, P.E., 2012. Rapid emplacement of the Central Atlantic Magmatic Province as a net sink for $CO_2$. *Earth and Planetary Science Letters* **323**, 27–39.

Seinfeld, J.H., Pandis, S.N., 1997. *Atmospheric Chemistry and Physics: From Air Pollution to Climate Change*. John Wiley & Sons Inc.

Self, S., Blake, S., Sharma, K., Widdowson, M., Sephton, S., 2008. Sulfur and chlorine in Late Cretaceous Deccan magmas and eruptive gas release. *Science* **319**, 1654–1657.

Shen, S.-Z., Crowley, J.L., Wang, Y. *et al.*, 2011. Calibrating the end-Permian mass extinction. *Science* **334**, 1367–1372.

Sobolev, S.V., Sobolev, A.V., Kuzmin, D.V. *et al.*, 2011. Linking mantle plumes, large igneous provinces and environmental catastrophes. *Nature* **477**, 312–U380.

Stothers, R.B., Wolff, J.A., Self, S., Rampino, M.R., 1986. Basaltic fissure eruptions, plume heights, and atmospheric aerosols. *Geophysical Research Letters* **13**, 725–728.

Sun, Y., Joachimski, M.M., Wignall, P.B. *et al.*, 2012. Lethally hot temperatures during the Early Triassic greenhouse. *Science* **338**, 366–370.

Svensen, H., Planke, S., Polozov, A.G. *et al.*, 2009. Siberian gas venting and the end-Permian environmental crisis. *Earth and Planetary Science Letters* **277**, 490–500.

Tabazadeh, A., Turco, R.P., 1993. Stratospheric chlorine injection by volcanic eruptions – HCl scavenging and implications for ozone. *Science* **260**, 1082–1086.

Textor, C., Graf, H.F., Herzog, M., Oberhuber, J.M., 2003. Injection of gases into the stratosphere by explosive volcanic eruptions. *Journal of Geophysical Research - Atmospheres* **108**.

Thordarson, T., Self, S., 1996. Sulfur, chlorine and fluorine degassing and atmospheric loading by the Roza eruption, Columbia River Basalt Group, Washington, USA. *Journal of Volcanology and Geothermal Research* **74**, 49–73.

Timmreck, C., Graf, H.F., Lorenz, S.J. *et al.*, 2010. Aerosol size confines climate response to volcanic super-eruptions. *Geophysical Research Letters* **37**.

Varekamp, J.C., Luhr, J.F., Prestegaard, K.L., 1984. The 1982 eruptions of El Chichón volcano (Chiapas, Mexico): character of the eruptions, ash-fall deposits, and gas-phase. *Journal of Volcanology and Geothermal Research* **23**, 39–68.

Visscher, H., Looy, C.V., Collinson, M.E. *et al.*, 2004. Environmental mutagenesis during the end-Permian ecological crisis. *Proceedings of the National Academy of Sciences of the United States of America* **101**, 12952–12956.

Ward, P.D., Montgomery, D.R., Smith, R., 2000. Altered river morphology in South Africa related to the Permian–Triassic extinction. *Science* **289**, 1740–1743.

Westrich, H., Gerlach, T., 1992. Magmatic gas source for the stratospheric $SO_2$ cloud from the June 15, 1991, eruption of Mount Pinatubo. *Geology* **20**, 867–870.

White, J., Bryan, S., Ross, P., Self, S., Thordarson, T., 2009. Physical volcanology of continental large igneous provinces: update and review. Studies in volcanology: the legacy of George Walker. *Special Publications of IAVCEI* **2**, 291–321.

Wignall, P.B., 2001. Large igneous provinces and mass extinctions. *Earth-Science Reviews* **53**, 1–33.

Wignall, P.B., Sun, Y., Bond, D.P. *et al.*, 2009. Volcanism, mass extinction, and carbon isotope fluctuations in the Middle Permian of China. *Science* **324**, 1179–1182.

# Index

Printed in the United States
By Bookmasters